Aerospace Engineering and Thermodynamics

Aerospace Engineering and Thermodynamics

Edited by **Russell Mikel**

Willford Press
New York

Published by Willford Press,
118-35 Queens Blvd., Suite 400,
Forest Hills, NY 11375, USA
www.willfordpress.com

Aerospace Engineering and Thermodynamics
Edited by Russell Mikel

International Standard Book Number: 978-1-68285-227-9 (Hardback)

Printed in China.

Contents

Preface

Aerospace engineering is a multidisciplinary field that deals with the manufacturing of spacecraft and aircraft. This book is a valuable compilation of topics, ranging from the basic to the most complex advancements in aerospace engineering, and the importance of thermodynamics in this field. It elucidates the concepts and innovative models around prospective developments with respect to thermodynamics and propulsion, fluid mechanics, pressure, power generation systems, combustion and solid mechanics, along with researches from experts around the globe. With state-of-the-art inputs by acclaimed experts of this field, this book targets students and professionals.

The researches compiled throughout the book are authentic and of high quality, combining several disciplines and from very diverse regions from around the world. Drawing on the contributions of many researchers from diverse countries, the book's objective is to provide the readers with the latest achievements in the area of research. This book will surely be a source of knowledge to all interested and researching the field.

In the end, I would like to express my deep sense of gratitude to all the authors for meeting the set deadlines in completing and submitting their research chapters. I would also like to thank the publisher for the support offered to us throughout the course of the book. Finally, I extend my sincere thanks to my family for being a constant source of inspiration and encouragement.

Editor

Experimental Study of Helical Coil Induction Water Heater using Induction Cooker

Tejas G. Patil[1], Atul A. Patil[2]

[1]*Deprtment of Mechanical Engineering, Godavari College of Engineering, India, tejas.patil@live.com*
[2]*Deprtment of Mechanical Engineering, Godavari College of Engineering, India, patilatul20@rediffmail.com*
Author Correspondence: Kolhe Nagar, Jalgaon

Abstract

This paper present a new phenomenon of heating of water flowing through helical coil using induction heating process is applied through induction cooker. The experimental setup is prepared according to proposed plan with helical coil configurations which is varying factor among the several has high heat transfer rate.Helical coil is designed as per limitations of induction cooker with some parameters and the values of water outlet temperature observed with variable flow rates of water during the experimental analysis is getting affected with the help of different helical coil configurations by changing number of turns with coil pitch, coil diameter and coil tube diameter maintainsconstant. The data obtained from experimentation is analyse to succeed highest heat transfer from induction cooker to water flowing through Helical coil with minimum time consumption, less pollution and safety aspect specially related to industrial applications.

Keywords: Induction Cooker, Helical coil, Induction Heating Process, Coil Pitch.

1. Introduction

Industry prerequisites of electricity and especially water at certain temperature of 85^0-90^0 for different processes to fulfill demands it required lot of electricity or fuel and also for hot water required at particular temperature using different techniques including Electric Heater, Gas Water Heater, Storage Heat Pump water heater, Heat Exchanger using burning of fuel etc. Due to degradable resources in environment it is necessary to find out alternative energy resources or use available resources effectively; so for water heating process it should aimed at having effective use over the supplied electricity, the new technique of water heating using induction heating process is being developed. Typical applications of induction heating are melting of metals, heating of metals for design, Brazing, Welding and all sorts of Surface Treatments.

The previous work done by Yuki Makimura,hideo tomita and shin-ichi Motegi (2010), on improvement of superheated steam generator by induction heating which is able to generate joule heat to metal effectively and this IH has been recently applied to small thermal power plants and home appliancesbecause of high thermal efficiency, safety , cleanliness and convenience for water heating . A green and renewable energy, high cost, thermal efficiency, safety and clean environmental are important issues that influenced the technology in home appliances recently such as in stove and water heating. Induction heating is a non-contact heating process and one of the new technologies in home appliance. There are some researches that discussed about induction heating process and implementation due to their clean and no pollution.Also previous work done by Nihar P Bara (2013), states the main advantages of using induction processes when compared to any other heating process (gas furnace) are among others, their fast heating rate, good reproducibility and low energy consumption and it is check with finite element analysis of induction furnace.

The previous work done byL. C. Meng, K.W.E. Cheng and K.W. Chan (2009) on heating performance improvement of the induction cooker studied magnetic field and heating performance with identifications of problems where it changes the induction coil with a Novel coil format is proposed. The conventional Induction heating has been extended to the catering industry as currently; the Induction Cooker draws more and more attentions as one of the popular kitchen appliances instead of ordinary stove plates andthese InductionCooker works with the suitable magnetic material pots such as Steel,Enameled Steel and Cast Iron etc. Instead of pot utilized in water heating process using Induction cooker for continuous hot water requirement; it is suggested that helical coil is being used for water flow through it which placed on Induction cooker where induction cooker acts as heat source to change the temperature of water. The previous work done by Ordoñez Flores Rafael, Reyes Castillo Fabiola and Carreño Hernández Carlos (2013) on water heater by magnetic induction shows new device that is proposed to heat the shower water as it obeys the excessive water wasting at start of knob of water, so it anticipates a water and gas saving. This water heater consumes less electrical energy than the resistance based boiler and it doesn't store hot water which signifies use of less energy than the gas based boilers. A review study done by N. D. Shirgire and P. Vishwanath Kumar (2013), on comparison of helical coil and straight tube heat exchanger heat transfer coefficient of helical coil is affected by geometry of heat exchanger. Helical coils offer advantageous over straight tubes due to their compactness and increased heat transfer coefficient.The effectiveness of heat exchanger greatly affected by hot water mass flow rate and cold water flow rate when cold water mass flow rate is constant and hot water mass flow rate increases the effectiveness decreases and increase in cold water mass flow rate for constant hot water mass flow rate resulted in increase in effectiveness hence flow rate of fluid plays vital role in experimenting with helical coil.

A review study by S.S. Pawar, V.K.Sunnapwar and B.A.Mujawar (2011), shows the heat transfer through helical coils for circular cross sections utilized either active (electric field, acoustic or surface vibrations etc.) or passive (Fluid additives or special surface geometries) techniques had been applied which can improve performance of heat exchanger in heat transfer. This shows that heat transfer in helical coil with circular cross section varied by altering basic geometry of the coil and different fluid types with different flow rates. The preceding work done byMofid Gorji Bandpy and Hasan Sajjadi (2010), accomplished about various aspect like coil diameter, coil tube diameter and coil step (coil pitch) of coil tube used in heat transfer process, considering coil step for coiled tube experimentally represented enhancement of coil step in a fixed tube diameter, shell side heat transfer coefficient increases. This also conclude shell side heat transfer coefficient raises with increases with heat flux rate of coil for fixed pipe diameter and heat transfer coefficient for coil inlet temperature increases with enhancement of coil step with fixed tube diameter at different mass flow rates. This shows coil step influence on heat transfer rate of coil for both laminar and turbulent flow. The previous work done by M.R. Salimpour (2008) on the experimental study shows that the shell-side heat transfer coefficients of the coils with larger pitches up to certain are higher than those for smaller pitches and the empirical correlation for constant temperature boundary condition is quite in agreement with the experimental data in low Dean number region. With experimental data indicates shortcoming of enhancement of coil pitch shows correspond to the lower values of helical number or loose-coiling conditions decreases the inner Nusselt number. These caused the heat transfer coefficients of coiled tubes are found closer to those of the straight pipes.

Helical coils are widely used in applications such as heat recovery system, chemical processing, food processing, nuclear reactors and high-temperature gas cooling reactors. Because of helical coils have a compact configuration, more heat transfer surface can be provided per unit of space than by the use of straight tubes, ease of manufacture and longer operating life. Hence helical coil is used instead of Spiral or Pancake Coil for heat exchange purpose. With the utilization of all these individual benefits of total system for helical coil water heater includes induction cooker and helical coil with measuring instruments which is developed to heat the water flows through it and is placed on Induction cooker. So there is continuous flow of water is maintained using rotameter attached in the system like in case of gas geysers.

2. Characteristics of Helical coil.

The schematic view of helical coil used as shown in figure 1. The pipe has outer diameter d_0. The coil has diameter of D (measured between the centresof the pipe), while the distance between the adjacent turns, called as pitch is P and the coil is diameter also called as pitch circle diameter (PCD). The ratio of pipe diameter to coil diameter (d_0/D) is called curvature ratio and number of turns of helical coil N.

Many researchers have identified the effect of all these parameters on the heat transfer rate to and from fluid flowing through helical coil by varying one or more parameter and keeping other constant. So in this present research work overall focus is given to number of turns of helical coil 'N' which three separate number of turns for helical coil selected arbitrary with N=3.25, N=4 and N=5 for experimental work keeping other parameters constant such as coil diameter D=109.5 mm, Pipe Diameter d_0= 9.5 mm and pitch of coil is 9.5 mm.

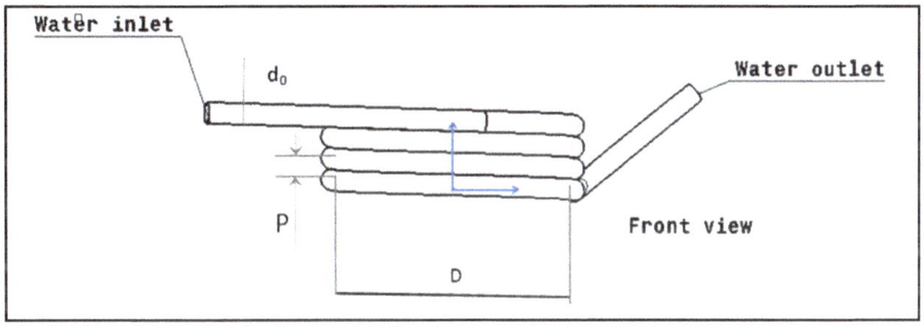

Figure 1: Basic Geometry of helical coil

So the previous work done by Ramchandra K. Patil, B.W. Shende and Prasanta K. Ghosh (1982) on design of helical coil heat exchanger where length of helical coil and volume occupied by coil is determined by analytical method as follows:

1. The length of coil L, needed to make N turns:

$$L = N\sqrt{((2*\Pi*r)^2 + p^2} \tag{1}$$

Where N= number of turns of helical coil
r = Average radius of helical coil, taken from centerline of helix to the centerline of coil
P = Pitch spacing between consecutive coil turns measured from center to center

Hence for number of turns (N) as 3.25, 4 and 5 turns for designing of helical coil

Length of coil For N=3.25,

$$L = 3.25\sqrt{((2*\Pi*54.75)^2 + 9.5^2}$$

L = 1118.44 mm.

Also extra length is added at inlet and outlet of helical coil for steady flow condition achieved for experimentation purposed as follows:

Extra length added to inlet of helical coil = 130 mm

Extra length added at outlet of coil varies with number of turn selected for experimentation purpose is as follows:

4. For N=3.25 Extra length added to outlet of helical coil = 75 mm
5. For N=4 Extra length added to outlet of helical coil = 95 mm
6. For N=5 Extra length added to outlet of helical coil = 120 mm

Length of coil with extra length L_{max} for N=3.25:-

$$L_{max} = L + 130 + 75 = 1118.44 + 130 + 74$$

$$L_{max} = 1322.44 \text{ mm} = 1.322 \text{ m.}$$

Similarly,
Length of coil for N=4 L_{max} = 1.601 m

Length of coil for N=5 L_{max} = 1.970 m

2. The volume occupied by coil, Vc:

$$Vc = \frac{(\Pi * d^2{}_0 * L\max)}{4} \tag{2}$$

Volume occupied by helical coil For N=3.25

$$Vc = \frac{(\Pi * 9.5^2 * 1.322)}{4}$$

$$Vc = 374824.9903 \text{ mm}^3 = 3.748349\text{e-}4 \text{ m}^3$$

Similarly,
Volume occupied by helical coil For N=4 is 1.1348e-4 m^3
Volume occupied by helical coil For N=5 is 1.39637e-7 m^3

3. Experimental Setup and Procedure

3.1 Experimental setup:-

It is necessary to evolve physical design of an experimental set up having provision of setting test points, adjusting test sequence, executing proposed experimental plan, provision for necessary experimentation for noting down the responses. The experimental set up is designed considering various physical aspects of its elements. The other sub system of the setup such as measuring instruments and environmental factors are considered. This generalized procedure of design of experimental set up however cannot be totally followed in the field of experimentation. This is because in the field of experimentation, we are carrying out the experimentation using the available ranges of the various independent variables to assess the value of the dependent variables. The experimental setup consists of Induction cooker, helical coil, Rotameter and thermocouple arrangement as shown in figure 2. The Bajaj Majesty ICX 6 induction cooker with rating 1600 w is used in test set up for this experimentation. Induction cooker is works at 1200 Watt as constant heat source for heat transfer to water flow through helical coil which is work on induction heating process for experimentation while helical coil is placed at center of induction cooker. Water is flow through helical coil from top to bottom of coil and coil is subjected to induction heating process. These helical coil shown in figure is replaced with other coil has differs in number of turns.Water at atmospheric temperature is supplied from tap to inlet of helical coil through rotameter where flow rate is maintaining forten different levels for each helical coil and flow from rotameter to insulated pipe where thermocouple is fixed which shows temperature on display unit provided with it to measure inlet water temperature. Water from insulated pipe is supplied to helical coilis placed on induction cooker for heat supply after passing through helical coil outlet it is passes from insulated pipe after helical coil where again thermocouple is attached to measure outlet water temperature and this water is collected in bucket for different application in house hold applications. Also portable thermometer is also used to measure temperature of water outlet temperature is shown in figure 2.

The experimental set up is designed with the following instrumentation.

- Induction cooker arrangement
- Helical coil arrangement
- Rotameter and thermocouple arrangement

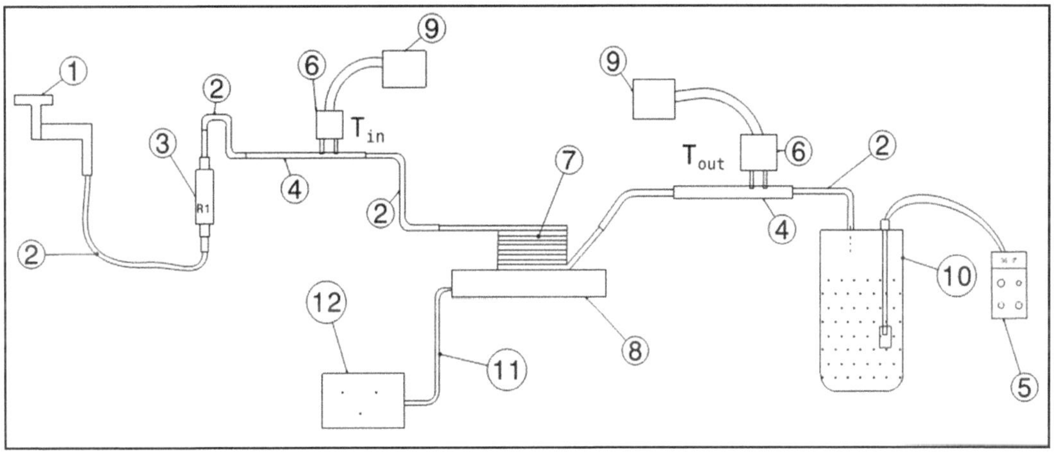

Figure: 2 Schematic View of Experimental setup

1. Water Tap
2. Flexible Pipe to continue water flow
3. Rotameter
4. Insulated Pipe for thermocouple
5. Portable thermometer
6. Thermocouple fixed at insulated pipe
7. Helical coil
8. Induction Cooker
9. Temperature Display unit
10. Plastic Bucket
11. Supply cable of induction cooker
12. Switch Board

3.2 Procedure for Test Run

Different Helical coils are used for performing the task of developing helical coil water heater using Induction cooker.Experiments are conducted for ten different water flow rates through the coil and for three different helical coil with N=3.25, N=4, N=5.The water flow rate in lph is noted down for different helical coils with number of turns 3.25, 4 and 5 in observation tables. As per the plan the flow rate is varied at ten levels. The readings for water inlet and outlet temperature for heat transfer analysis of helical coils are noted down.Measurements are taken only after the temperatures attainsteady values.During the courseof each set of experiments, the heat supply from induction cooker at 1200 w is kept constant, which ensures a constant heat transferin induction heating process.

4. Results and discussion

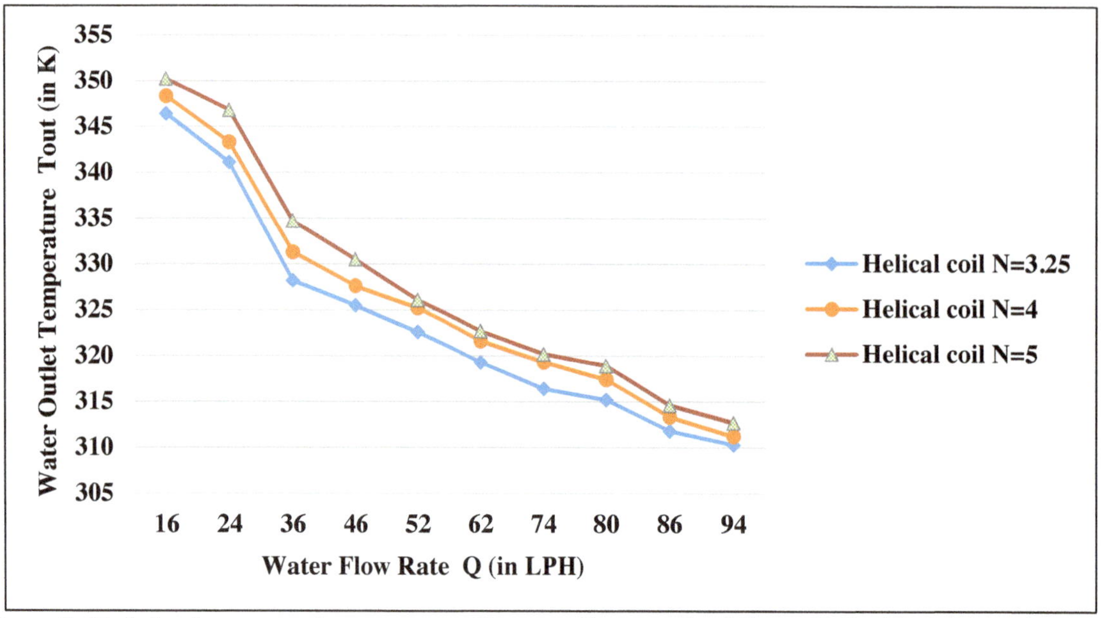

Figure 3: Variation in water outlet temperature Vs. water flow rate for different Helical N=3.25, N=4 and N=5

It is observed from the figure3 variations in water outlet temperature Vs. water flow rate for different Helical coil using number of turns N=3.25, N=4 and N=5. It is found that when one helical coil is selected at different water flow rate viz. 15 lph to 95 lph the water outlet temperature decreases with increase in water flow rates. During the change of Helical coil as number of turns increases form N=3.25 to N=5 value of water outlet temperature is also increased at same water flow rate. It also clarify from the results that as number of turns is increases which causes increase in length of pipe required to form helical coil indicates that higher the length of pipe used will increase the water outlet temperature of helical coil.

5. Conclusions

Helical coil water heater with the help of induction cooker is non-contactless, pollution free and safety setup is used to heat water up to certain temperature with variations in flow rates. It is concluded from experimentation as number of turns of helical coil is increases the water outlet temperature also increases at constant flow rates keeping other parameters constant.This experimentation set up is utilised in industry where hot water requirement is necessary for different process which leads to optimize helical coil design to get maximum output water temperature using induction cooker with least energy consumption due to faster process rate.

References

1. L. C. Meng, K.W.E. Cheng and K.W. Chan (2009), Heating Performance Improvement and Field Study of the Induction Cooker, 2009 3rd International Conference on Power Electronics Systems and Applications
2. N. D. Shirgire, and P. Vishwanath Kumar (2013), Review on Comparative Study between Helical Coil and Straight Tube Heat Exchanger, IOSR Journal of Mechanical and Civil Engineering (IOSR-JMCE). Volume 8, Issue 2 (Jul. - Aug. 2013), PP 55-59

3. Nihar P Bara (2013),Finite Element Analysis of Induction Furnace for Optimum Heat Transfer, International Journal of Innovative Research in Science, Engineering and Technology, Vol. 2, Issue 5, May 2013. PP 1313-1319.
4. Ordoñez Flores Rafael, Reyes Castillo Fabiola and Carreño Hernández Carlos (2013), Water Heater by Magnetic Induction, 978-1-4673-6155-2/13 2013 IEEE Journal
5. Ramchandra K. Patil, B.W. Shende and Prasanta K. Ghosh (1982), Designing a Helical-Coil Heat Exchangers , Chemical Engineering, 13 December 1982
6. Yuki Makimura, hideotomita and shin-ichi Motegi (2010), Improvement of Superheated Steam Generator by Induction Heating, International conference on computer applications and Industrial Electronics (ICCAIE), December 5-7, 2010, Kuala Lumpur, Malaysia
7. S.S. Pawar, V.K.Sunnapwar and B.A.Mujawar (2011), A Critical Review of Heat Transfer through Helical Coils of Circular Cross Section, Journal of Scientific and Industrial Research Vol. 70, October 2011, PP 835-843.
8. Mofid Gorji Bandpy and Hasan Sajjadi (2010), An Experimental Study of the Effect of CoilStep on Heat Transfer Coefficient in Shell-Side of Shell-and-Coil Heat Exchanger, International Journal of Mechanical and Materials Engineering 1:2 2010. PP 80-85.
9. R. Salimpour (2008), Heat transfer coefficients of shell and coiled tube heat exchangers, Experimental Thermal and Fluid Science 33 (2009) 203–207.

A Brief Author Biography

Tejas G. Patil is Lecturer in GF's Godavari College of Engineering, Jalgaon, India. He has qualifications of Licentiate in Mechanical Engineering (LME) from VJTI, MUMBAI in 2008 and Graduation degree in Mechanical engineering from university of Pune in 2011 and PerusingPost-Graduation degree ME specialization in Thermal. special interest of field is related to heat transfer improvement in existing system or develop new concepts.

Atul A. Patil is Assistant Professor in GF's Godavari College of Engineering, Jalgaon, India. He received degree in BE Mechanical in 2002, Post-Graduation ME with specialization in Designin 2005 from Shivaji University, Kolhapur and submitted Ph.D. Thesis on Design and Development of Energy Efficient Exhaust System for Four Stroke Compression Ignition Engine to North Maharashtra University, special interest of field is related to heat transfer analysis using software simulation to validate with the experimental results.

2

Comparative Study of Different Passages over Dimpled Plate Evaporator for a Small Water Cooler

Mukund Y Pande[1], Atul Patil[2]

[1]*Department of Mechanical Engineering, Godavari College of Engineering, India,*
pande21mukund@gmail.com
[12]*Department of Mechanical Engineering, Godavari College of Engineering, India,*
patilatul20@rediffmail.com
Author Correspondence: [1]GF'S Godavari college of Engineering Jalgaon Maharashtra. India

Abstract

Bonded coil construction have surface area coming in contact with tank after brazing is small compared with the total surface area of the coil and also there is indirect contact between water to be cooled and the evaporator coils. Due to this type of construction most of the portion of the evaporator coil remains ineffective. Plate coil exchanger has direct surface area connect between two fluids. The whole heat exchanger surface is dipped in one of the fluid so effective area of contact is total surface area. They can be made from stainless steel material to maintain hygienic condition of drinking the water. Cleaning and maintenance of this heat exchanger is so easier. But milk cooler is a good example of dimpled plate heat exchanger so it is proposed to analyze performance of this heat exchanger os an evaporator in small capacity water cooler.

Keywords: Evaporator, Dimpled plate.

1. Introduction

Water cooler is a common example of refrigeration system observed in day to day life. The purpose of water cooler is to make water available at a constant temperature irrespective of ambient temperature. They are meant to produce cold water at about 7°C to 13°C. It has the primary function of cooling potable water and dispensing it to integral and/or remote areas, by means of a complete mechanical refrigeration system here are three types of water cooler such as storage type, bottler type and instantaneous type water cooler. Out of which storage type water cooler is most popular among all other types. Refrigeration system of water cooler consists of different components as follows Hermetic compressors are commonly used for both 50 and 60 Hz AC applications. Compressors are similar to those used in household refrigerators. Forced air-cooled condensers are most commonly used. Capillary tubes are used almost exclusively for refrigerant flow control in hermetically sealed systems. This water cooler has the evaporator formed by refrigerant tubing bonded to the outside of a storage tank; such evaporator is called bonded coil evaporator. This bonded coil construction has very small effective surface area of evaporator. As coils used are of circular cross section, there is line contact or small area contact between coils and storage tank. Remaining area of the evaporator coils is covered with insulation and remains ineffective, so effective surface area is very less compared with total surface area of the coil. When a dimpled wall is used in channel flow, dimple geometry play a key role in the heat transfer and drag coefficients. The optimum configuration should provide the higher heat transfer improvement with less drag for a specific application. As of today, there exists no road map to determine the optimum dimple configuration.

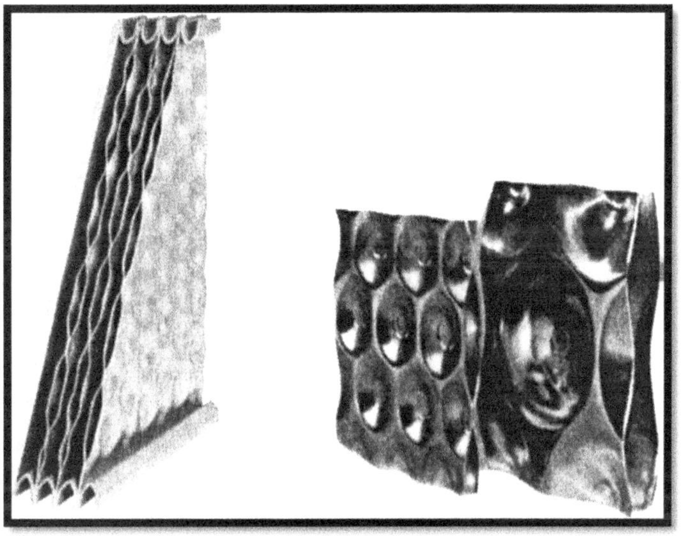

figure.1: Dimple on plate

During the last few years, studies on turbulent flow over dimpled walls have provided some insight into the application of this technology for turbine blade and jet impingement cooling. However, application of dimples in laminar flow was scarcely explored. Microelectronic cooling and micro fluids, where the flow regime is mostly laminar, are two potential fields of application for dimple technology. Most existing archival papers which utilize dimpled surfaces consider turbulent flows at higher Reynolds numbers. Existing thermal performance data for dimpled surfaces in laminar flows are rare, even though such data are useful for applications such as electronic cooling, where lower Reynolds numbers and lower speeds are present. The present study is aimed at partially remedying this deficiency by providing a systematic set of data which illustrate the effects of dimples on local and spatially-averaged surface Nusselt number distributions, as well as on friction factors in channels with laminar flow. Of particular interest are the changes which occur because of the absence of turbulence transport. In some cases, vastly different trends are present as flow conditions are altered, or as some aspect of channel geometry is changed. Two different dimple depths and three channel heights are employed at Reynolds numbers based on channel height ranging from 260 to 1030. Also included are laminar flow experimental results, which illustrate the combined effects of dimples and protrusions as they are located on opposite channel walls.

2. Literature review

Bulk milk cooler is a stainless steel insulated tank fitted with agitator and a refrigeration system. Milk when pumped into this tank is circulated till it reaches less than 4°C. Because of insulation the temperature is maintained. Whenever the temperature exceeds the refrigeration system operates. Bulk milk coolers are used extensively for cooling milk at the collection point, thus maintaining the quality of milk. These are hygienic, easy to use and available in various sizes.

N. **Xiao et al. [10]** The present study provides the systematic set of data which illustrate the effects of an array of dimples on local and spatially averaged surface Nusselt number distributions, as well as on friction factors in channels with laminar flow. Trends of spatially-averaged Nusselt numbers and friction factors are provided as they vary with dimple depth, channel height, Reynolds number and the use of protrusions on the opposite channel wall. When compared with turbulent flow results, the present laminar data illustrate changes due to the absence of turbulence transport. For example, in contrast to turbulent flows, the present laminar flow data show that there is no overall benefit from the use of a top wall with protrusions. In addition, spatially-averaged Nusselt number ratios and friction factor ratios measured on a deep dimpled surface with a smooth top wall show trends which are opposite from ones observed in turbulent flows, since lower laminar heat transfer augmentations are present for smaller channel heights when compared at the same Reynolds number.
Derya Burcu Ozkan et al [6] examine parameter affecting the frost formation on the evaporator of a refrigerator and the structure of frost. Air velocity both at the air inlet and outlet channels of the evaporator were performed, and the effect of the air velocity on frost formation was examine. In this experiment parameter affecting on frost formation on the evaporator and structure of the frost were examined and frost thickness on the chosen fin was measured. In this experiment the compressor of the refrigerator operated at 100% load and cooling was carried out for 5 Hr.

S. D. Hwang et al. [7] in present study, heat transfer and thermal performance of a periodically dimple-protrusion patterned surface have been investigated to enhance energy-efficiency in compact heat exchangers. The local heat transfer coefficients on the dimple/protrusion walls are derived using a transient TLC (Thermo-chromic Liquid Crystal) technique at low Reynolds number. The periodically patterned surface is applied to the bottom wall only or both the bottom and top walls in the test duct. On the single-side patterned walls, various secondary flows generated from the dimple/protrusion coexist. The vortices induced from the upstream affect strongly on the downstream pattern. For the double-side patterned wall case, vortex interaction affected by the opposite wall enhances highly the heat transfer. The heat transfer augmentation is higher in the lower Reynolds number due to the effective vortex interactions. Therefore, the performance factor considering both heat transfer enhancement and pressure loss increases with decreasing the Reynolds number.
O. The capacity of a water cooler is expressed in liters per hour and is the quantity of water cooled in one hour from a specified inlet temperature to a specified dispensing temperature. Normal standard capacities of water coolers range from 3.8 to 114 L/h.
Water coolers are rated on the basis of continuous flow capacity under specified water temperature and ambient conditions. ARI standard 1010-84, Drinking –Fountain and self –contained Mechanically – Refrigerated Drinking –Water Cooler, gives the generally accepted rating conditions and references test methods as prescribed in ASHRAE standard 18-1987, Methods of Testing for Rating Drinking – Water Coolers with Self-Contained Mechanical Refrigeration Systems.

	Temperature 0C				
Type of Cooler	Ambient (o C)	Inlet Water (o C)	Cooled Water (o C)	Heated Portable Water (o C)	Spill %
Bottle Cooler	32	32	10	74	NONE
Pressure Types using pre cooler or Non-pre cooler Drain(bubbler service)	32	27	10	74	60
Not using pre-cooler or Other heat transferring device	32	27	10	74	NONE

Compartment coolers	Standard rating conditions for water cooling noted above apply. During the standard rating test, no ice shall melts, nor shall the average temperature exceed 7.8 (°C) in the refrigerated compartment.

Table.1: Rating Of cooler

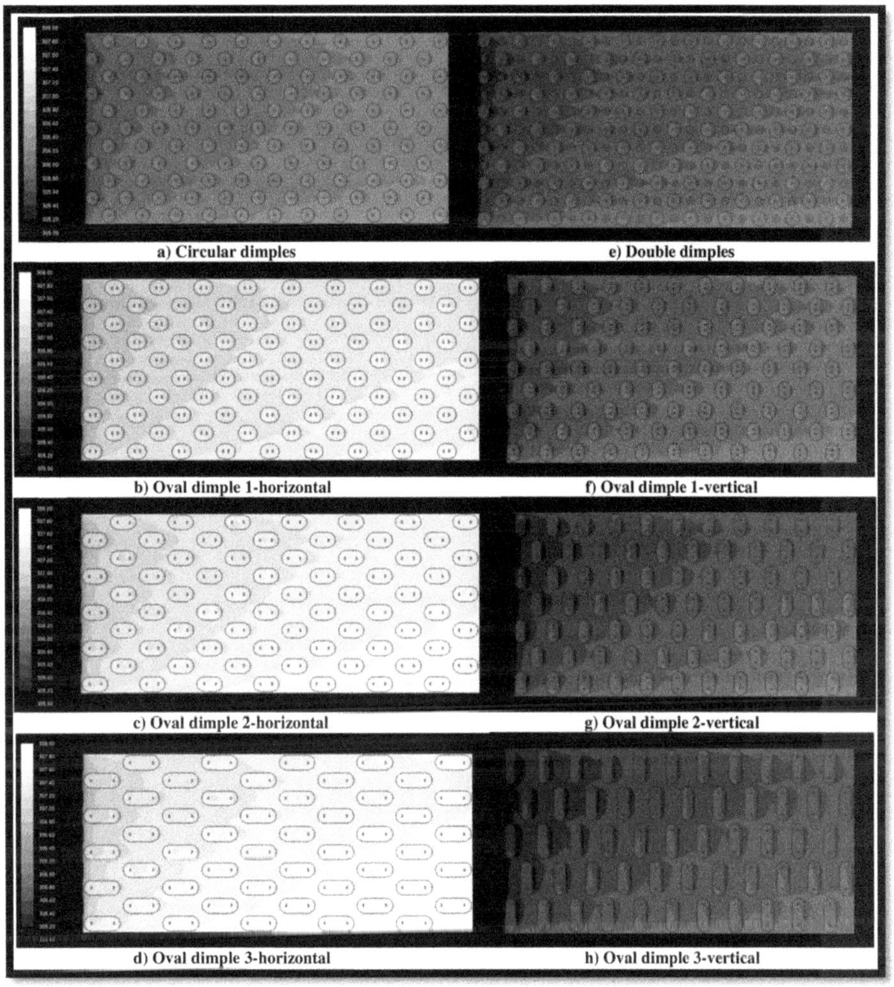

Figure.2 Temperature distribution over the dimpled surfaces. Fluid flow is from the left to right. Heat flow is from the bottom to the top. *a*. Circular dimples, *b*. Oval dimple 1-horizontal, *c*. Oval dimple 2-horizontal, *d*. Oval dimple 3-horizontal, *e*. Double dimples, *f*. Oval dimple 1-vertical, *g*. Oval dimple 2-vertical, and *h*. Oval dimple 3-vertical

3. Experimental Setup and Procedure

3.1 Experimental procedure

i. Water level in tank is at 40 liter is ensured.
ii. Hand operated valve for water supply is open is ensured.
iii. Hand operated valves H_2 and H_4 kept at open position and H_1 and H_3 at closed position for operating system with bonded coil evaporator.
iv. Power supply switched "ON".
v. Initial reading of temperatures noted down.
vi. Set up run for about 1 hr till temperature drop for 17°C is observed.
vii. Faucet opened and flow of water adjusted to maintain temperature drop of 17°C.
viii. Set up is allowed to run for about 1 hr till temperature drop observed is stable.

ix. Temperatures at all points noted down.
x. Time required for 10 flashes in energy meter noted down.
xi. Time required for flow of 1 liter of water noted down.
xii. Condenser and evaporator pressure in pressure gauges noted down.

Procedure was repeated by operating system for dimpled evaporator. Hand operated valves H_1 and H_3 kept at open position and H_2 and H_4 at closed position for that purpose.

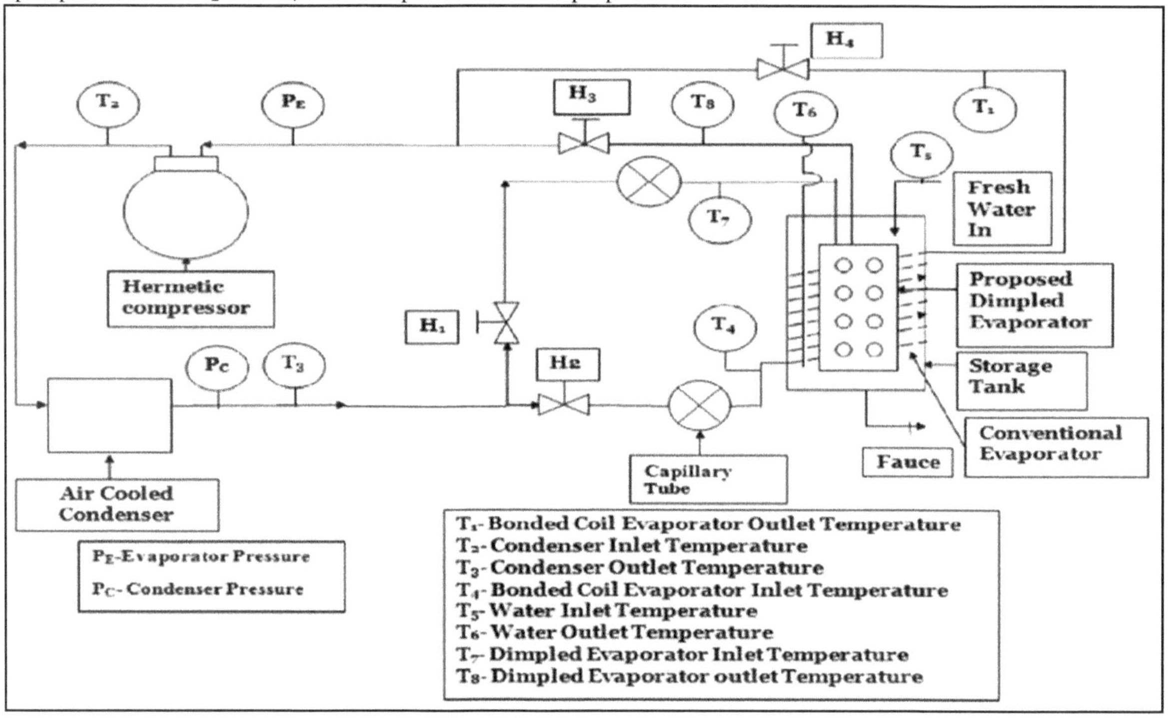

Figure.3: Layout of system

Assumptions

a) There is no heat loss from suction line.
b) There is no heat loss from storage tank.

Calculations

A). Carnot COP = $\frac{T_L}{(T_H - T_L)}$

Where TL= Lowest Temperature in cycle i.e. saturation Temperature Corresponds to evaporator pressure (°K),

TH= Highest Temperature in cycle i.e. saturation Temperature Corresponds to condenser pressure (°K)

B). Theoretical COP = $\frac{Theoretical\ Refrigerating\ Effect}{Theoretical\ Compressor\ Work}$

Enthalpy values according to P-h Chart of refrigerant R134a Corresponding to pressure and temperature conditions of evaporator and condenser inlet and outlet.

$h_1 =$ kJ/kg,

$h_2 =$ kJ/kg,

$h_3 =$ kJ/kg,

$h_4 =$ kJ/kg,

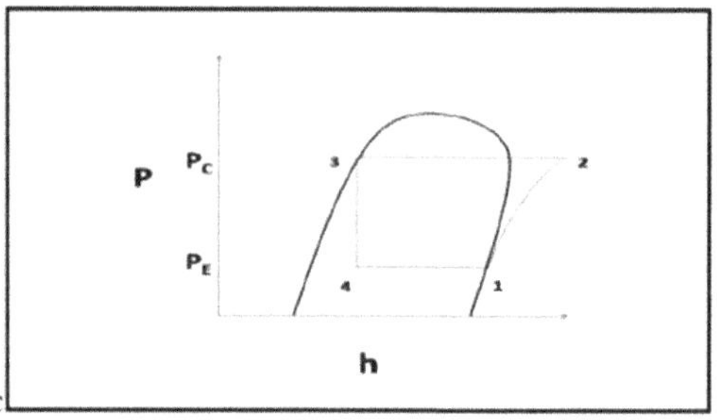

Figure.4: P-h Chart

Theoretical refrigerating effect = $(h_1 - h_4)$ kJ/kg

Theoretical Compressor Work = $(h_2 - h_1)$ kJ/kg

C). Actual COP = $\frac{Actual\ Refrigerating\ Effect}{Actual\ Compressor\ Work}$

Actual Refrigerating Effect = **mw × Cpw × (T5-T6) kW**

$m_w = \frac{1}{t_w}\frac{kg}{s}$

Actual Compressor work = Energy supplied / Time

Actual Compressor work = $\frac{10}{N} \times \frac{3600}{t_{10}}$ kW Where, N=Energy meter constant in rev/kW-hr.

N=320 rev/kW-hr

D). Relative COP = $\frac{\text{Actual COP}}{\text{Theoretical COP}}$

E). Capacity of water cooler = $\frac{3600}{Time\ to\ floe\ 1\ li\ water}\ \frac{li}{Hr}$

F). Evaporator Capacity = $\frac{Actual\ Refrigeraring\ effect}{Area\ of\ Evaporator}\ \frac{w}{m^2}$

4. Results and discussion

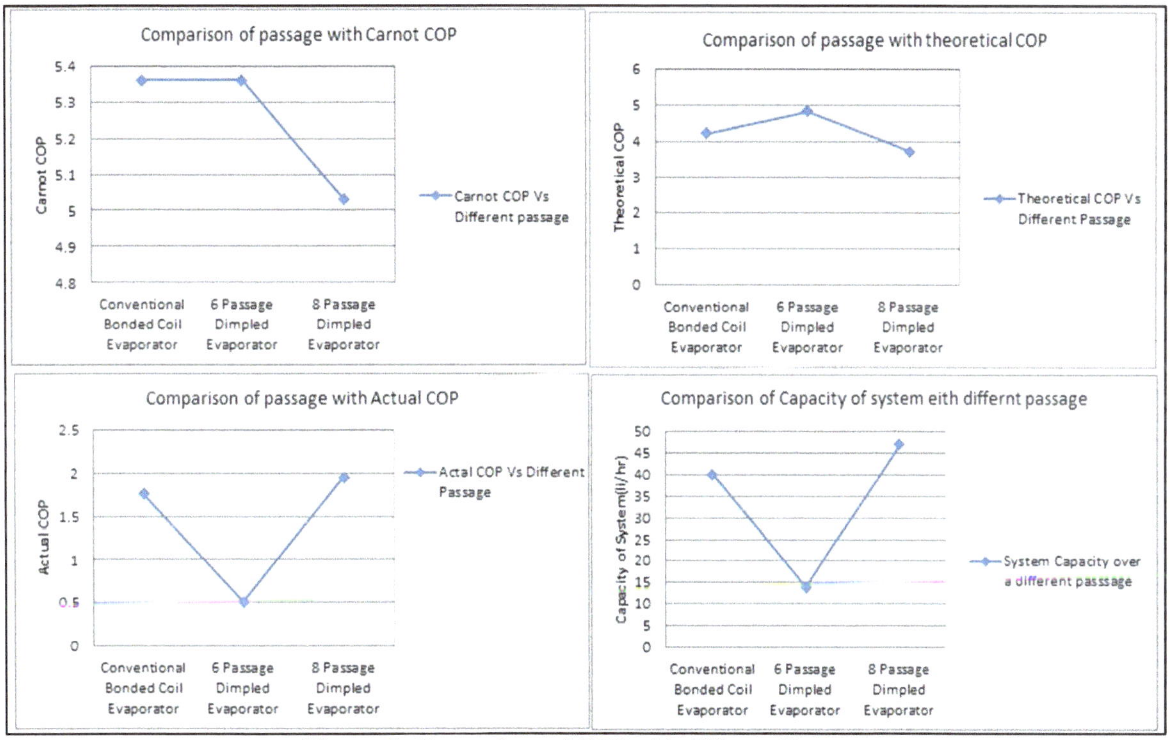

Figure.5: Comparison of COP with different Passage

Capacity observed was nearly same of the capacity specified by the company; system was operating as per design conditions. Results obtain for bonded coil evaporator can be used to compare performance of the system with proposed dimpled evaporator. To increase refrigerating effect, it was decided to narrowing the width of the passages. As width of passages going to reduce, number of passages for given width of the plate were going to increase, increasing length of the path followed by the refrigerant. Passages formed of dimpled plate evaporator with 8 passages were observed better than previous dimpled evaporator. Refrigerant was following expected flow path and get evaporated. As length of path was more vapor leaving evaporator was observed superheated. System balancing observed difficult for dimpled evaporator.

5. Conclusion

Based on the experiments performed and analysis of the results following conclusions were drawn System operating with 8 passages dimpled evaporator outperforms both bonded coil evaporator and 6 passages dimpled evaporator. Evaporator having 8 passages was more effective due to increased length of path for the refrigerant. It was causing effective utilization of surface area and increasing time for which refrigerant was in the evaporator. Refrigerating effect was increased by 17.61% for 8 passages evaporator with relatively less increase in compressor work (5.8%). It was observed that COP improved from 1.76 to 1.95. As coils were soldered to the drum for bonded coil evaporator, surface area and hence evaporator capacity was not effectively utilized.

6. References

1. R. K. Shah, D. P. Sekulic, "Fundamentals of Heat Exchanger Design," John Wiley & Sons, Inc. 2003.
2. Transfer The Heat Transfer People, Texas, brochure
3. T. Kuppan, "Heat Exchanger Design Handbook," CRC Press, 2000.
4. ASHRAE Refrigeration Handbook, 1994, Chapter 49, pp.49.1-5.
5. Plate Coil prime surfaces, Sweden, brochure
6. S. K. Group, Slovenia, brochure
7. S. D. Hwang, H. G. Kwon, H. H. Cho, "Local heat transfer and thermal performance on periodically dimple-protrusion patterned walls for compact heat exchangers," Energy, 2010, vol. 35, pp. 5357-5364.
8. N. Xiao, Q. Zhang, P. M. Ligrani, R. Mongia, "Thermal performance of dimpled surfaces in laminar flows," International Journal of Heat and Mass Transfer, 2009, vol. 52, pp. 2009–2017.

A Brief Author Biography

Mukund Y. Pande is a Post-graduation student of GF's Godavari College of Engineering, India. and I recived degree in BE mechanical in 2011. Special interest of field is related to refrigeration System improvement in existing system or develop new concepts

Atul A. Patil is Assistant Professor GF's Godavari College of Engineering, India, India. He received degree in BE Mechanical in 2002, Post-Graduation ME with specialization in Design in 2005 from Shivaji University, Kolhapur and submitted Ph.D. Thesis on Design and Development of Energy Efficient Exhaust System for Four Stroke Compression Ignition Engine to North Maharashtra University, special interest of field is related to heat transfer analysis using software simulation to validate with the experimental results.

INVESTIGATION OF CHATTER STABILITY IN BORING OPERATIONS USING POLYMERS AS IMPACT DAMPERS

K.Esakkiraja[1], B.Sam Antony[2], A.Kannan[3]

[1] *Assistant Professor, Adithya Institute of Technology, Coimbatore, e-mail: esakkiraja89@gmail.com*
[2] *Assistant Professor, Ranganathan Engineering College, Coimbatore, e-mail: bsamantony@gmail.com*
[3] *Assistant Professor, SVS College of Engineering, Coimbatore, e-mail: kannankalam@yahoo.co.in*

Abstract

Chatter is a problem of instability in the metal cutting process. The phenomenon is characterized by violent vibration, poor surface finish and causes reduction of the life of the tool. In all cutting operations vibrations are induced due to the deformation of the work piece. The main objective of this proposed work is to introduce the impact dampers in between the tool insert and tool and reduce the chatter. In this investigation, improvement of damping capability of boring tool and reduction of chatter vibration using dampers are analyzed. The polymers selected for the dampers are acrylate, epoxy and polyamides. The main contributions of the current study are observed that the natural frequency of the boring tool has been enhanced with dampers. From harmonic and transient analysis it is observed that there is a reduction in amplitude of vibration with respect to frequency and time taken to decay the amplitude of boring tool with dampers.

Keywords: chatter, boring, polymer impact dampers.

1. Introduction

In the boring operation the tool holder shank are subjected to a dynamic excitation due to the deformation of the work material during the cutting operations. The chip formation process usually induces vibration in the machine-tool system. Energy from chip formation process excites the machine tool system. Modes of the work piece may also influence the vibration. The relative dynamic motion between cutting tool and work piece will affect the result of machining, in particular the surface finish. Further more, the tool life is correlated with the amount of vibration, with the modern trend of machine tool development: accuracy and reliability are gradually becoming more prominent features.

The main objective of the proposed work is to introduce the impact damper in between the tool insert and tool holder. In this investigation, improvement of the damping capability of boring tool and reduce the chatter vibration using three types of polymers which are polyamides, acrylate, and epoxy. These materials having high density and so that inertia mass of boring tool is increased to reduce the chatter vibration in boring operation. From the result suggesting which damping materials will be suitable for reduce the chatter vibration of boring operations comparing with each other

.

2. Properties of tool and polymers

Tool property

Material : EN31

Density :7.84×10^{-6} kg/mm^3

Young's modulus : 2.84×10^5 N/mm^2

Poisons ratio :0.3

Insert property

Material : cemented carbide

Density :14.95×10^{-6} kg/mm^3

Young's modulus : 6.25×10^5 N/mm^2

Poisons ratio :0.22

Acrylate property

Density :1.4×10^{-6} kg/mm^3

Young's modulus : 7.5 Gpa

Poisons ratio :0.35

Epoxy property

Density :2.2×10^{-6} kg/mm^3

Young's modulus : 6.9 Gpa

Poisons ratio :0.4

Polyamide property

Density :1.1×10^{-6} kg/mm^3

Young's modulus : 2.8 Gpa

Poisons ratio :0.41

3. Investigation of dynamic behavior of boring tool using ansys

3.1 Modal analysis

Modal analysis is used to determine the natural frequencies and mode shapes of the structure. The natural frequencies and mode shapes are very important parameters in the design of a structure for dynamic loading conditions. They are also required for to do a spectrum analysis or a mode superposition harmonic or transient analysis. Modal analysis in the ansys program is a linear analysis any nonlinear ties such as plasticity and contact element, will be ignored even if are defined. We can choose from three mode extraction methods-reduced, full and damped. The damped method allows you to include damping structure.

3.2 Results of modal analysis

Figure 1 first mode shape of boring tool without dampers

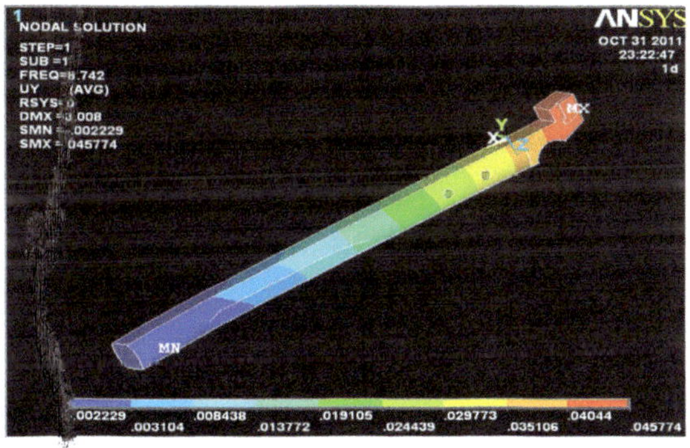

Figure 2 first mode shape of boring tool with acrylate impact damper

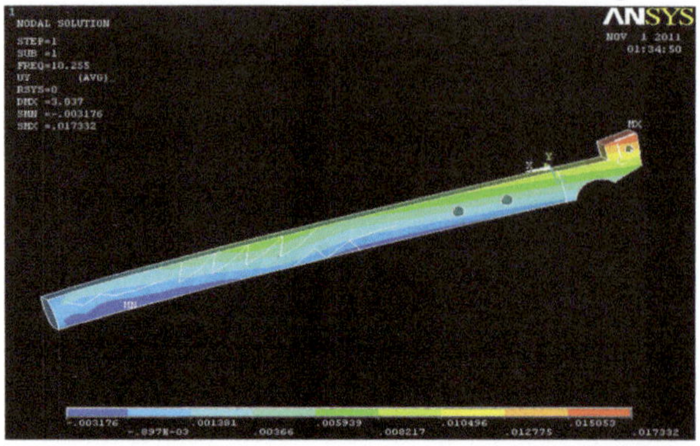

Figure 3 first mode shape of boring tool with Epoxy impact damper

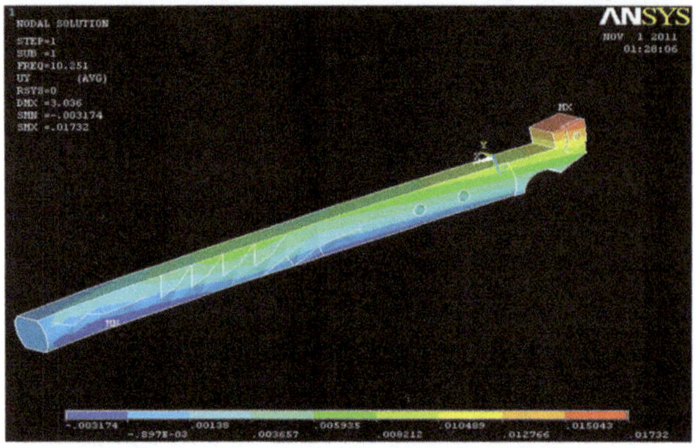

Figure 4 first mode shape of boring tool with polyamide impact damper

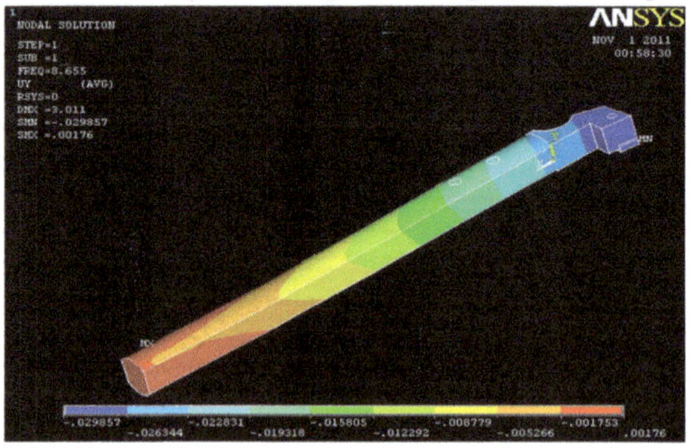

Figure 5 second mode shape of boring tool without dampers

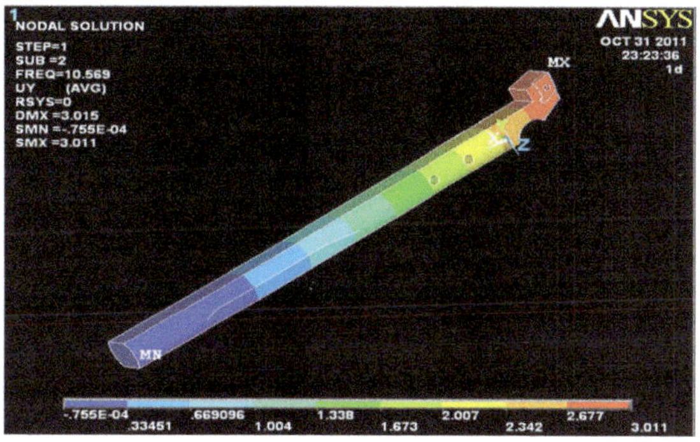

Figure 6 second mode shape of boring tool with acrylate impact damper

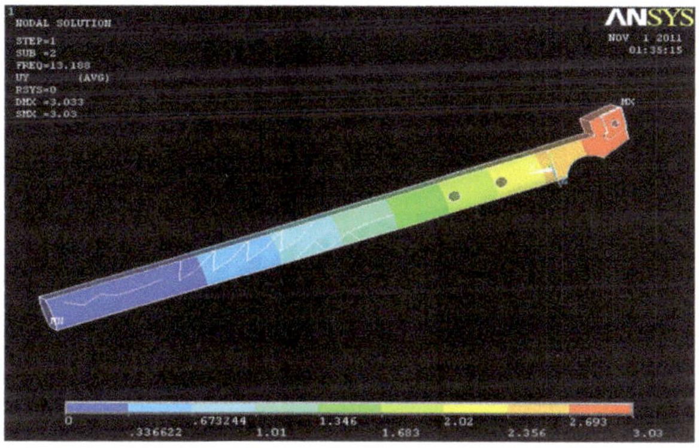

Figure 7 second mode shape of boring tool with epoxy impact damper

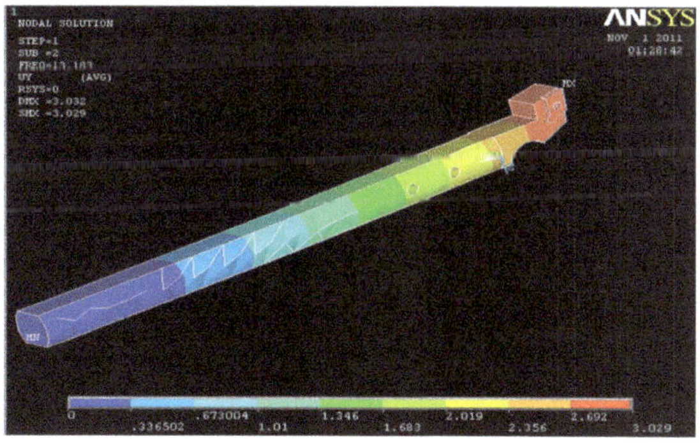

Figure 8 second mode shape of boring tool with polyamide impact damper

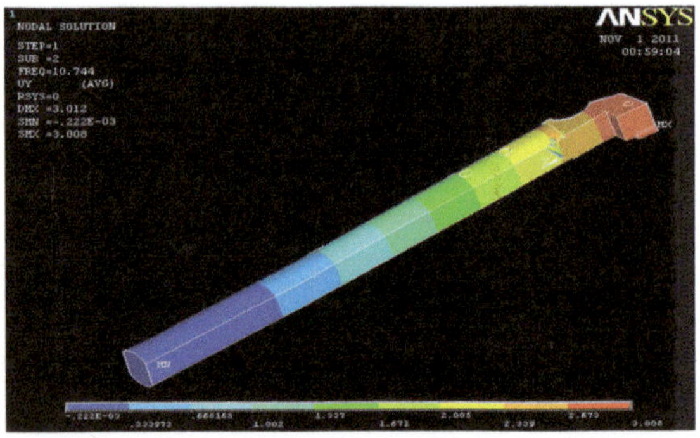

Figure 9 third mode shape of boring tool without damper

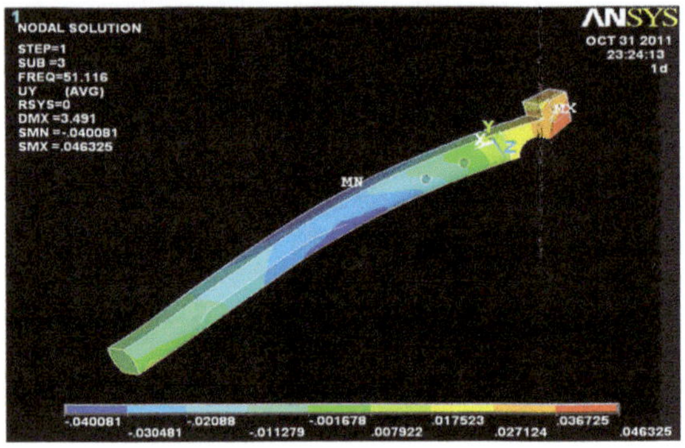

Figure 10 third mode shape of boring tool with acrylate impact damper

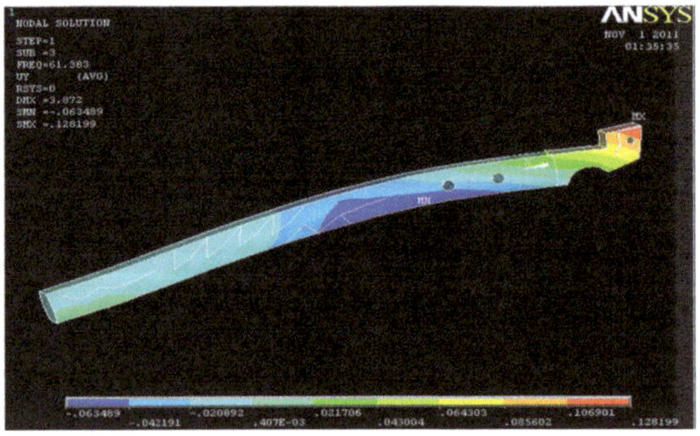

Figure 11 third mode shape of boring tool with epoxy impact damper

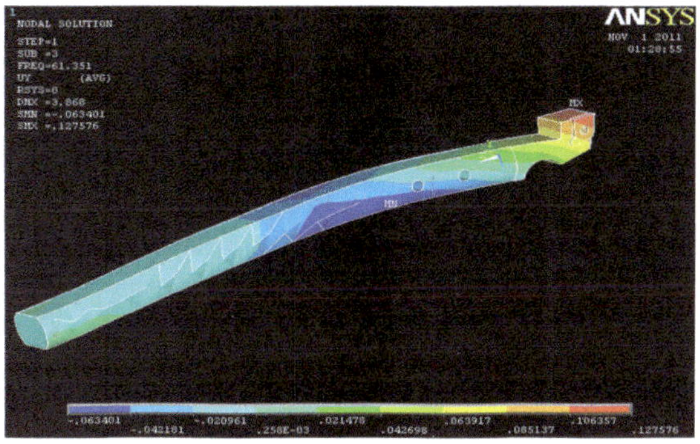

Figure 12 third mode shape of boring tool with polyamide impact damper

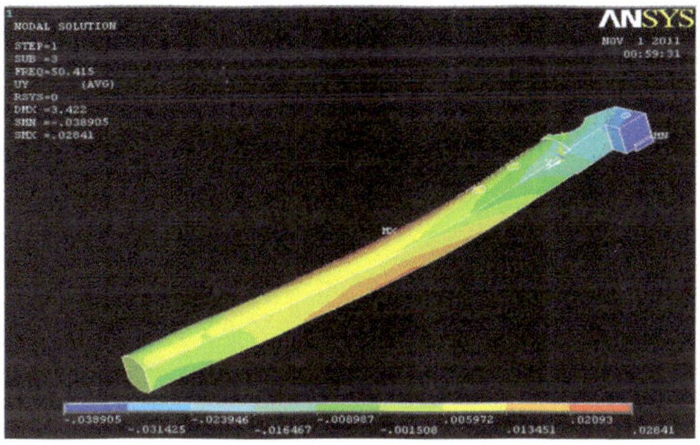

Table 1 Natural frequencies of the boring tool

Mode	thout damp (Hz)	Acrylate dai (Hz)	Epoxy dai (Hz)	polyamide dai (Hz)
1	8.7422	10.255	10.251	8.6554
2	10.569	13.188	13.183	10.744
3	51.116	61.383	61.351	50.415

From these results on the modal analysis, three modes of extraction have been carried out such mode shapes are vertical bending, horizontal bending, and twisting. The frequencies of modal analysis of boring tools with and without dampers are not coinciding with each other in different mode shapes of the boring tool because the impact dampers with high stiffness is introduced between the tool and the tool insert.

3.3 Harmonic response analysis

Harmonic response analysis is a technique used to determine the steady-state response of a linear structures response at several load that vary sinusoidal with time. The idea is to calculate the structure response at several frequencies and obtain graph of some response quality verse frequency. Peak response are then identified on the graph and stressed reviewed at those peak frequencies.

3.4 Results of harmonic analysis

Figure 13 harmonic analysis of boring tool without damper

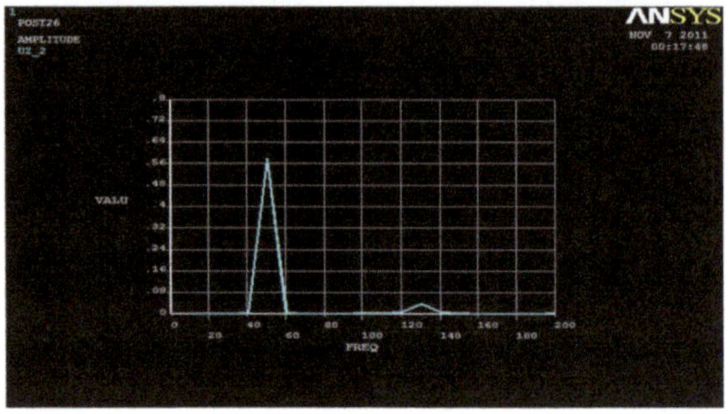

Figure 14 harmonic analysis of boring tool with acrylate damper

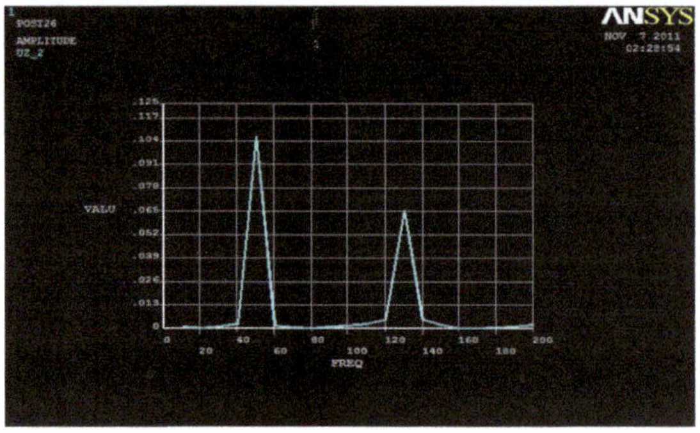

Figure 15 harmonic analysis of boring tool with epoxy damper

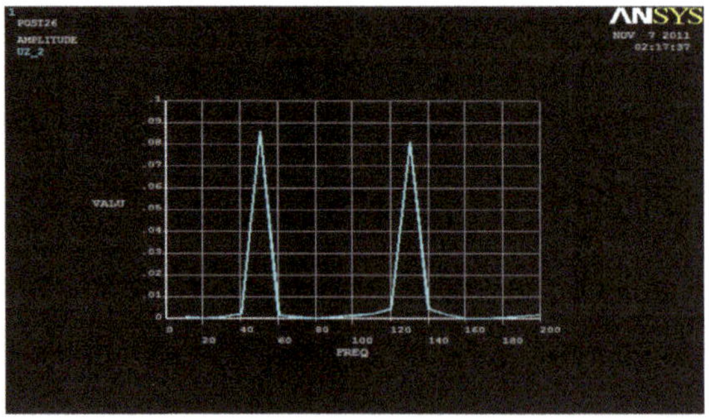

Figure 16 harmonic analysis of boring tool with polyamide damper

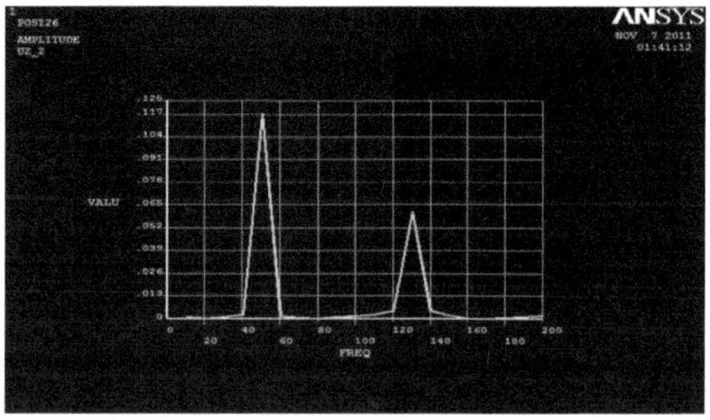

The results obtained from the harmonic analysis of boring tool with and with out damper plotted as below. The peak vale of the amplitude has been reduced with respect to the frequencies in this analysis. The results are shown from figure 13 To 16

Table 2 Comparison of peak value

Damping materials	Peak value(mm)
Without damper	0.60
Vith Acrylate impact damper	0.107
With Epoxy impact damper	0.088
ith polyamide impact dampe	0.117

3.5 Transient dynamic analysis

Transient dynamic analysis is a technique used to determine the dynamic response of structure under the action of any general time-dependent loads. You can use these types analysis to determine the time-varying displacement, strain, stress and forces in a structure as it response to any combination of static, transient, harmonic load. The time scale of the loading is such that the inertia or damping effect are consider important.

3.6 Results of transient dynamic analysys using adams

Figure 17 transient analysis of boring tool without damper

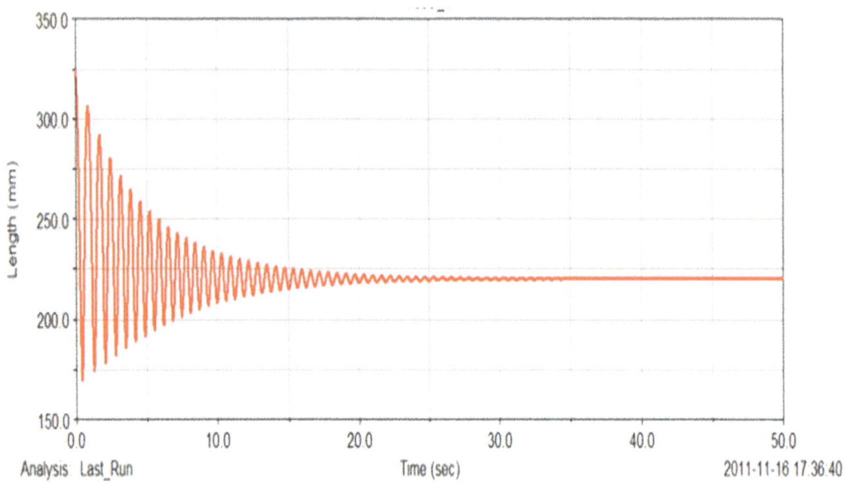

Figure 18 transient analysis of boring tool with acrylate impact damper

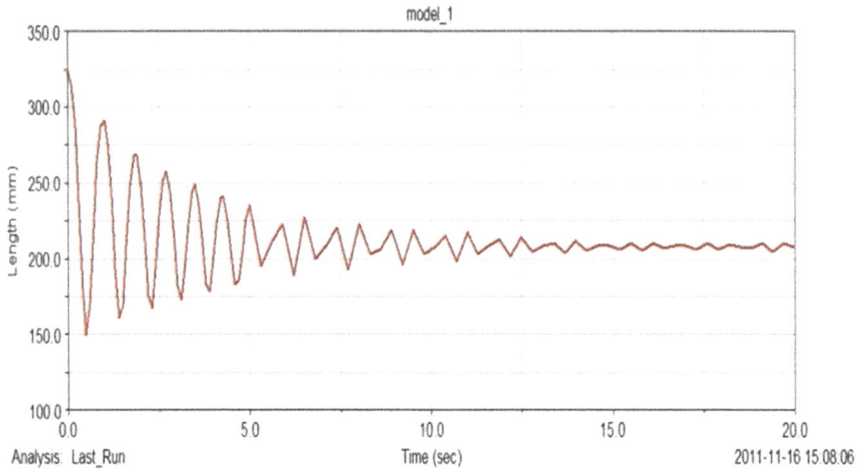

Figure 19 transient analysis of boring tool with epoxy impact damper

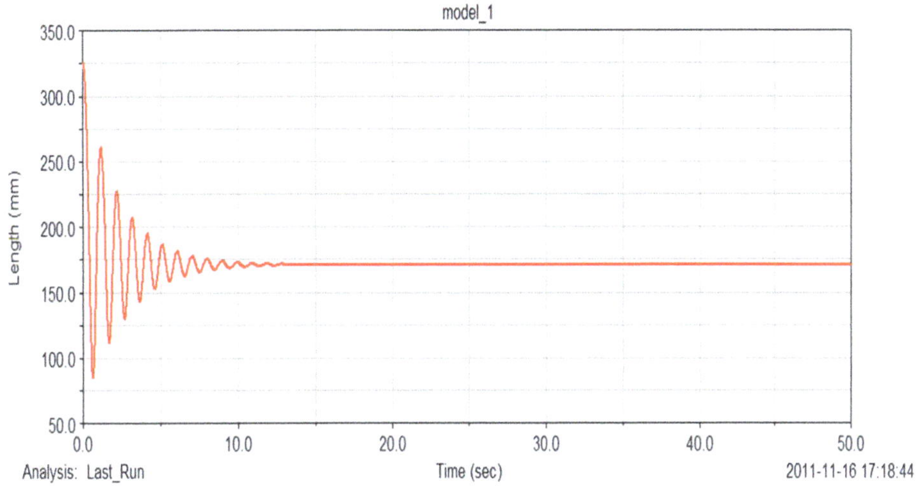

Figure 20 transient analysis of boring tool with polyamide impact damper

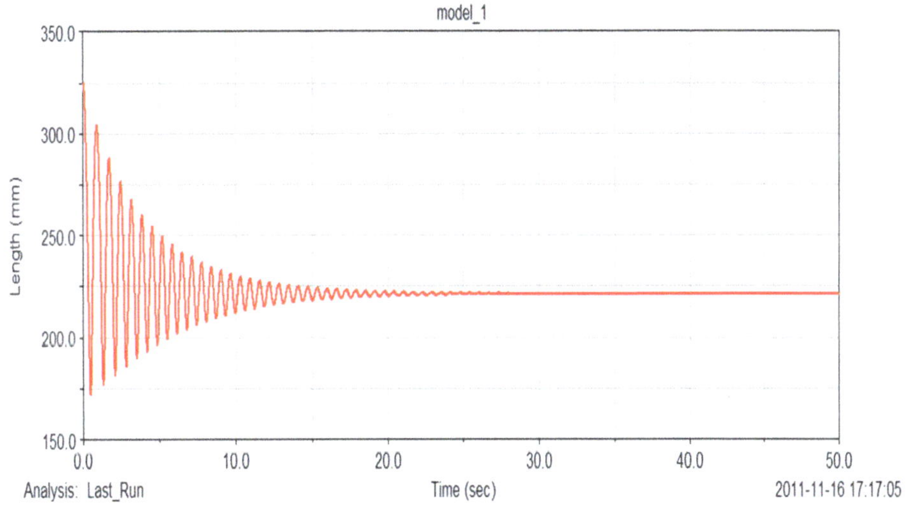

Table 3 Time to decay the amplitude of boring tool

Damping materials	Time taken to decay (sec)
Without damper	35
Vith acrylate impact damper	20
With Epoxy impact damper	12
ith polyamide impact dampe	25

The amplitude of the boring tool for without damping materials is reduced with epoxy damper which is very less reduction but the time taken to decay the amplitude is reduced considerable amount. Time taken to decay the amplitude of boring tool without damping materials starts from 35 seconds is reduced to 12 seconds for with epoxy.

3.7 Comparison of damping ratio

The amplitude values are taken from adams transient analysis results for calculations of damping ratio. From the transient analysis, we know the amplitude decay with respect to the time has been reduced with damped boring tools when compared with undamped boring tool.So the damping ratio can be calculated from the transient analysis by using logarithmic decrement which is ratio between the successive amplitudes(x1,x2).

Logarithmic decrement $\delta = \ln(x1/x2)$

Damping ratio $\xi = \delta/2\pi$

For boring tool without damper, the damping ratio is 0.0075. from the comparisons the damping ratio has been increased with damped boring tool when compared without damper. Epoxy damper only gives the significant improvement in damping ratio.

4. Conclusion

The conclusions drawn from the investigations are the results were obtained from modal analysis of boring tool with and without dampers. It is observed that the natural frequency of boring tool has been enhanced with damper.

From the harmonic analysis, it is observed that amplitude of the vibration has been reduced with damper. In that analysis with Epoxy damper the amplitude of vibration of boring tool has been reduced by 82.6% when compared with the boring tool without damper.

From the transient analysis it is observed that there is reduction in time taken to decay the amplitude of boring tool with dampers.

From the comparison of damping ratio of different damper the damper gives the increased damping ratio by 96.7% when compared with boring tool without damper.

Table 4 Comparison of damping ratio

Damping material	Without damper	With acrylate impact damper	With Epoxy impact damper	With poly-amide impact damper
Damping ratio	0.0075	0.1243	0.2308	0.0116

Hence the results of the finite element analysis of boring tool with epoxy impact damper shows that there is significant improvement in the dynamic behavior of structural analysis, when compared with other dampers.

5. References

1.D.D.L .Chung, 2001, " Review materials for vibration damping", journal of material science, vol 36

2.Emre ozlu erhan budak, 2007, "analytical modeling of chatter stability in turning and boring operations-part I: Model development, ASME, vol.129.

3.M.H. Migue´ lez, L.Rubio, J.A.Loya, J.Ferna´ndez-Sa´ez, 2010 "Improvement of chatter stability in boring operations with passive vibration absorbers" Department of Continuum Mechanics and Structural Analysis, University of Madrid, Spain.

4.Evita Edhi, Tetsutaro Hoshi 2001, "Stabilization of high frequency chatter vibration in fine boring by friction damper". Toyohashi University of Technology, Japan

5.Guillem Quintana, Joaquim Ciurana, 2011, "Chatter in machining processes", University of Girona, Spain

6.Jon Robert Pratt, 1997, "vibration control of chatter suppression with application of boring bar", Blacksburg, Virginia

7.Lonnie houck, tony l.schmitz, and k.scott smith, 2006," tunable holder for boring bars", university of Floride, USA

8.Dongki won, 2004, "Numerical analysis and parameter study of mechanical damper in machine tool". University of Florida, USA

Experimental Verification of Damping Coefficient by Half Power Band Width Method

Prof.Dr.S.H.Sawant

Professor, Mechanical Engg.Dept, Dr.J. J. Magdum college of Engineering,Jaysingpur, Kolhapur, India
sanjaysawant2010@gmail.com

Abstract

Modal analysis is a worldwide used methodology that allows fast and reliable identification of system dynamics in complex structures. In the last decades several methods have been developed in quest to improve accuracy of modal models extracted from test data and to enlarge the applicability of modal analysis in industrial context. Damping in composite materials is an important parameter affecting the dynamic behavior of structures, controlling the resonant and near-resonant vibration levels.

In this paper vibration characteristics of a cantilever beam made with different materials such as Steel, Spring Steel, Wood, Rubber, Fiber Reinforced Plastic (FRP) are studied experimentally.

Keywords: Fibre Reinforced Plastic (FRP), Half Power Band Width, Modal Analysis, Resonant.

1. Introduction

Damping properties are of significant importance in determining the dynamic response of structures, and accurate prediction of them at the design stage, especially in the case of light-weight structures [1]. Damping in composite materials is an important parameter affecting the dynamic behavior of structures, controlling the resonant and near-resonant vibration levels. For the solution of variety of noise and vibration problems, especially those associated with vibrations of structures made of sheet metal, surface damping treatments are often used [2]. Hence in order to design composite structure determination of damping coefficients is necessary.

2. Theoretical Background

Structures vibrate in special shapes called mode shapes when excited at their resonant frequencies. A mode shape is the characteristics deformation shape defined by relative amplitudes of the extreme positions of vibration of a system at a single natural frequency. The modal parameters are the natural frequencies, damping ratios and modal masses associated with each of the mode shapes.

Experimental modal analysis is a supportive tools to study the various vibration problems. In addition to verifying the theoretical results, it helps to study the actual dynamic behavior of a structure away from idealization. Modal analysis is the representation of the dynamic properties of an elastic structure in terms of its modes of vibration. The elastic structure tends to bend, twist and vibrate when subjected to an external dynamic force. The dynamic properties are also known as the modal parameters which are unique for the structure [3]. The modal parameters are:

Natural Frequency

Damping Factor

The damping in a system can be obtained from free vibration decay curve. Where the free vibration test is not practical, the damping may be obtained from the frequency response curve of forced vibration test. The damping ratio determination method comes originally from the electrical engineering domain and is called half power bandwidth method. On the response curve at the harmonic forced vibrations shown in Fig.2.1 one identifies the points A and B, corresponding to the maximum response, multiplie by $1/\sqrt{2}$ (in electrical engineering this value matches to a point where the electric power amplifier is at half of the maximum value).

The damping ratio is given by the equation [4]:

$$2\xi = \frac{\omega_2 - \omega_1}{\omega_n} \qquad (1)$$

$$C = \zeta^* C \qquad (2)$$

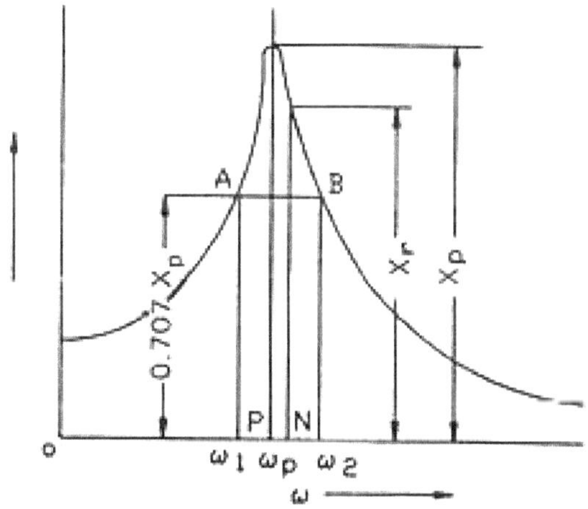

Fig. 1. Determination of Equivalent Viscous Damping from Resonant Amplitude.

Hence an accurate value of ξ and C can be found by construction and measurements.

3. Design and Experimentation

3.1Elements of Experimental Setup:

1. FFT (Fast Fourier Transform) Analyzer.
2. Magnetic Exciter
3. Impact Hammer
4. Accelerometer
5. Apparatus

Fig. 2 shows the developed experimental setup used for determination of damping coefficients and Fig.3 shows different types of composite beams used for determination of damping coefficient.

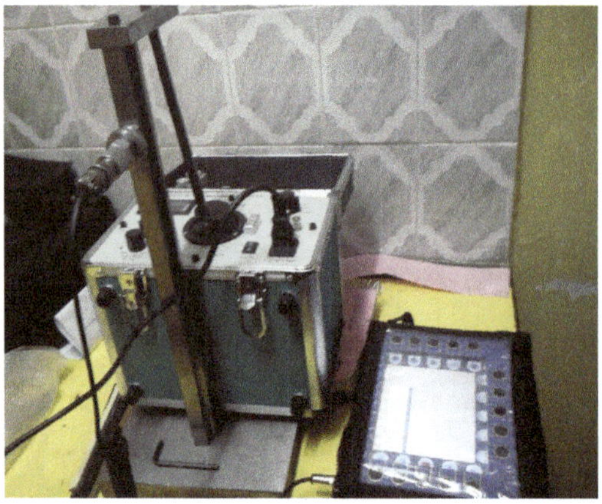

Fig.2 Experimental Setup

Fig. 3 Beams of Different Materials

The procedure used for the determination for damping coefficient by this method is given below

1. Fix the beam with the help of clamp fixed on the table.
2. Make proper connections of the vibscanner and accelerometers.
3. Place accelerometer at the free end of the cantilever beamt to measure the vibration response.
4. Struck the free end of a cantilever beam with a wooden mallet.
5. Record all the data obtained from the vibrating beam with the help of vibscanner as accelerometer is attached to it.
6. Repeat the whole experiment for different material by changing the parameters like length & thickness.
7. Record the whole set of data and then import it into the PC.

4. Results and Discussions

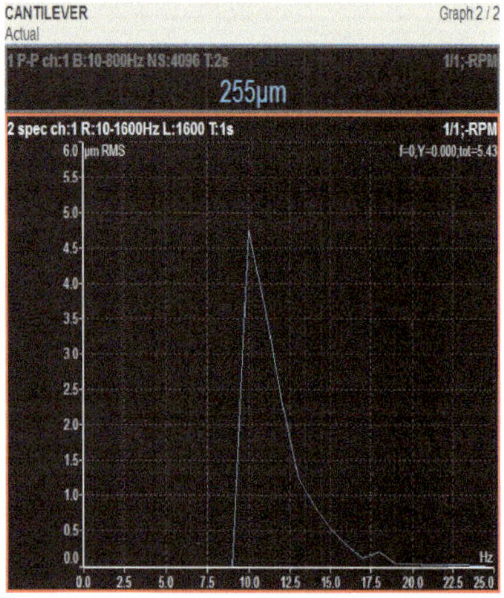

Graph 1Displacement Vs Frequency for Steel.

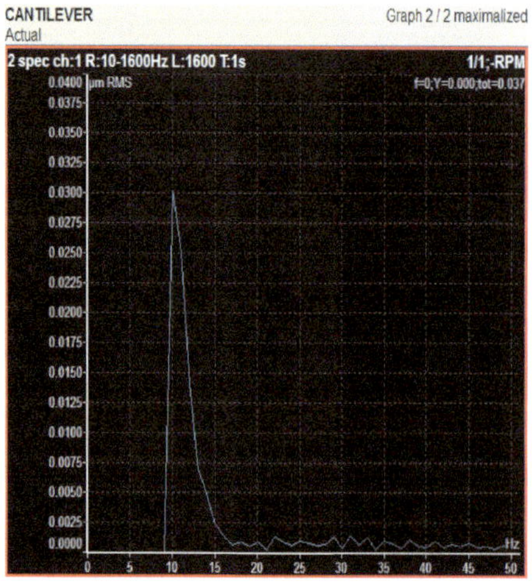

Graph 2 Displacement Vs Frequency for Steel on Steel.

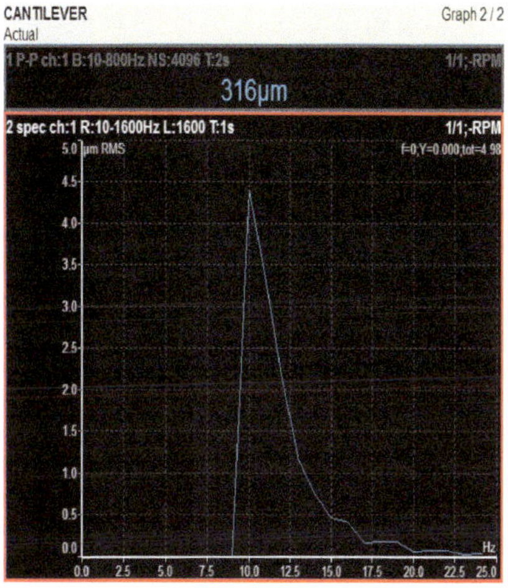

Graph 3 Displacement Vs Frequency for Wood on Steel.

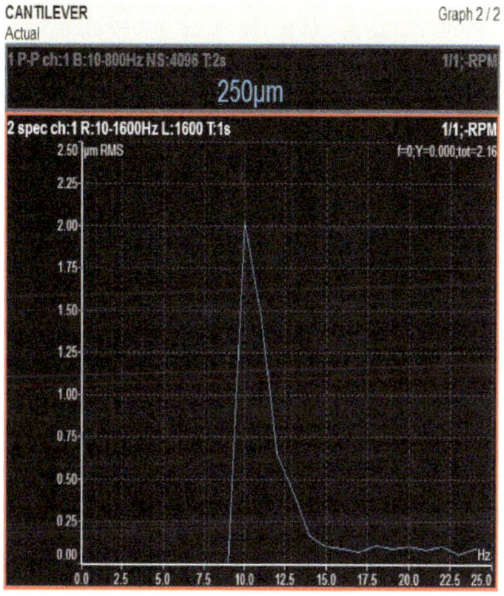

Graph 4 Displacement Vs Frequency for Rubber on Steel.

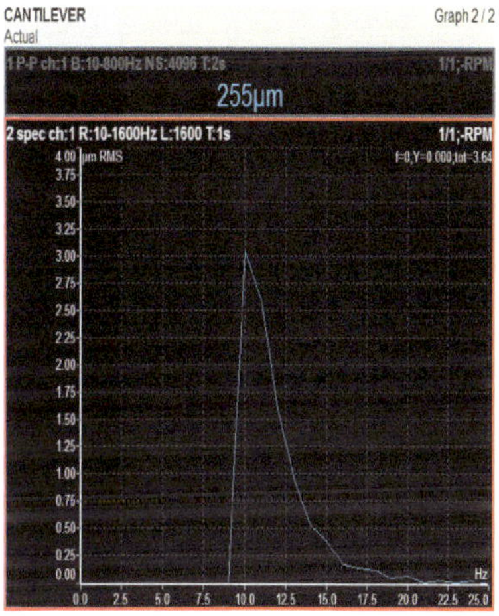

Graph 5 Displacement Vs Frequency Spring Steel on Steel

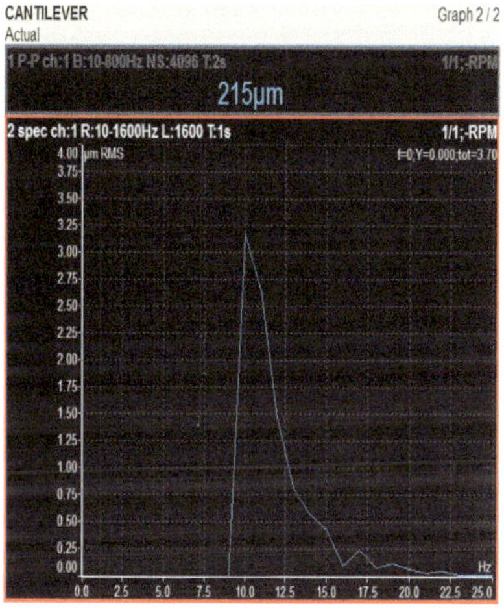

Graph 6 Displacement Vs Frequency for for FRP on Steel

The variation in Displacement with respect to Frequency is tabulated in Table 4.1

Table 4.1 Variation of Displacement with respect to Frequency for Steel Beam.

Sr. No.	Beam	Peak Amplitude (Xp)	Half Power Amplitude (X)
1	Steel	4.7083	3.3287
2	Steel on Steel Beam	0.03	0.2121
1	Wood on steel	4.3666	3.087
2	Rubber on steel	2	1.41
3	Spring steel on steel	2	2.16027
4	FRP on steel	3.1388	2.2191

The variation in damping coefficient with respect to c/s area is tabulated in Table 4.2

Table 4.2: Variation in Damping Coefficient with Respect to c/s Area

Sr. No.	Damping coefficient Ns/m		Percentage Variation
	Steel	Steel on Steel	
1	39.089	167.74	76.69 %

The variation in Equivalent Damping Coefficient of bonded beams with Respect to Steel on Steel is tabulated in Table 4.3.

Table 4.3 Variation in Equivalent Damping Coefficient of Bonded Beams with Respect to Steel on Steel

Sr. No.	Beam	Damping Coefficient N-s/m	Percentage Variation
1	Wood on Steel	108.99	35.024 %
2	Rubber on Steel	253.61	33.859 %
3	Spring steel on Steel	171.83	2.83 %
4	FRP on Steel	197.25	14.96 %

From Table 4.3 it is observed that

The Percentage Variation of Damping Coefficient of Wood on Steel with respect to Steel on Steel is 35.024 %

The Percentage Variation of Damping Coefficient of Rubber on Steel with respect to Steel on Steel is 33.859 %

The Percentage Variation of Damping Coefficient of Spring Steel on Steel with respect to Steel on Steel is 2.83 %

The Percentage Variation of Damping Coefficient of FRP on Steel with respect to Steel on Steel is 14.96 %

From above result it can be concluded that,
Equivalent damping behavior of spring steel is nearly equal to that of steel

As the variation of equivalent damping coefficient is maximum for wood on steel, wood has more damping property as compared to rubber and FRP.

From Table 4.1 it is observed that when c/s of beam doubled, the damping coefficient for steel varies by 76.69%.

The variation in equivalent damping coefficient of bonded beams with respect to steel on steel is tabulated in Table 4.3.

5. Conclusion

Half Power Band Width method is useful for determination of damping coefficient in case of solid damping and equivalent damping coefficient in case of solid and slip damping of various materials. The damping coefficient of Fiber Reinforced Plastic (FRP) is more than Spring Steel (Spring Steel is a Steel used to make leaf spring in a 3 wheeler automobile, which are used to dampen the vibrations generated in an automobile) So we can make the leaf spring of a 3 wheeler automobile with Fiber Reinforced Plastic (FRP).

References

1. John Butterworth, Jin Hee Lee, Barry Davidson, "Experimental Determination of Modal Damping from Full Scale Testing", 13th World Conference on Earthquake Engineering, Vancouver, B.C., Canada, August 1-6, 2004.
2. Radu Cruciat, Cristian Ghindea, "Experimental Determination of Dynamic Characteristics of Structures", Mathematical Modelling in Civil Engineering, No.4 – 2012, pp. 51-59.
3. Muszynska A., "Half-power Bandwidth method for the Evaluation of Synchronous and Non-Synchronous Quadrature Stiffness", Orbit Magazine, June 1994, pp. 7-10.
4. S. S. Rao, Mechanical Vibrations, 3rd Edition, Addison-Wesley Publishing Company (1995).

5

ENERGY MODELLING ON CYLINDRICAL TOOL GEOMETRY OF ALUMINUM ALLOYS USING "C" PROGRAM

D.Muruganandam[1*], D.Raguraman[1] , and S.Briolin Nayagam
[1]Assistant Professor, Dept. of production engineering, Sairam Engineering college, Chennai, India
[2]Assistant Professor, Dept. of Instrumentation and Control Engineering, production engineering, Sairam Engineering college, Chennai, India
[3] Student ,Department of Computer Science, Jeppiaar Institute of Technology, India

*Corresponding author

Abstract

Friction stir welding is competing technology used in many engineering applications. Many researchers showed their interest on developing numerical modelling on friction stri welding. The friction stir welding process is dependent mainly by controlling the major weld parameters. The major weld parameters are rotational speed, transverse speed and axial load.In this study,a energy model for taper tool geometry is developed numerically using these weld parameters which estimates the energy generated due to friction and plastic deformation. The generated energy model is programmed using "C" program. By using the program we can reduce the time of calculating the effective energy and maximum temperature using the developed numerical equations.

Keywords: Friction Stir Welding, C-Program, Energy, Temperature model.

1. Introduction

Friction Stir Welding (FSW) is a process of joining two similar and dis-similar metals by the non-consumable tool. The major weld parameters involved are rotational speed, transverse speed and axial load. The joining of two metals is made by bonding the two similar and dis-similar metals. The creation of bond develops energy in between two metals by the application of direct heat. The major sources for direct heat is from friction and chemical reaction between group of atoms while colliding and cohesion. The atomic mechanism and chemical reactions are neglected for the simplicity on the creation of numerical equations.

Friction-stir welding (FSW) enables welding similar and dissimilar metals such as high-strength aluminum alloys. Friction-stir welding was developed and patented by The Welding Institute (TWI) in 1991 [1].

Friction-stir welding is carried out using a rotating tool that is attached to a shoulder piece and the whole unit is translating over the line of welding. The rotation and translation of the pin within and on top of the line of welding generates heat, which is used to weld the workpieces. Heat is generated due to plastic deformation of the workpiece and the effect of the friction between the surfaces of the tool and the workpiece [2,3]. Figure 1 is a schematic of the friction stir welding process.

Y. K. Yousif et al., shows the possibility of the use of neural networks for the calculation of the mechanical properties of welded Al plates using FSW. The Levenberg – Marquardt algorithm shows better performance than gradient descent[4].

M Selvaraj et al., developed a temperature dependent slip factor based on three-dimensional thermal model for friction stir welding with capability of predicting thermal cycle, temperature distribution, power required and heat generated per unit length for friction stir welding of stainless steel[5].

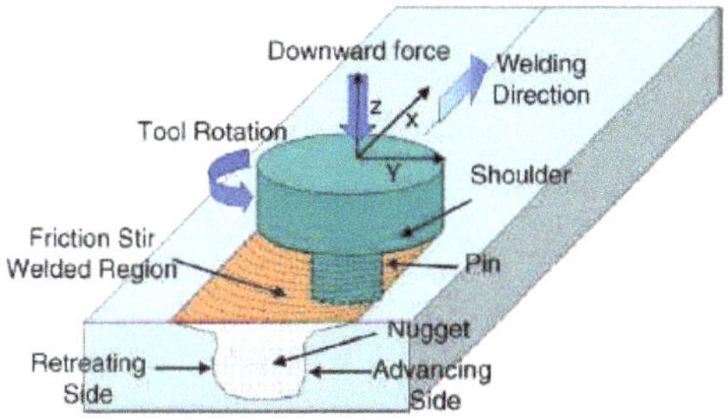

Figure 1. Schematic illustration of Friction Stir Welding

Samir A. Emam et al., created a simple model that estimates the energy generated in friction-stir welding is presented. The model accounts for the heat generated due to friction between the weld tool and the surface of the workpiece and heat generated due to plastic deformation. The heat due to plastic deformation is found to have a significant effect on the resulting temperature especially at low-energy levels[6].

Darko M. VELJI et al., is developed a 3-D finite element model in the commercial code ABAQUS/Explicit using the arbitrary Lagrangian-Eulerian formulation. Through the numerical modelling he found that the amounts of heat generated by friction and plastic deformation suddenly increase after contact between the workpiece and the tool shoulder is established. Also found that the heat generated by friction between the workpiece and the tool can account for the largest percentage of the generated heat[7].

S. Sulaiman et al., studied a heat input equation for FSW and found the main factors in heat generation and heat input issues as rotational and traveling speeds. If the heat input is too low, suitable plastic conditions will be prevented and will cause voids during welding and eventually, in extreme cases, the tool may break[8].

P. Edwards et al., proposed a energy model, energy input values below 1.4 kW lead to 'cold' welding conditions, while values greater than 2.5 kW produce 'hot' welding conditions. Optimal processing conditions were found to have energy input values between 1.4 and 2.5 kW for the given material thickness and tooling configuration[9].

Yuh J. Chao et al., studied the numercal equation that determines the heat flux from the friction which is applied at the lower end of the tool . Only about 5% of the heat generated by the friction process flows to the tool and the rest flows to the workpiece. The ''heat efficiency'' in FSW is thus 95%, which is very high relative to the traditional fusion welding where the heat efficiency is typically 60 to 80%[10].

With the proposed analytical approach of Vijay Shivaji Gadakh et al., one can directly see the peak temperature for respective taper probe angle under given process conditions which will be helpful for predicting the mechanical properties for that Al alloy and hence elimination of post weld testing cost and time[11].

Binnur Goren Kiral et al., used the moving heatb resource model and found the maximum temperature near the weld increases as the tool holding time and rotational speed are increased[12].

Muruganandam et al. used Tauguchi technique for the identification of rotational speed for higher strength using cylindrical and taper model[13]. In the present work, a simple energy-based model for the friction stir weld is proposed for taper pin. The model aims at estimating the heat generated due to plastic deformation within the work pieces and friction between the tool surfaces and the work pieces. This generated energy and the associated maximum temperature are compared to the results available in the literature to verify the proposed model.

2. Proposed Energy Model

Previous studies [14] assume that heat generated due to friction of the pin shoulder on the workpiece surface is dominant and the heat generated due to the plastic deformation within the workpiece and the friction of the pin of the material is negligible. However, other authors e.g. Heurtier et al. [14] and Hamilton et al. [15] consider the heat generated from both the friction of the pin shoulder and plastic flow. In fact, the energy due to plastic deformation and friction of the shoulder with the surface of the workpiece are related and competing each other. As the heat generated by the shoulder is low, the flow stress is higher and hence the resulting plastic deformation energy increases. On the other hand, as the heat generated by the shoulder is high, the flow stress reduces and as a result the plastic strain contribution decreases. In this work, heat is modeled to be generated by the friction of the shoulder and plastic deformation.

Figure 2 presents a cylindrical geometry of the friction-stir welding tool. It is assumed that the tool rotates with an angular speed of ω and transversely translates along the line of welding with a speed of vo . The tool is acted upon by a compressive force F.

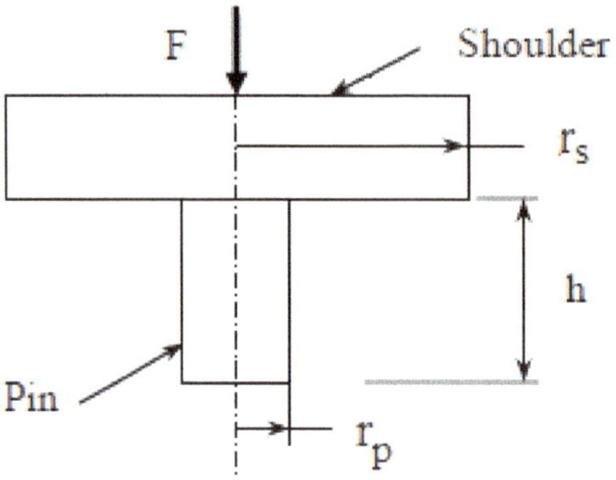

Figure 2. Geometry of the FSW tool

Following, Hamilton et al. [15], the energy generated per unit length of the weld E is given by

$$Ef = Pf / vo \qquad (1)$$

where P is the total power generated by friction. Assuming that the total torque due to friction of the pin, shoulder, and pin circumference with the workpiece surfaces is Tf , the frictional power is then given by

$$Pf = Tf\ \omega \qquad (2)$$

where ω is the pin angular speed. To find an expression for the total frictional torque Tf , we let

$$Tf = Ts + Tp \qquad (3)$$

where Ts is the torque generated by the shoulder and Tp is torque generated by the pin. Assuming that a uniform shear stress τ occurs during welding, we obtain

$$\int_{r_p}^{r_s} \tau(2\pi r) r\,dr + \int_0^{r_p} \tau(2\pi r) r\,dr + 2\pi r_p^{\,2} h \tau \qquad (4)$$

where rs is the radius of the shoulder, r_p is the radius of the pin, and h is the height of the pin, as shown in Fig. 2. Equation 3 reduces to

$$T_f = 2\pi\tau r_s^2 [\frac{1}{3} r_s + \frac{r_p^2}{r_s^2} h] \quad (5)$$

Assuming a coefficient of friction is μ , the total friction force Ff due to the compressive force F is given by

Ff = μF (6)

Noting that $\pi \tau r_s^2$ is the total friction force, Eq. 3 yields

$$T_f = 2\mu F [\frac{1}{3} r_s + \frac{r_p^2}{r_s^2} h] \quad (7)$$

Substituting Eq. 5 into Eqs. 1 and 2, we obtain

$$E_f = 2\mu F [\frac{1}{3} r_s + \frac{r_p^2}{r_s^2} h] \frac{\omega}{v_o} \quad (8)$$

Equation (8) defines the energy per unit length of the weld due to friction between the tool and the workpiece. For a given tool geometry, tool speed, and workpiece material, this energy can be easily identified. As can be noted from Eqn. (1), the power generated due to friction is given by

$$P_f = 2\mu F [\frac{1}{3} r_s + \frac{r_p^2}{r_s^2} h]\, \omega \quad (9)$$

The other source of heat is due to the plastic deformation within the workpiece. Provided that the plastic deformation within the workpiece is totally transformed into heat, the heat generated due to plastic deformation per unit weld length can be expressed as follows:

Ep =σε V (10)

where σ and ε are the stress and strain, respectively, V = is the volume per unit length of the base material. The stress σ is given by

$$\sigma = K\, \varepsilon^n \exp\left(\frac{mQ}{R_G T}\right) \quad (11)$$

where K is the strength coefficient, n is the strain hardening exponent, m is the strain rate sensitivity, Q is the apparent activation energy, RG is a constant equals 8.32 J $mol^{-1}K^{-1}$, and T is the absolute temperature. As a result, the energy generated due to plastic deformation per unit length of the weld is expressed as follows:

$$E = K\, \varepsilon^{n+1}\, blh \exp\left(\frac{mQ}{R_G T}\right) \quad (12)$$

The stress-strain relationship in the room temperature for AA2024, AA7075 and AA 6061 are shown in this figure 3.

Because the temperature is already unknown, the accurate calculation of the plastic strain energy needs an iterative process. Since the effect of the energy due to plastic deformation is much smaller than that due to friction, a simple model was proposed as follows:

$$E_p = \sigma_e \varepsilon_e\, blh \quad (13)$$

where σe is the equivalent (effective) stress and εe is the effective strain. Using the finite element method, Heurtier et al. [3] found that the effective strain is around six. In this case, the effective stress is assumed to be constant, and hence the area under the curve becomes a rectangle. The power generated due to the plastic deformation is given by

$$Pp = \sigma e \varepsilon e^{blh} \, vo \qquad (14)$$

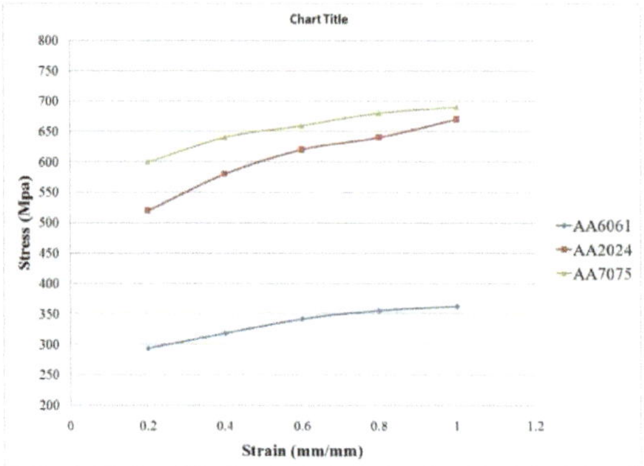

Figure 3. True-stress true-strain curve for AA 2024,AA7075 and AA6061.

The total energy generated per unit length of the wild is the sum of the energy generated due to friction between the tool and the workpiece surface and the plastic deformation within the workpiece.

The term effective energy is introduced to take into account the case where the height of the FSW tool pin h is different from the thickness of the workpiece t. The effective energy is defined as follows:

$$E_{eff} = \frac{h}{t} E = \beta E \qquad (15)$$

where β is the coefficient of transfer efficiency.

3. Result and discussion

The energy model proposed in the previous section accounts for both the frictional heating and the heating results from the plastic deformation. In order to validate the model, the total energy presented in taper model is adopted into the empirical formula developed by Hamilton et al. The formula is obtained from the experimental results available in the literature. The empirical formula is given by

$$\frac{T_{max}}{T_s} = 1.56 \; X \; 10^{-4} \; E_{eff+0.54} \qquad (16)$$

where Tmax is the maximum temperature generated within the weld, Ts is the solidus temperature in Kelvin, Eeff is the effective energy generated per unit length of weld in J/mm as given in Eq. (18). The non dimensionlized maximum temperature is obtained so that the deduced formula can serve for the AA6000 and AA7000 series considered in the study.

A good agreement is obtained except in the low-energy level region. The reason is that Hamilton et al. neglected the energy generated by the plastic deformation, which is significant when the frictional heating is low. In the present taper model, the heating due to plastic deformation has been taken into consideration through the scaling factor as given by Eq. (15). To validate this taper model, we find the total energy generated for given welding parameters and solved for the maximum temperature. The obtained maximum temperature is then compared to both the cylindrical model and the taper model. It is found out that considering the heat due to plastic deformation enabled this model to better fits the experimental results at all energy levels[16].

Three aluminum alloys are considered using different welding parameters such as tool geometry and welding speed. Table 1 presents the three alloys with the material characteristics and tool geometry used for each case.

Welding parameters including the tool rotational speed, tool translational speed, and the acting normal force are given in Table 2.

A comparison between the heat energy obtained by cylindrical model and the heat energy obtained using the proposed taper model in this study is given in Table 3. Investigating these results, one notes that at low-energy levels, the model of plastic deformation estimates the generated energy. This is expected since the plastic deformation is significant in this region, which is ignored in the model at higher energy level due to scaling factor close to zero.

Table 1.Material characteristics and tool geometry of the aluminum Alloys used

Aluminium alloy		6061 - T6	2024 - T6	7075 - T6
Ref.		L1	L2	L3
t (mm)		6	6	6
ρ (kg/m3)		2700	2780	2810
Cp (J/Kg K)		896	875	960
K (W/m K)		167	177	130
α (x 10-6 cm/cm/C)		23.4	22.8	23.2
Ts (K)		855	910	748
Tool geometry	ri (mm)	12.0	12.0	12.0
	ro(mm)	9.5	9.5	9.5
	h (mm)	6.0	6.0	6.0

On the other hand, the energy predicted using present taper model seems much greater than the cylindrical model, which reflects the enhancement of the model by including the plastic deformation. Moreover, in the high energy level, up to 2000 J/mm, the results are very close.

The graphs (Fig.5 and Fig.6) are plotted between the rotational speed and effective energy for both the cylindrical and taper model. By data interpretation it is clear that the change in tool geometry from cylindrical shape to taper shape increases the effective energy. However, the model is mainly based on geometry so the change in material is not much effected the effective energy.

Table 2:Welding rotational and translational speeds and the applied normal force.

Alloy	Case	R.P.M.	v_o (mm/s)	F (KN)
AA 6061 – T6	1	200	1.5	10
	2	400	1.5	10
	3	600	1.5	10
	4	800	1.5	10
	5	1000	1.5	10
	6	1200	1.5	10
AA 2024 – T6	7	200	1.5	10
	8	400	1.5	10
	9	600	1.5	10
	10	800	1.5	10
	11	1000	1.5	10
	12	1200	1.5	10
AA 7075 – T6	13	200	1.5	10
	14	400	1.5	10
	15	600	1.5	10
	16	800	1.5	10
	17	1000	1.5	10
	18	1200	1.5	10

"C" Program for cylidrical model:

```
#include<stdio.h>
#include<conio.h>
void main()
{
int F,w,ro,ri,h,s;
float v,Hs,i,j,k,l,m,n,o,p,q,r,t;
printf("Enter value of Axial Load(4 KN to 12 KN):");
scanf("%d",&F);
printf("Enter value of Transverse speed(1.5 to 5.5 mm/min):");
scanf("%f",&v);
```

```
printf("Enter value of Rotational speed(200 r.p.m to 1200 r.p.m):");
scanf("%d",&w);
printf("Enter value of Shoulder diameter - ro(10 mm to 20 mm):");
scanf("%d",&ro);
printf("Enter value of Pin diameter(6 mm to 12 mm):");
scanf("%d",&ri);
printf("Enter value of pin height(6 mm to 10 mm):");
scanf("%d",&h);
printf("Enter value of solidus temperature of metal as integer values(E.g.AA6061->167,AA2024-
>177,AA7075->130):");
scanf("%d",&s);
Hs=(ro/3);
i=ri*ri;
j=ro*ro;
k=(i/j)*h;
l=k+Hs;
m=2*0.5*F;
n=((2*(3.14/60))*w);
o=n/v;
p=m*l*o;
printf("The energy generated is: %f",p);
q=(5.7/6)*p;
r=0.54+((1.564/10000)*q);
t=s*r;
printf("\n the r value is:%f",r);
printf("\n The maximum temperature generated is:%f",t);
getch();
clrscr();
}
```

Output: Output of the program is shown in Figure.4

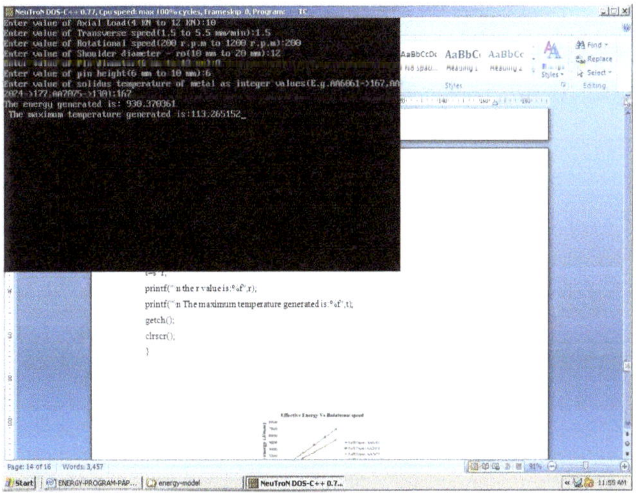

Figure 4. Calculated values of energy generated and maximum temperture from the "C" program for the given weld parameters

Table 3. Energy and maximum temperature obtained using taper Model and cylindrical model[16]

Case #	E (J/mm) – Cylindrical model	E (J/mm) – Taper model	Tmax – cylindrical model	Tmax – Taper model
1	975	1574	114	129
2	1881	2467	137	150
3	2824	3703	160	181
4	3760	4932	183	211
5	4702	6168	207	241
6	5643	7401	230	271
7	988	1587	121	137
8	1883	2469	145	159
9	2826	3705	170	191
10	3772	4944	195	223
11	4716	6181	220	255
12	5659	7418	244	287
13	1019	1618	90	101
14	1894	2480	107	117
15	2835	3714	125	141
16	3778	4950	143	164
17	4718	6184	161	188
18	5659	7417	179	211

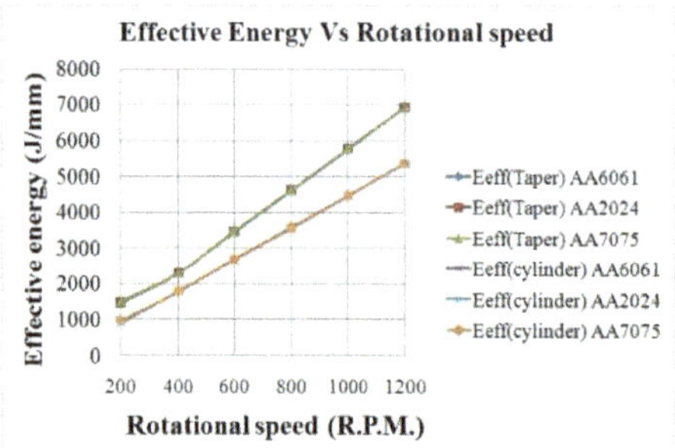

Figure 5. Variation of the effective energy with rotational speed for cylindrical and taper model

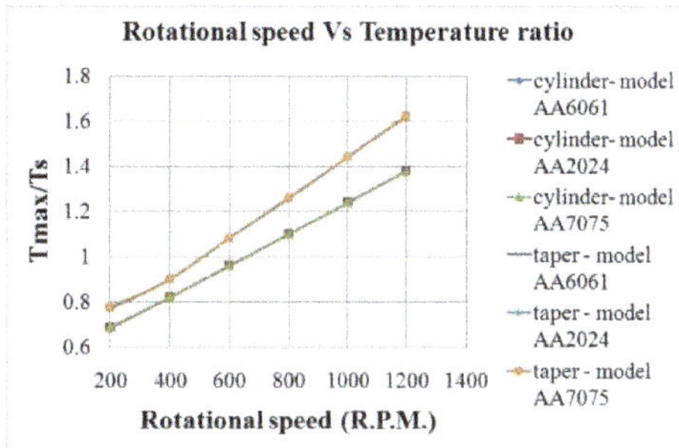

Figure 6. Variation of the temperature ratio with rotational speed for cylindrical and taper model

The present work is focussed on writing "C" program for cylindrical model. It is evident from the manual calculations that scaling factor consideration have less effect except the low energy levels. So it is neglected in "C' programming and the coding is written for the identification of energy generated and maximum temperature for the known parameters.

4. Conclusions

In this study, a simple cylindrical model that estimates the energy and maximum temperature generated in friction-stir welding is identified using "C" programing. The usage of "C" program is minimised to calculate the number of equation substitution and solving.

The model accounts for the heat generated due to friction between the weld tool and the surface of the workpiece and heat generated due to plastic deformation. However the scaling factor is neglected as it have the effects only at minor energy levels. This model can be enhanced by considering an accurate model for the plastic deformation that will automatically predicts the heat generation due to plastic deformation at different energy levels. An accurate thermal model that simulates the heat transfer within and around the weld tool could be used.

5. References

[1] W. M Thomas, "*Friction Stir Butt Welding*", International Patent Application PCT/GB92, Patent Application GB125978.8, 1991.

[2] P. Heurtier, M. J. Jones, C. Desrayaud, J. H. Driver, F. Fontheillet, and D. Allehaux, "*Mechanical and Thermal Modeling of Friction Stir Welding*", Journal of Materials Processing Technology, Vol. 171, pp. 348-357, 2006.

[3] R. S. Mishra, and Z. Y. Ma, "*Friction Stir Welding and Processing*", Materials Sciences and Engineering, R 50, pp. 1-78, 2005.

[4] Y. K. Yousif, K. M. Daws , B. I. Kazem, "*Prediction of Friction Stir Welding Characteristic Using Neural Network*", Jordan Journal of Mechanical and Industrial Engineering, Volume 2, Number 3, 151 – 155, 2008.

[5] M.Selvaraj, *"A temperature dependent slip factor based thermal model for friction stir welding of stainless steel"*, Indian Academy of Sciences, Vol. 38, Part 6, pp. 1393–1405, 2013.

[6] Samir A. Emam, Ali El Domiaty, *"A Refined Energy-Based Model for Friction-Stir Welding"*, World Academy of Science, Engineering and Technology, Vol:3, 05-23, 2009.

[7] Darko M. Velji, Marko P. Rakin, Milenko M. Perovic,Bojan I. Medjo ,Zoran J. Radakovi,Petar M. Todorovi, and Mirko N. Pavisi, *"Heat Generation during Plunge Stage in Friction Stir Welding"*, Thermal Science, Vol. 17, No. 2, pp. 489-496,2013.

[8] S. Sulaiman, S. Emamian, M. N. Sheikholeslam, and M. Mehrpouya, *"Review of the Effects of Friction Stir Welding Speed on Stainless Steel Type 304L"*, International Journal of Materials, Mechanics and Manufacturing, Vol. 1, No. 185-87, 2013.

[9] P. Edwards and M. Ramulu, *"Effect of process conditions on superplastic forming behaviour in Ti–6Al–4V friction stir welds"*, Science and Technology of Welding and Joining, VOL 14, NO 7,669-680,2009.

[10] Yuh J. Chao,X. Qi,W. Tang, *"Heat Transfer in Friction Stir Welding -Experimental and Numerical Studies"*, Transactions of the ASME, Vol.125, 138-145, 2003.
[11] Vijay Shivaji Gadakh, Kumar Adepu, *"Heat generation model for taper cylindrical pin profile in FSW"*, J. Mater. Res. Technol. 02:370-375, 2013.
[12] Binnur Goren Kiral, Mustafa Tabanoglu, H. Tarik Serindag, *" Finite Element Modeling Of Friction Stir Welding In Aluminum Alloys Joint"*, Mathematical and Computational Applications, Vol. 18, No. 2, pp.122-131, 2013.
[13] D.Muruganandam, D.Raguraman, and Sushil Lal Das, *"Analysis of mechanical strength of dissimilar aluminium alloys AA7075 and AA2024 in friction stir welding using Taguchi's technique"*, International Journal of Research in Aeronautical and Mechanical Engineering, Vol.1,Issue 6, pp. 13-22, 2013.
[14] P. Heurtier, M. J. Jones, C. Desrayaud, J. H. Driver, F. Fontheillet, and D. Allehaux, *"Mechanical and Thermal Modeling of Friction Stir Welding"*, Journal of Materials Processing Technology, Vol. 171, pp. 348-357, 2006.
[15] Hamilton, S. Dymek, and A. Sommers, *"A Thermal Model of Friction Stir Welding in Aluminum Alloys"*, International Journal of Machine Tools and Manufacture, Vol. 48, pp. 1120-1130, 2008.
[16] D.Muruganandam, D.Raguraman, B.Senthilkumar and Susil Lal Das, *"FSW of Aluminium alloys – Energy model for taper tool geometry"*, Elixir Mech. Engg. 65, 19864-19872, 2013.

6

Friction Stir Welding

[1]Karan Singla, [2]Dr. Deepak Bhardwaj, [3]Vikas Burman

[1]Department of Mechanical Engineering MIET, Shahabad India, Email: karan.singla1@gmail.com
[2]Department of Mechanical Engineering, BITS, Bhiwani India
[3]Department of Mechanical Engineering , RPIIT, Karnal India

Abstract

Welding is a fabrication process used to join materials, usually metals or thermoplastics, together. During welding, the work pieces to be joined are melted at the joining interface and usually a filler material is added to form a weld pool of molten material that solidifies to become a strong joint. In contrast, Soldering and Brazing do not involve melting the work piece but rather a lower melting point material is melted between the work pieces to bond them together. Friction Stir Welding (FSW) was invented by Wayne Thomas at TWI (The Welding Institute), and the first patent applications were filed in the UK in December 1991. Initially, the process was regarded as a "laboratory" curiosity, but it soon became clear that FSW offers numerous benefits in the fabrication of aluminum products. Friction Stir Welding is a solid-state process, which means that the objects are joined without reaching melting point. This opens up whole new areas in welding technology. Using FSW, rapid and high quality welds of 2xxx and 7xxx series alloys, traditionally considered un-weld able, are now possible.

Keywords: Friction; Welding; Aluminum; fabrication

1. Introduction

1.1 Literature survey

Following literature survey has been summarized here under

Ericsson and Sandstrom (2002) investigated that the fatigue strength of friction stir (FS) welds is influenced by the welding speed, and also compare the fatigue results with results for conventional arc-welding methods: MIG-pulse and TIG. The Al–Mg–Si alloy 6082 was FS welded in the T6 and T4 temper conditions, and MIG-pulse and TIG welded in T6. The T4- welded material was subjected to a post-weld ageing treatment. According to the results, welding speed in the tested range, representing low and high commercial welding speed, has no major influence on the mechanical and fatigue properties of the FS welds. At a significantly lower welding speed, however, the fatigue performance was improved possibly due to the increased amount of heat supplied to the weld per unit length.

The MIG-pulse and TIG welds showed lower static and dynamic strength than the FS welds. This is in accordance with previous comparative examinations in the literature on the fatigue strength of fusion (MIG) and FS welds. The TIG welds had better fatigue performance than the MIG pulse welds. The softening of the alloy around the weld line has been modelled. Using a model without adjustable parameters, a fair description of the hardness profiles across the weld as a function of welding speed was obtained. The softening in front of the Friction Stir Welding tool was also estimated. At the low and high welding speeds a full and partial softening is predicted, respectively.

Liu et al. (2003) in their research paper discussed the friction stir weld ability of the 2017-T351 aluminium alloy and determine optimum welding parameters, the relations between welding parameters and tensile properties of the joints. Researchers found that the tensile properties and fracture locations of the joints are significantly affected by the welding process parameters. When the optimum revolutionary pitch is 0.07 mm/rev corresponding to the rotation speed of 1500 rpm and the welding speed of 100 mm/min, the maximum ultimate strength of the joints is equivalent to 82% that of the base material. Though the voids-free joints are fractured near or at the interface between the weld nugget and the thermo-mechanically affected zone (TMAZ) on the advancing side, the fracture occurs at the weld centre when the void defects exist in the joints.
Kovacevic (2003) In their research friction stir welding (FSW) is a relatively new welding process that may have significant advantages compared to the fusion processes as follow: joining of conventionally non-fusion weld able alloys, reduced distortion and improved mechanical properties of weld able alloys joints due to the pure solid-state joining of metals. In this paper, a three-dimensional model based on finite element analysis is used to study the thermal history and thermo mechanical process in the butt-welding of aluminium alloy 6061-T6.

Huseyin Uzun et al. (2004) investigated that the joining of dissimilar Al 6013-T4 alloy and X5CrNi18-10 stainless steel was carried out using friction stir welding (FSR) technique. The microstructure, hardness and fatigue properties of fiction stir welded 6013 aluminium alloy to stainless steel have been investigated. Optical microscopy was used to characterize the microstructures of the weld nugget, the heat affected zone (HAZ), thermo-mechanical affected zone (TMAZ) and the base materials.

Cavaliere et al. (2005) investigated the mechanical and micro structural properties of dissimilar 2024 and 7075 aluminium sheets joined by friction stir welding (FSW). The two sheets, aligned with perpendicular rolling directions, have been successfully welded; successively, the welded sheets have been tested under tension at room temperature in order to analyze the mechanical response with respect to the parent materials.
Kovacevic (2005) In their research thermo-mechanical simulation of friction stir welding can predict the transient temperature field, active stresses developed, forces in all the three dimensions and may be extended to determine the residual stress. The thermal stresses constitute a major portion of the total stress developed during the process. Boundary conditions in the thermal modelling of process play a vital role in the final temperature profile.

Driver a (2005) In the present paper, a three-dimensional thermo mechanical model for Friction Stir Welding (FSW) is presented. Based on the velocity fields classically used in fluid mechanics and incorporating heat input from the tool shoulder and the plastic strain of the bulk material, the semi-analytical model can be used to obtain the strains, strain rates, and estimations of the temperatures and micro-hardness in the various weld zones. The calculated results are in good agreement with experimental measurements performed on a AA2024-T351 alloy friction stir welded joint.

Marzol et al. (2006) established a friction stir welding (FSW) process parameters envelope for an AA 6061 alloy reinforced with 20% of Al2O3 particles, and determine properties of the obtained joints. After a brief description of the FSW technique, and the difficulties in joining MMCs, experimental procedure is illustrated. Microstructure has been observed with optical microscope, and images have been analyzed with image analysis software. Micro hardness and tensile tests have been also carried out. The tool's stirring effect has a substantial influence on the reinforcement particles distribution and shape. Tensile testing revealed joint efficiencies over 80% for the Rp0, 2 and of slightly more than 70% for the Rm, with failure outside the stir zone. The parameter envelope determined in the present study resulted in defect free, high strength welds.

Watanabe et al. (2006) tried to butt-weld an aluminium alloy plate to a mild steel plate by friction stir welding, and investigated the effects of a pin rotation speed, the position for the pin axis to be inserted on the tensile strength and the microstructure of the joint. The behaviour of the oxide film on the faying surface of the steel during welding also was examined. The main results obtained are as follows. Butt-welding of an aluminium alloy plate to a steel plate was easily and successfully achieved by friction stir welding. The maximum tensile strength of the joint was about 86% of that of the aluminium alloy base metal. A small amount of inter metallic compounds was formed at the upper part of the steel/aluminium interface, while no inter metallic compounds were observed in the middle and bottom parts of the interface. The regions where the inter-metallic compounds formed seemed to be fracture paths in the joint. Many fragments of the steel were scattered in the aluminium alloy matrix and the oxide film removed from the faying surface of the steel by the rubbing motion of a rotating pin was observed at the interface between the steel fragments and the aluminium alloy matrix.

Scialpi et al. (2006) studied the effect of different shoulder geometries on the mechanical and micro structural properties of a friction stir welded joints have been studied in the present paper. The process was used on 6082 T6 aluminium alloy in the thickness of 1.5 mm. The three studied tools differed from shoulders with scroll and fillet, cavity and fillet, and only fillet. The effect of the three shoulder geometries has been analyzed by visual inspection, macrograph, HV micro hardness, bending test and transverse and longitudinal room temperature tensile test. The investigation results showed that, for thin sheets, the best joint has been welded by a shoulder with fillet and cavity.

Ceschini et al. (2006) investigated that the application of this solid state welding technique to particles reinforced composites seems very attractive, since it should eliminate some typical defects induced by the traditional fusion welding techniques, such as: gas occlusion, undesired interfacial chemical reactions between the reinforcement and the molten matrix alloy, inhomogeneous reinforcement distribution after welding. The present work describes the effect of the FSW process on the microstructure and, consequently, on the tensile and low-cycle fatigue behaviour, of an aluminium matrix (AA7005) composite reinforced with 10 vol. % of Al2O3 particles (W7A10A).

Zhang et al. (2007) represent the 3D material flows and mechanical features under different process parameters by using the finite element method based on solid mechanics. Experimental results are also given to study the effect of process parameters on joining properties of the friction stir welds. Numerical results indicate that the tangent flow constitutes the major part in the material flow. The shoulder can accelerate the material flow on the top half of the friction stir weld.

Villegas (2007) studied the macro cystis integrifolia and Lessonia trabeculata form vast kelp beds providing a three-dimensional habitat for a diverse invertebrate and Wsh fauna oV northern chile. Habitat modiWcations caused by the El Niño Southern Oscillation (ENSO) are likely to alter the inhabiting communities. The aim of this study was to reveal relationships between distinct habitat structures of a M. integrifolia kelp bed, a dense L. trabeculata kelp bed and L. Trabeculata patches colonizing a barren ground, and the associated dominant macrobenthic key species.

Amancio-Filhoa et al. (2007) described that aircraft aluminium alloys generally present low weld ability by traditional fusion welding process. The development of the friction stir welding has provided an alternative improved way of satisfactory producing aluminium joints, in a faster and reliable manner.

Cavalierea et al. (2007) analyzed the effects of processing parameters on mechanical and micro structural properties of AA6082 joints produced by friction stir welding. Different welded specimens were produced by employing fixed rotating speeds of 1600rpm and by varying welding speeds from 40 to 460 mm/min. The SEM observations of the fatigue specimens, welded at 115 mm/min, showed that at higher stress amplitude levels the cracks initiate at the surface of the welds. By decreasing the stress amplitude the cracks initiate by the internal defects.

Elangovan et al. (2007) studied the influences of tool pin profile and welding speed on the formation of friction stir processing zone in AA2219 aluminium alloy.AA2219 aluminium alloy has gathered wide acceptance in the fabrication of light weight structures requiring a high strength to weight ratio. In this investigation an attempt has been made to study the effect of tool pin profiles and welding speed on the formation of friction stir processing zone in AA2219 aluminium alloy. From this investigation, the following important conclusions are derived.

- Of the five tool pin profiles used to fabricate the joints, square pin profiled tool produced defect free FSP region, irrespective of welding speeds.

- Of the three welding speeds used to fabricate the joints, the joints fabricated at a welding speed of 0.76mm/showed superior tensile properties, irrespective of tool pin profiles.
- Of the 15 joints, the joint fabricated using square pin profiled tool at a welding speed 0.76mm/s exhibited maximum tensile strength, higher hardness and finer grains in the FSP region.

Muthukumaran et al. (2008) states that Electromagnetic radiation is emitted during the transient stage of elastic to plastic deformation of metals and alloys. In the present work, aluminium plates were welded by friction stir welding (FSW) at different process parameters, such as tool rotational speed, traverse speed and rake angle. The EMR fundamental frequencies emitted during the tensile failure of the welds were measured and recorded. The variation in the fundamental frequency was analyzed by fuzzy modelling using MATLAB and it was observed that an increase in the first mode of metal transfer decreases the fundamental frequency. Further, the fundamental frequency of a weld was estimated from the obtained model and found to be closer to the experimental results. It will be more useful for metal flow analysis as well as online condition monitoring of the welds which are used in critical applications.

Chen et al. (2008) Al–Si alloy and pure titanium were lap joined using friction stir welding technology. Microstructure and tensile properties of joints were examined. The maximum failure load of joints reached 62% of Al–Si alloy base metal with the joints fractured at the interface. X-ray diffraction results showed that new phase of TiAl3 formed at the interface. The microstructure evolution and the joining mechanism of aluminium–titanium joints were systematically discussed.

Moreira et al. (2008) Studied that mechanical and metallurgical characterization of friction stir welded butt joints of aluminium alloy 6061-T6 with 6082-T6 was carried out. For comparison, similar material joints made from each one of the two alloys were used. The work included microstructure examination, micro hardness, tensile and bending tests of all joints. An approximate finite element model of the joint, taking into account the spatial dependence of the tensile strength properties, was made, modelling a bending test of the weldments. This study shows that the friction stir welded dissimilar joint present intermediate mechanical properties when compared with each base material. In tensile tests the dissimilar joint displayed intermediate properties. For instance in the hardness profile the lowest values were obtained in the AA6082-T6 alloy plate side where rupture occurred, and in the nugget all type of joints present similar values.

Rodrigues a (2008) In this research paper present work friction stir welds produced in 1 mm thick plates of AA 6016-T4 aluminium alloy, with two different tools, were analyzed and compared concerning the microstructure and mechanical properties. For each tool, the welding parameters were optimized in order to achieve non-defective welds. The differences in mechanical properties between the two types of welds are explained based in TEM micro structural analysis. Despite the under matched characteristics of the ''cold'' welds relative to the base material, formability tests demonstrated that these welds improve the drawing performance of the welded sheets.

Jai KWON et al. (2009) performed the experiment of Friction stir welding between 5052 aluminium alloy plates with a thickness of 2 mm was performed. The tool for welding was rotated at speeds ranging from 500 to 3 000 r/min under a constant traverse speed of 100 mm/min. In all tool rotation speeds, the SZ exhibits higher average hardness than the base metal. Especially at 500 r/min, the average hardness of the SZ reaches a level about 33% greater than that of the base metal. At 500, 1 000 and 2 000 r/min, the tensile strength of the friction stir welded (Friction stir welded) plates is similar to that of the base metal (about 204 MPa). The elongation of the Friction stir welded plates is lower than that of the base metal (about 22%). However, it is noticeable that the maximum elongation of about 21% is obtained at 1 000 r/min.

Sandra Zimmer et al. (2009) presents the results of an experimental investigation, done on the friction stir welding (FSW) plunging stage. Previous research works showed that the axial force and torque generated during this stage were characteristic for a static qualification of a FSW machine. Therefore, the investigation objectives are to better understand the relation between the processing parameters and the forces and torque generated. It is an interesting way to present the experimental results. This kind of representation can be useful for the processing parameters choice. They can be chosen according to the force and torque responses and consequently to the FSW machine capacities.

Hwang (2010) This study aimed to experimentally explore the thermal history of a work piece undergoing Friction Stir Welding (FSW) involving butt joining with pure copper C11000. In the FSW experiments, The appropriate temperatures for a successful FSW process were found to be between 460 ◦C and 530 ◦C. These experimental results and the process control of temperature histories can offer useful knowledge for a FSW based process of copper butt joining.
Hattel a (2010) Studied that the post-welding stress state, strain history and material conditions of friction stir welded joints are often strongly idealized when used in subsequent modelling analyses, typically by neglecting one or more of the features above. But, it is obvious that the conditions after welding do influence the weld performance. The objective of this paper is to discuss some of the main conflicts that arise when taking both the post-welding material conditions and stress strain state into account in a subsequent structural analysis

Kanwer S. Arora et al. (2010) in this research, successful friction stir welding of aluminium alloy 2219 using an adapted milling machine is reported. The downward or forging force was found to be dependent upon shoulder diameter and rotational speed whereas longitudinal or welding force on welding speed and pin diameter. Tensile strength of welds was significantly affected by welding speed and shoulder diameter whereas welding speed strongly affected percentage elongation.

Hwang et al. (2010) experimentally explore the thermal history of a work piece undergoing Friction Stir Welding (FSW) involving butt joining with pure copper C11000. In the FSW experiments, K-type thermocouples were used to record the temperature history at different locations on work piece. This data, combined with the preheating temperature, tool rotation speeds and tool moving speeds allowed parameters for a successful weld to be determined.

Riahi (2010) In this research residual stress is lower in friction stir welding (FSW) compared with other melting weldment processes. This is due to being solid-state process in its nature. There are several advantages in utilizing stir welding process. Lower fluctuation and shrinkage in weldment metal enhanced mechanical characteristics, less defects, and ability to weld certain metals otherwise impractical by other welding processes are to name just a few of these advantages. In the prediction of results of residual stress, only heat impact was studied. This was recognized as the main element causing minor difference in results obtained for simulation in comparison with that of actual experiment.

Tozak et al. (2010) newly developed tool for friction stir spot welding (FSSW) has been proposed, which has no probe, but a scroll groove on its shoulder surface (scroll tool). By use of this tool, FSSW has been performed on aluminium alloy 6061-T4 sheets and the potential of the tool was discussed in terms of weld structure and static strength of welds. The experimental observations showed that the scroll tool had comparable or superior performance to a conventional probe tool. The shear fracture took place at smaller shoulder plunge depths or at shorter tool holding times, while the plug fracture occurred at larger shoulder plunge depths or at longer tool holding times. It was indicated that the tensile-shear strength and associated fracture modes were determined by two geometrical parameters in the weld zone.

S. Rajakumar et al. (2011) observed that AA6061 aluminium alloy has gathered wide acceptance in the fabrication of light weight structures requiring high strength-to-weight ratio and good corrosion resistance. Friction-stir welding (FSW) process is an emerging solid state joining process in which the material that is being welded does not melt and recast. This process uses a non-consumable tool to generate frictional heat in the abutting surfaces. The FSW process and tool parameters play a major role in deciding the joint strength. Joint strength is influenced by grain size and hardness of the weld nugget region. Hence, in this investigation an attempt was made to develop empirical relationships to predict grain size and hardness of weld nugget of friction-stir-welded AA6061 aluminium alloy joints. The empirical relationships are developed by response surface methodology incorporating FSW tool and process parameters. A linear regression relationship was also established between grain size and hardness of the weld nugget of FSW joints.

Aval (2011) In this work, the thermo-mechanical responses during the dissimilar friction stir welding of aluminium alloys have been evaluated employing a three-dimensional model and the finite element software ABAQUS. Also, both experimental and simulated results indicate that welding fixtures significantly affect the residual stress profiles as well as their magnitudes.

Kumaran et al.(2011) In this research numerous advancements have been occurring in the field of materials processing. Friction welding is an important solid-state joining technique. In this research project, friction welding of tube-to-tube plate using an external tool (FWTPET) has been performed, and the process parameters have been prioritized using Taguchi's L27 orthogonal array. Genetic algorithm (GA) is used to optimize the welding process parameters. The practical significance of applying GA to FWTPET process has been validated by means of computing the deviation between predicted and experimentally obtained welding process parameters.

Elangovan et al.(2012) The researchers **in** this paper focuses on the development of an effective methodology to determine the optimum welding conditions that maximize the strength of joints produced by ultrasonic welding using response surface methodology (RSM) coupled with genetic algorithm (GA). RSM is utilized to create an efficient analytical model for welding strength in terms of welding parameters namely pressure, weld time, and amplitude. Experiments were conducted as per central composite design of experiments for spot and seam welding of 0.3- and 0.4-mm-thick Al specimens. An effective second-order response surface model is developed utilizing experimental measurements. Response surface model is further interfaced with GA to optimize the welding conditions for desired weld strength. Optimum welding conditions produced from GA are verified with experimental results and are found to be in good agreement.

Mariano et al. (2012) presents a literature review on friction stir welding (FSW) modelling with a special focus on the heat generation due to the contact conditions between the FSW tool and the work piece. A reliable FSW process modelling depends on the fine tuning of some process and material parameters. Usually, these parameters are achieved with base on experimental data. The numerical modelling of the FSW process can help to achieve such parameters with less effort and with economic advantages.

ZHANG (2012) studied that, the thermal modelling of underwater friction stir welding (FSW) was conducted with a three-dimensional heat transfer model. The vaporizing characteristics of water were analyzed to illuminate the boundary conditions of underwater FSW. Temperature dependent properties of the material were considered for the modelling. For underwater joint, the high-temperature distributing area is dramatically narrowed and the welding thermal cycles in different zones are effectively controlled in contrast to the normal joint.

Bhatt (2013) In this research we observed that Friction stir welding (FSW) of AA6061-T6 aluminium alloy has been attempted to overcome limitations of fusion welding of the same. The FSW tool, by not being consumed, produces a joint with predominant advantages of high joint strength, lower distortion and absence of metallurgical defects. Process parameters such as tool rotational speed, tool traverse speed and axial force and tool dimensions play an important role in obtaining a specific temperature distribution and subsequent viscosity distribution within the material being welded; the former controlling the mechanical properties and later the flow stresses within the material in turn.

Guo (2013) Studied that the Dissimilar AA6061 and AA7075 alloy have been friction stir welded with a variety of different process parameters. In particular, the effects of materials position and welding speed on the material flow, microstructure, micro hardness distribution and tensile property of the joints were investigated. It was revealed that the material mixing is much more effective when AA6061 alloy was located on the advancing side and multiple vortexes centres formed vertically in the nugget. l computational model.

KEIVANI (2013) have studied in their work, friction stir welding (FSW) is applied extensively in industry for joining of nonferrous metals especially aluminium. A three-dimensional model based on finite element analysis was used to study the thermal characteristic of copper C11000 during the FSW process.

Liu a (2013) In their research, the 4 mm thick 6061-T6 aluminium alloy was self-reacting friction stir welded at a constant tool rotation speed of 600 r/min. The specially designed self-reacting tool was characterized by the two different shoulder diameters. The results of transverse tensile test indicated that the elongation and tensile strength of joints increased with increasing welding speed. The defect-free joints were obtained at lower welding speeds and the tensile fracture was located at the heat affected zone adjacent to the thermal mechanically affected zone on the advancing side.

Shen a (2013) Studied that, the friction stir welding (FSW) technique is considered to offer advantages over fusion welding in terms of dissimilar jointing. However, some challenges still exist in the butt FSW of dissimilar Ti and Al metals. The present research employed a modified butt joint configuration into the FSW of Ti–6Al–4V alloy to Al–6Mg alloy with a special pin plunge setup, aiming to obtain a high-quality Ti-to-Al joint, avoid butt flaw or Al melting, and reduce the tool shoulder attrition. Under different FSW process conditions, the examinations and analyses of macro/micro-structures, mechanical tensile properties and fractographies of the dissimilar joints were conducted.

Simoes a, (2013) their work describes the thermo-mechanical conditions during Friction Stir Welding (FSW) of metals have already been subject of extensive analysis and thoroughly discussed in literature, in which concerns the FSW of polymers,
From the study it was possible to conclude that, due to the polymers rheological and physical properties, the thermo-mechanical conditions during FSW are very different from that registered during welding of metals, leading to completely different material flow mechanisms and weld defect morphologies.

Pan (2013) In their research the Friction Stir Welding (FSW) is a complex thermal-mechanical process. Numerical models have been used to calculate the thermal field, distortion and residual stress in welded components but some modelling parameters such as film coefficient and thermal radiation of the work pieces may be technically difficult and/or expensive to measure experimentally.
By comparing the FEM numerical results with experimental results, the FSW process thermal parameters have been successfully identified. This automatic parameters characterization procedure could be used for the FSW process optimization.

Ni (2014) observed that the Thin sheets of aluminium alloy 6061-T6 and one type of Advanced high strength steel, transformation induced plasticity (TRIP) steel have been successfully butt joined using friction stir welding (FSW) technique. Tensile tests and scanning electron microscopy (SEM) results indicate that the weld nugget can be considered as aluminium matrix composite, which is enhanced by dispersed sheared-off steel fragments encompassed by a thin inter-metallic layer or simply inter-metallic particles. Effects of process parameters on the joint microstructure evolution were analyzed based on mechanical welding force and temperature that have been measured during the welding process.

He (2014) Have done a relatively new solid-state joining technique which is widely adopted in different industry fields to join different metallic alloys that are hard to weld by conventional fusion welding. Numerical analysis of friction stir welding will allow many different welding processes to be simulated in order to understand the effects of changes in different system parameters before physical testing, which would be time-consuming or prohibitively expensive in practice. The main methods used in numerical analysis of friction stir welding are discussed and illustrated with brief case studies. In addition, several important key problems and issues remain to be addressed about the numerical analysis of friction stir welding and opportunities for further research are identified.

REFERNCES

[1] A. górka1 and D. kocańda1,2, Mass transport in a high-gradient thermal field in the course of friction stir welding and modification of the upper layer, Materials Science, Vol. 47, No. 2, September, 2011 (Ukrainian Original Vol. 47, No. 2, March–April, 2011).

[2] Bo Li a, Zhenhua Zhang a, Yifu Shen a,⇑, Weiye Hub, Lei Luo a, Dissimilar friction stir welding of Ti–6Al–4V alloy and aluminum alloy employing a modified butt joint configuration: Influences of process variables on the weld interfaces and tensile properties, Materials and Design 53 (2014) 838–848, Received 26 May 2013,Received in revised form 5 July 2013,Accepted 7 July 2013,Available online 18 July 2013.

[3] C.M. Chen, R. Kovacevic *,Finite element modeling of friction stir welding—thermal and thermomechanical analysis, International Journal of Machine Tools & Manufacture 43 (2003) 1319–1326, Received 12 May 2003; accepted 10 June 2003.

[4] Diogo Mariano Neto & Pedro Neto. Numerical modellings of friction stir welding process: a literature review.

[5] D.M. Rodrigues a,*, A. Loureiro a, C. Leitao a, R.M. Leal a,b, B.M. Chaparro a,c, P. Vilaça d, Influence of friction stir welding parameters on the microstructural and mechanical properties of AA 6016-T4 thin welds, Materials and Design 30 (2009) 1913–1921, Received 21 July 2008,Accepted 9 September 2008,Available online 21 September 2008.

[6] F. Simoes a,b, D.M. Rodrigues b,⇑, Material flow and thermo-mechanical conditions during Friction Stir Welding of polymers: Literature review, experimental results

and empirical analysis, Materials and Design 59 (2014) 344–351, Received 27 October 2013,Accepted 16 December 2013,Available online 18 February 2014.

[7] H. Jamshidi Aval & S. Serajzadeh & A. H. Kokabi, Experimental and theoretical evaluations of thermal histories and residual stresses in dissimilar friction stir welding of AA5086-AA6061, Int J Adv Manuf Technol (2012) 61:149–160, Received: 7 April 2011 / Accepted: 24 October 2011 / Published online: 13 November 2011.

[8] Hui-jie ZHANG, Hui-jie LIU, Lei YU, Thermal modeling of underwater friction stir welding of high strength aluminum alloy, Trans. Nonferrous Met. Soc. China 23(2013) 1114_1122, Received 23 February 2012; accepted 26 June 2012.

[9] H.J. Liu a,⇑, J.C. Hou a,b, H. Guo a, Effect of welding speed on microstructure and mechanical properties of self-reacting friction stir welded 6061-T6 aluminum alloy, Materials and Design 50 (2013) 872–878, Received 22 January 2013,Accepted 30 March 2013,Available online 11 April 2013.

[10] H.J. Liu, H. Fujii, M. Maeda, K. Nogi. Tensile properties and fracture locations of friction-stir-welded joints of 2017-T351 aluminium alloy. Journal of Materials Processing Technology 142 (2003) 692–696.

[11] J.F. Guo ⇑, H.C. Chen, C.N. Sun, G. Bi, Z. Sun, J. Wei, Friction stir welding of dissimilar materials between AA6061 and AA7075 Al alloys effects of process parameters, Materials and Design 56 (2014) 185–192, Received 12 August 2013,Accepted 29 October 2013,Available online 9 November 2013.

[12] J.H. Hattel a,*, K.L. Nielsen b, C.C. Tutum a, The effect of post-welding conditions in friction stir welds: From weld simulation to ductile failure, European Journal of Mechanics A/Solids 33 (2012) 67e74, Received 28 September 2010,Accepted 22 October 2011,Available online 23 November 2011.

[13] K. Elangovan, V. Balasubramanian.Influences of tool pin profile and welding speed on the formation of friction stir processing zone in AA2219aluminium alloy. Journal of materials processing technology 2 0 0 (2 0 0 8) 163–175.

[14] Kanwer S. Arora & Sunil Pandey & Michael Schaper &Rajneesh Kumar Effect of process parameters on friction stir welding of aluminium alloy 2219-T87, Springer-Verlag London Limited 2010, Received: 2 December 2009 / Accepted: 31 January 2010 / Published online: 20 February 2010.

[15]M. Ericsson, R. Sandstro¨m.Influence of welding speed on the fatigue of friction stir welds, and comparison with MIG and TIG, Received 4 March 002; received in revised form 14 June 2002; accepted 1 July 2002.

[16] Mohammad Riahi & Hamidreza Nazari, Analysis of transient temperature and residual thermal stresses in friction stir welding of aluminum alloy 6061-T6 via numerical simulation, Int J Adv Manuf Technol (2011) 55:143–152, Received: 5 December 2009 / Accepted: 16 November 2010 / Published online: 1 December 2010

[17] Mario J. Villegas · Jürgen Laudien ·Walter Sielfeld · Wolf E. Arntz, Macrocystis integrifolia and Lessonia trabeculata (Laminariales; Phaeophyceae) kelp habitat structures and associated macrobenthic community oV northern Chile, Helgol Mar Res (2008) 62 (Suppl 1):S33–S43, Received: 6 June 2007 / Revised: 2 November 2007 / Accepted: 9 November 2007 / Published online: 20 December 2007.

[18] M. GHOSH, R.K. GUPTA, and M.M. HUSAIN, Friction Stir Welding of Stainless Steel to Al Alloy: Effect of Thermal Condition on Weld Nugget Microstructure, The Minerals, Metals & Materials Society and ASM International 2013

[19] Nuran Bradley, The Response Surface Methodology, A thesis for award of Master of Science in Applied Mathematics & Computer Science, 2007.

[20] P. Cavalierea, R. Nobilea, F.W. Panellaa, A. Squillace. Mechanical and micro structural behaviour of 2024–7075 aluminium alloy sheets joined by friction stir welding. International Journal of Machine Tools &Manufacture 46 (2006) 588–594.

[21] P. Cavaliere, G. Campanile, F. Panella, A. Squillace. Effect of welding parameters on mechanical and microstructuralproperties of AA6056 joints produced by Friction Stir Welding. Journal of Materials Processing Technology 180 (2006) 263–270.

[22] P. Heurtier a, M.J. Jones b, C. Desrayaud a, J.H. Driver a,∗, F. Montheillet a, D. Allehaux c, Mechanical and thermal modelling of Friction Stir Welding, Journal of Materials Processing Technology 171 (2006) 348–357, Received 16 April 2004; received in revised form 1 January 2005; accepted 6 July 2005.

[23] R. KEIVANI1, B. BAGHERI2, F. SHARIFI2, M. KETABCHI2, M. ABBASI3, Effects of pin angle and preheating on temperature distribution during friction stir welding operation, Trans. Nonferrous Met. Soc. China 23(2013) 2708−2713, Received 31 October 2012; accepted 14 January 2013.

[24] Scialpi, L.A.C. De Filippis, P. Cavaliere. Influence of shoulder geometry on microstructure andmechanical properties of friction stir welded 6082 aluminium alloy. Materials and Design 28 (2007) 1124–1129

[25] Sandra Zimmer & Laurent Langlois & Julien Laye & Regis Bigot. Experimental investigation of the influence of the FSWplunge processing parameters on the maximum generated force and torque, Springer-Verlag London Limited 2009, Received: 22 January 2009 / Accepted: 22 June 2009 / Published online: 21 July 2009.

[26] S. Muthukumaran & Kumar Pallav &Vikas Kumar Pandey & S. K. Mukherjee. A study on electromagnetic property during friction stir weld failure, Springer-Verlag London Limited 2006, Received: 19 March 2006 / Accepted: 10 October 2006 / Published online: 12 December 2006. [15] S. Rajakumar & C. Muralidharan & V. Balasubramanian. Statistical analysis to predict grain size and hardness of the weld nugget of friction-stir-welded AA6061-T6aluminium alloy joints, Springer-Verlag London Limited 2011Received: 29 October 2009 / Accepted: 17 March 2011 / Published online: 20 April 2011.

[27] S. Senthil Kumaran & S. Muthukumaran & S. Vinodh. Optimization of friction welding of tube-to-tube plate using an external tool by Taguchi method and genetic algorithm.

[28] S. Elangovan & K. Anand & K. Prakasan. Parametric optimization of ultrasonic metal welding using response surface methodology and genetic algorithm

[29] Takehiko Watanabe, Hirofumi Takayama, Atsushi Yanagisawa. Joining of aluminium alloy to steel by friction stir welding. Journal of Materials Processing Technology 178 (2006) 342–349.

[30] T4 S.T. Amancio-Filho, S. Sheikhi,J.F. dos Santos, C. Bolfarini. Preliminary study on the microstructure and mechanical properties of dissimilar friction stir welds in aircraft aluminium alloys 2024-T351 and 6056-journal of materials processing technology 2 0 6 (2 0 0 8) 132–142.

[31] Vijay Soundararajan, Srdja Zekovic, Radovan Kovacevic*,Thermo-mechanical model with adaptive boundary conditions for friction stir welding of Al 6061, International Journal of Machine Tools & Manufacture 45 (2005) 1577–1587, Received 5 January 2005; accepted 15 February 2005, Available online 21 April 2005.

[32] Xun Liu, Shuhuai Lan, Jun Ni ⇑,Analysis of process parameters effects on friction stir welding of dissimilar aluminum alloy to advanced high strength steel, Materials and Design 59 (2014) 50–62, Received 22 October 2013, Accepted 1 February 2014, Available online 7 February 2014.

[33] Xiaocong He, Fengshou Gu, Andrew Ball, A Review of Numerical Analysis of Friction Stir Welding, Progress in Materials Science (2014), Received Date: 2 September 2013, Revised Date: 29 November 2013, Accepted Date: 6 March 2014.

[34] Xingguo Zhou, Wenke Pan*, Donald MacKenzie, Identifying friction stir welding process parameters through coupled numerical and experimental analysis, International Journal of Pressure Vessels and Piping 108-109 (2013) 2e6,

[35] Y.C. Chen, K. Nakata. Microstructural characterization and mechanical properties in friction stir welding of aluminium and titanium dissimilar alloys. Materials and Design 30 (2009) 469–474

[36] Y.M. Hwang*, P.L. Fan, C.H. Lin, Experimental study on Friction Stir Welding of copper metals, Journal of Materials Processing Technology 210 (2010) 1667–1672, Received 15 January 2010, Received in revised form 8 April 2010, Accepted 31 May 2010.

[37] Yong-Jai KWON, Seong-Beom SHIM, Dong-Hwan PARK. Frictions stir welding of 5052 aluminium alloy plates. Trans. Nonferrous Met. Soc.China 19(2009) s23-s27.

[38] Z. Zhang & H. W. Zhang. Springer-Verlag London Limited 2007 Material behaviours and mechanical features in friction stir welding process, Received: 8 February 2006 / Accepted: 19 June 2006 / Published online: 3 February 2007.

7

DESIGN AND OPTIMIZATION OF AUTOMOTIVE MULTI-LEAF SPRING BY FINITE ELEMENT METHOD

Shahrukh Shamim[1], Jamil Anwer [2]

[1]B-Tech. Scholar, 121101151_shahrukh@manit.ac.in
[2]B-Tech. Scholar, jamilanwer10@gmail.com
Author Correspondence: Department of Material Science and Metallurgical Engineering, Maulana Azad National Institute of Technology, Bhopal, M.P., India.-462051, 121101151_shahrukh@manit.ac.in

Abstract

The automobile industry has shown increased interest in the replacement of steel spring with fibre glass composite leaf spring due to high strength to weight ratio. The present study describes design and comparative analysis of multi-leaf spring made of three different materials. The modelling of the leaf spring has been done in commercially used FEM software ANSYS 14.0, and for finite element analysis the model was imported in the Static Structural analysis and Harmonic Response workbench of ANSYS 14.0. The Materials used for the multi-leaf spring is AISI 6150 Steel (Oil Quenched 845°C and 650°C tempered), Ti-6Al-4V alloy and S-Glass fiber Composite. Harmonic analysis for vibrations due to road irregularity is carried out for all the materials. von-Mises stress and deformation is the output parameters in static structural analysis..

Keywords: Multi Leaf Spring, FEM, ANSYS, Static Structural Analysis, Harmonic Analysis.

1. Introduction

Leaf springs are one of the oldest suspension components that are still frequently used, especially in commercial vehicles. Leaf spring is one of the key components of vehicle suspension system. Increasing competition and innovations in automobile sector tends to modify the existing products or replacing old products by new and advanced materials. Leaf springs are mainly used in suspension systems to absorb shock loads in automobiles like light motor vehicles, heavy duty trucks and in rail systems. To meet the needs of the natural resource conservation, energy and economy, recently the automobile manufacturers have been attempting to minimize the weight of the vehicle.

More efforts are taken in order to increase the comfort of the user. Appropriate balance of comfort riding qualities and economy in the manufacturing of leaf springs has become an essential requirement. Multi-Leaf springs are vital suspension elements used on mini loader trucks necessary to minimize the vertical vibrations, impacts and jolts due to road inconsistency and to ensure safety of the passengers and cargo. Many past recorded data shows that steel leaf springs are manufactured by EN45, EN45A, 60Si7, EN47, and 50CrMoCV etc. these materials are widely used for the manufacture of the conventional multi leaf springs.The introduction of the composite materials made it possible to reduce the weight of the leaf springs without any diminution of load carrying capacity and stiffness. Studies were conducted on the application of the composite materials for automobile suspension system (leaf springs).

Non-ferrous metals are also used including phosphorous and titanium for parts necessitating corrosion resistance and beryllium for springs carrying electrical current. Multi leaf springs used in the automotive vehicles normally consist of full length leaves and graduated length leaves. The specimen under this study consists of seven leaves. Finite element analysis using ANSYS 14.0 software has been carried on multi-leaf springs to obtain approximate solutions to the boundary value problems in engineering. Multi-Leaf springs were modelled in typical way and simulated for the kinematic and dynamic comparatives. In the present study, leaf spring has been analysed for static strength and harmonic response using 3D finite element analysis. ANSYS 14.0 has been utilized in the creation of the three dimensional model and its static structural and harmonic response for analysis when subjected to vertical loads.

2. Problem Formulations

Multi-Leaf Spring for commercial vehicle needs to be of high strength in order to minimize the vertical vibrations, impacts and bumps due to road irregularities and to ensure safety of the passengers and cargo. The main objective of this work is:

- To develop structural modelling of Multi-Leaf Spring using ANSYS Design 14.0 software.
- Static structural and Harmonic response analysis of Multi-Leaf Spring using ANSYS v14.0.
- Comparison between AISI 6150 Steel (Oil Quenched 845 °C and 650 °C tempered), Ti-6Al-4V alloy (Annealed) and S-Glass Fiber Composite Multi-Leaf spring in terms of deformation, von-Mises stress and Amplitude (mm) versus frequency curves.

3. Materials for Multi-Leaf Spring

The materials which are widely used for production of the parabolic leaf springs and conventional multi leaf springs are EN45, EN45A, 60Si7, EN47, 50Cr4V2, 55SiCr7 and 50CrMoCV4 etc. The leaves are heat treated after the forming process. The heat treatment of spring steel products imparts greater strength and therefore greater load bearing capacity. In general terms higher alloy content is mandatory to ensure adequate harden ability when the thick leaf sections are used. The materials used in this work are AISI 6150 Steel, Ti-6Al-4V alloy and S-Glass Fiber Composite. Compositions of these materials are listed in table 1a-c.

Table 1a: Composition of AISI 6150 Steel used in this study

Elements	C	Cr	Fe	Mn	P	Si	S	V
%Composition	0.48%	0.8%	97.09%	0.7%	0.035%	0.15%	0.040%	0.15%

Table 1b: Composition of Ti-6Al-4V alloy used in this study

Elements	C	Fe	N	V	Al	Ti	H	O
%Composition	0.080%	0.4%	0.050%	3.5%	5.5%	87.6%	0.015%	0.20%

Table 1c: Composition of S-Glass Fiber Composite used in present study

Elements	Al_2O_3	B_2O_3	BaO	CaO	FeO	MgO	NaO_2	Other	SiO_2
%Composition	24.8%	0.01%	0.2%	0.01%	0.21%	10.27%	0.27%	0.03%	64.2%

Selected mechanical property of AISI 6150 Steel, Ti-6Al-4V alloy and S-Glass Fiber composite is discussed in table 2.

Table 2: Mechanical Property of Materials used in present study

Materials→ / Parameters↓	AISI 6150 Steel	Ti-6Al-4V alloy	S-Glass Fiber Composite
Density	7.85 g/cc	4.42 g/cc	2.48 g/cc
Young's Modulus, E	205 GPa	105GPa	86.9GPa
Poisson's Ratio	0.29	0.31	0.22
Ultimate Tensile Strength	1015 MPa	895MPa	4585MPa
Tensile strength, Yield	979 MPa	828MPa	--

4. Geometrical Parameters of Multi-Leaf Spring

Modern application of old leaf springs is the parabolic leaf spring for automobiles. The new innovative design is distinguished by the use of less leaves whose thickness varies from the centre to the outer side following a parabolic pattern. In addition to shocks, the leaf springs may carry loads, brake torque, driving torque, etc. Parabolic leaf springs for automobiles has better load bearing capacity with less weight. In the present work multi-leaf spring of commercial vehicles is considered for analysis. The modelling of the Leaf Spring has been carried out in ANSYS Design 14.0 and the geometrical parameters are shown in table 3.

Table 3: Geometrical Parameters of Multi-Leaf Spring

Parameters	**Value** (all dimensions are in mm)
Distance between eyes	1291
Thickness of all leaves	10
Width of each leaves	60
Total number of leaves	7 Leaves
Free Camber	150

The basic view of Multi-Leaf Spring is shown in Fig. 1a&b.

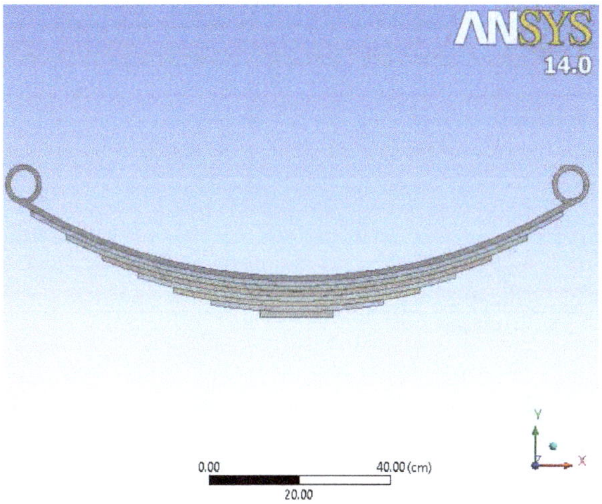

Fig.1a Front View of Multi-Leaf Spring

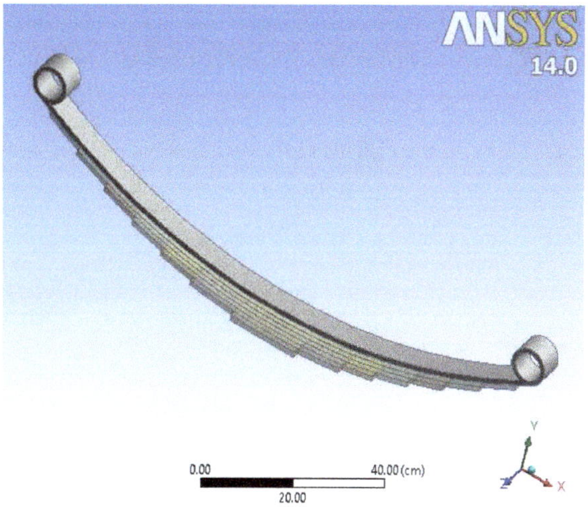

Fig. 1b Top view of Multi-Leaf Spring

5. Finite Element Analysis

The study has been carried out on a multi-leaf spring consisting of seven leaves used by commercial vehicle. Multi-leaf spring of given specification is subjected to Static structural and Harmonic response analysis using ANSYS to find the stresses and deformation. The main objective of this analysis is to study the multi-leaf steel leaf spring and verification of the results within the desirable limits. The general process FEA is divided into three phases, Pre-Processor, Solution and Post-Processor. Virtual modelling of the leaf spring is done and is imported to ANSYS 14.0 workbench. Same model is used for the static analysis as well as for harmonic response analysis with three different materials, namely AISI 6150 Steel, Ti-6Al-4V alloy and S-Glass Fibre Composite.

5.1.Meshing

Mesh Generation is one of the most critical aspects of engineering simulations. Meshing involves division of the entire model into small pieces called elements. It is also known as piecewise approximation. In present case, number and type of elements taken is shown in table 3. Meshed Surface is shown in Fig. 2.

Table 3: Total number of Nodes and Elements

Entity	Size
No. of Nodes	28861
No. of Elements	12410

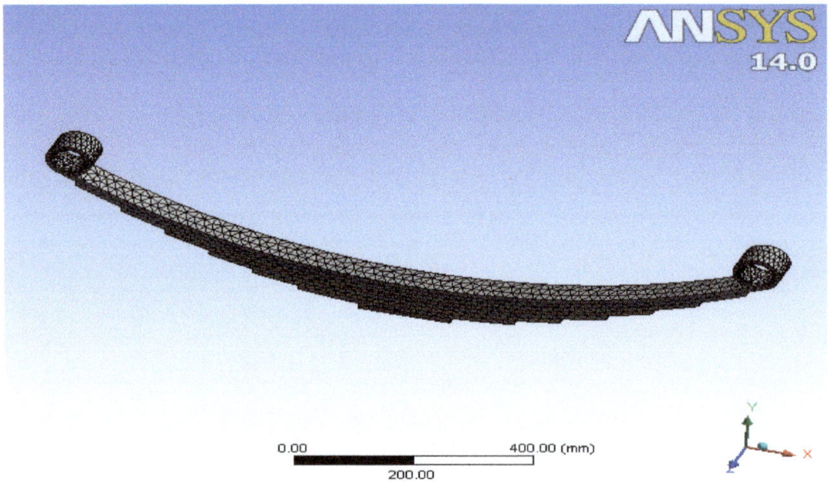

Fig.2: Meshed Surface of Multi-Leaf Spring

6. Results and Discussions

6.1. Static Analysis

Using ANSYS, Static analysis is carried out after applying the boundary conditions. The maximum von-mises stress and maximum deformation in AISI 6150 steel, Ti-6Al-4V alloy and S-Glass fibre composite is shown in Fig 4a-f. A graph is plotted as shown in Fig. 5 between Load and Von-mises stress. It is evident from the graph that there is a linear relation between load and bending stress.

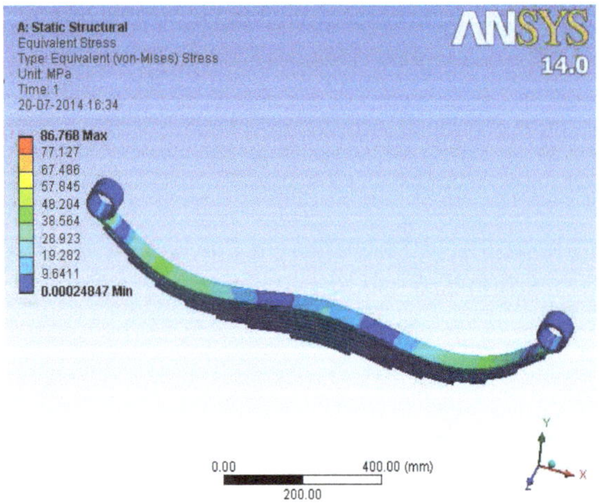

Fig. 4a: Von-Mises stress in AISI 6150 Steel

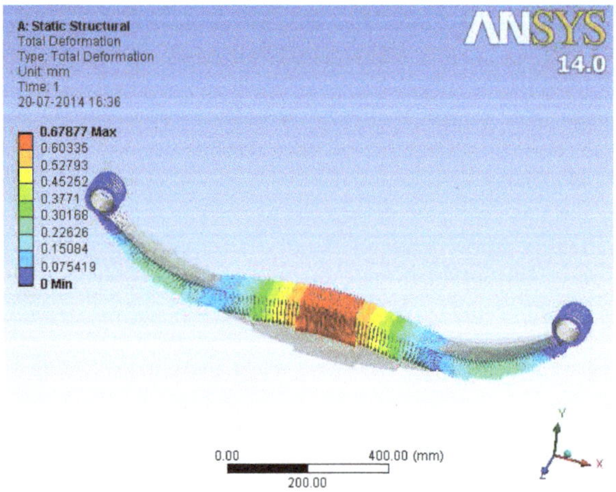

4b: Deformation in AISI 6150 Steel (Vector Display)

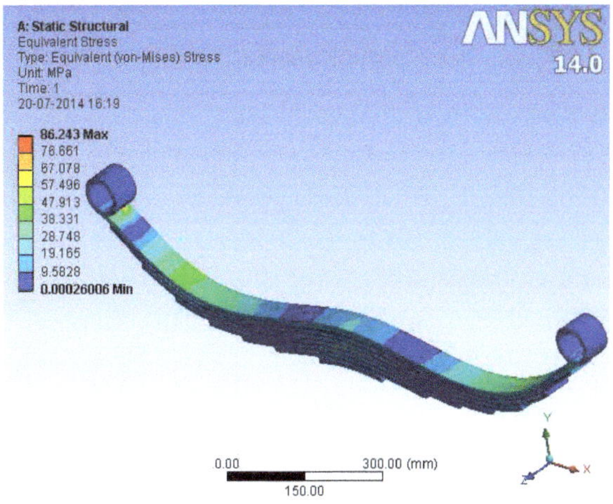

4c: Von-Mises stress in Ti-6Al-4V Alloy

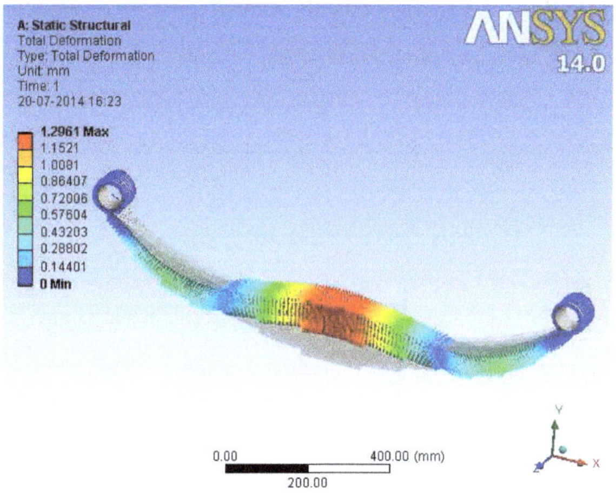

4d: Deformation in Ti-6Al-4V alloy

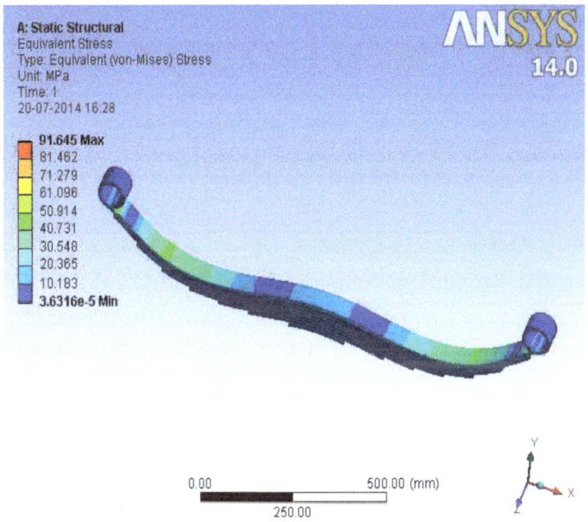

4e: Von-Mises stress in S-Glass Fiber Composite

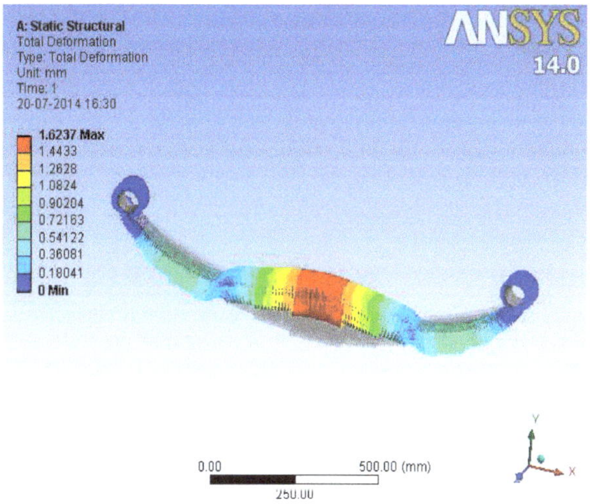

4f: Deformation in S-Glass Fiber Composites

Vertical load, ranging from 1000 N to 25000 N, is applied. Parameters under considerations are Maximum von-Mises Stress and weight of the multi-spring materials.

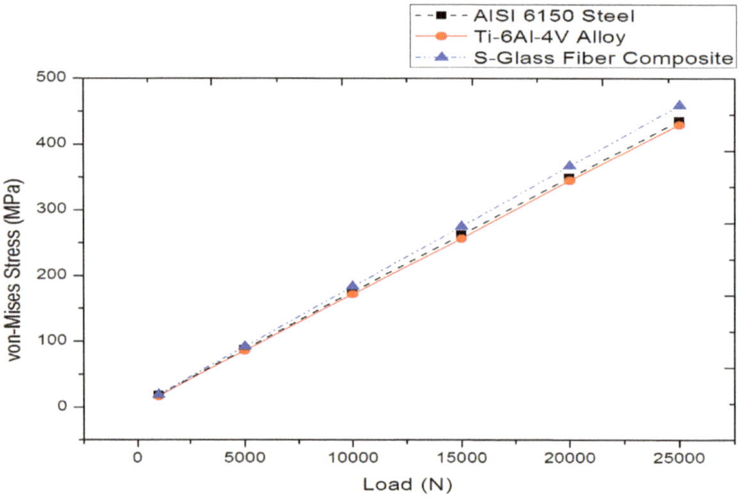

Fig. 5: Liner relationship between applied load and von-Mises stress

6.2. Harmonic Analysis

Harmonic response analysis is a technique used to determine the steady-state response of a linear structure to loads that vary harmonically with time. Main motive is to calculate the structure's response at several frequencies and obtain a graph of displacement versus frequency. Plot of amplitude versus frequency for AISI 6150 steel, Ti-6Al-4V alloy and S-Glass Fiber Composite is shown in Fig. 6a-c.

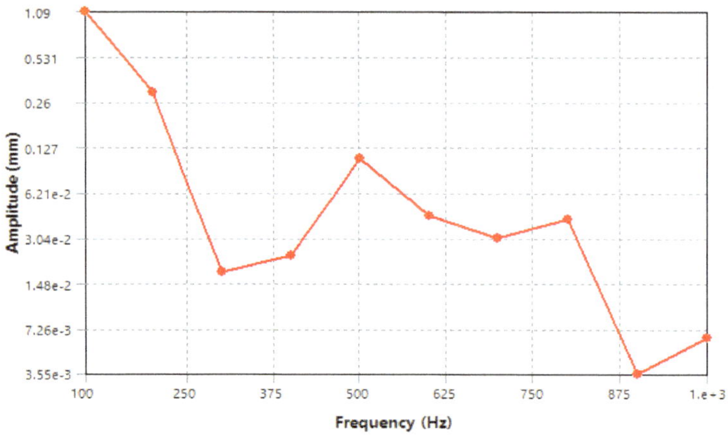

Fig. 6a: Amplitude vs. Frequency curve for AISI 6150 steel leaf spring

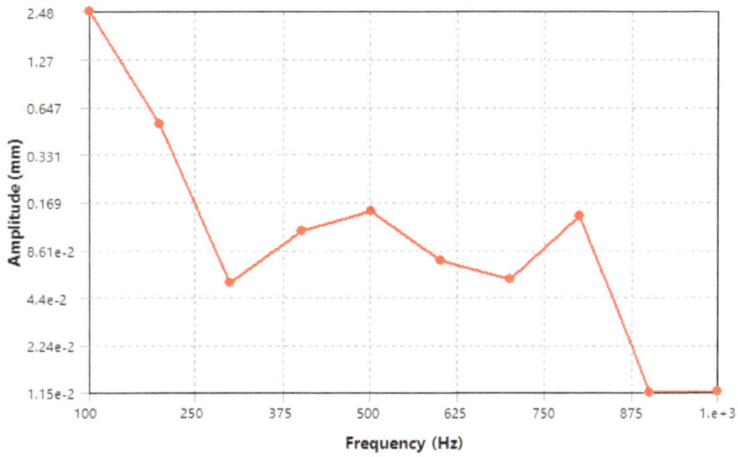

Fig. 6b: Amplitude vs. Frequency curve for Ti-6Al-4V Alloy leaf spring

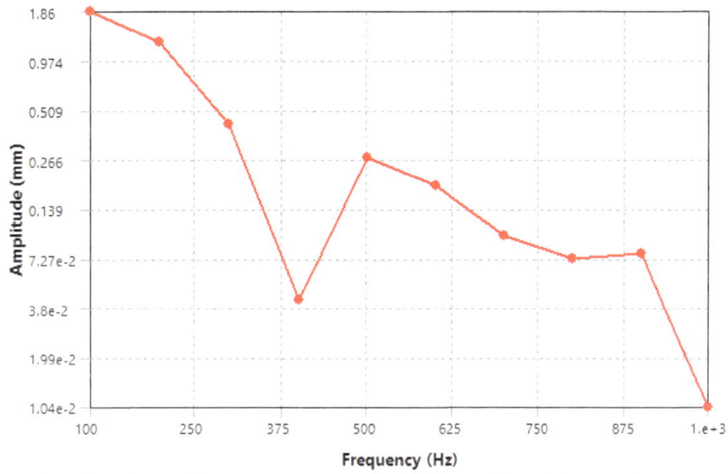

Fig. 6c: Amplitude vs. Frequency curve for S-Glass Fiber Composite leaf spring

7. Conclusion

The leaf spring based on S-Glass Fiber Composite and Ti-6Al-4Valloy has a lower mass compared to AISI 6150 Steel. Total mass of S-Glass Fiber Composite, Ti-6Al-4V alloy and AISI 6150 steel based multi-leaf spring is 10.278 Kg, 18.318 Kg and 32.532 Kg respectively. Reducing the leaf spring mass in automobiles, we can achieve better riding comfort against hard braking and acceleration. Under the same static load conditions the stresses in leaf springs are found with great difference. Stress in S-Glass Fiber Composite is found to be more as compared to conventional AISI 6150 steel. Titanium alloy based leaf spring, under same static loading conditions, shows lesser stress than S-Glass Fiber composite and AISI 6150 steel based leaf spring.

References

1. M. Venkatesan, Jan-Feb 2012, "Design And Analysis Of Composite Leaf Spring In Light Vehicle", *International Journal of Modern Engineering Research (IJMER)* Vol.2, Issue.1, pp-213-218 ISSN: 2249-6645.

2. H.A.Al-Qureshi, 2001, "Automobile leaf spring from composite materials", *Journal of Material Processing Technology*, 118, 58-61.

3. U.S. Ramakanth & K. Sowjanya. Mar 2013, "Design and Analysis of Automotive Multi-leaf Spring using Composite Materials", *International Journal of Mechanical Production Engineering Research and Development (IJMPERD),* ISSN 2249-6890, Vol. 3, Issue 1, 155-162. 4. Senthil Kumar,

SabapathyVijayarangan, 2001, "Analytical and experimental studies on fatigue life prediction of steel and composite multi-leaf springs for light passenger vehicles using life data analysis", *Journal of Material Processing Technology*

5. Daugherty. R.L, 1981, "Composite leaf springs in heavy truck applications", *International conference on composite material proceedings of japan US conference, Tokyo*, pp. 529-538.

6. Rajendra Prasad Sahu, Dask Raj Kothari and A.K. Jain, March 2013, "Design and comparative analysis of Multi Leaf Spring using Non-Conventional Materials", *VSRD International Journal of Mechanical, Civil, Automobile and Production Engineering*, Vol. 3 No. 3 /39, e-ISSN : 2249-8303, p-ISSN : 2319-2208.

7. Nikolas Philipson and Modelan A B, "Leaf spring modelling", science park S E 22307, Lund Sweden.

8. ASM Metals Reference Book, Third edition, Michael Bauccio, Ed. ASM International, Materials Park, OH, 1993.

9. http://www.alloywire.com/titanium_alloy_grade_5.html

10. Bhagwan D. Agarwal and Lawrence J. Broutman, 1990, "Analysis and Performance of Fiber Composites", *2nd ed., John Wiley & Sons,* Inc., NY, (1990)

11. Manas Patnaik, Narendra Yadav and Ritesh Dewangan, July-Aug 2012, "Study of a Parabolic Leaf Spring by Finite Element Method & Design of Experiments", *International Journal of Modern Engineering Research (IJMER)* Vol.2, Issue 4, pp-1920-1922 ISSN: 2249-6645.

Biography

Shahrukh Shamim- Shahrukh Shamim is a B-Tech. Scholar, majoring in Material Science and Metallurgy Engineering from Maulana Azad National Institute of Technology- Bhopal, M.P., India. His research interests include Mechanical Testing of Materials, Finite Element Analysis, Computational Fluid Dynamics and Advanced Materials.

Jamil Anwer- Jamil Anwer is a B-Tech. Scholar, majoring in Mechanical Engineering from Maulana Azad National Institute of Technology- Bhopal, M.P., India. His research interests include Mechanical Testing and Finite Element Analysis.

8

Heat Transfer Enhancement Techniques in Two Phase Closed Thermosyphon: A Review

Faddas Nikhil A.[1]**, Prof. Mali Kundlik V.**[2]

[1]*M.E. (Heat Power Engineering) Student, Sinhgad College of Engineering, pune, nafaddas@gmail.com*

[2]*Asst. Professor, Department of Mechanical Engineering, Sinhgad college of Engineering pune, kvmali.scoe@sinhgad.edu*

Abstract

The performance of heat transfer is one of the most important research areas in the field of thermal engineering. Due to the high heat transfer effectiveness, thermosyphon has its own importance in the low temperature heat transfer. Researchers observed that geometrical factors and working solutions have significant influence on the performance of thermosyphon. Therefore the experimental study of thermosyphon is essential to find out the factors affecting the performance of thermosyphon. In this review paper main focus is given to parameters like filling ratio, aspect ratio, heat load, mass flow rate and inclination angle, which affects the thermal performance of thermosyphon. Also the new heat transfer enhancement techniques like use of binary mixture, use of ultrasonic wave, resurfacing and CFD analysis are enlighten. From the literature it seems to be need of binary solution, new efficient and minimum ODP and GWP refrigerants and mathematical modelling of thermal performance of thermosyphon.

Keywords: *Binary Mixture, Controllable and uncontrollable factors, Heat transfer limitations, Mathematical and computational modelling, Mechanical and surface modifications.*

1. Introduction

Due to the human need for energy, a more efficient way of using it is a major challenge in the scientific community. The thermal performance of thermosyphon is one the most important part of these types of investigation in the field of heat transfer.

1.1. *Thermosyphon*

Thermosyphon is an enclosed two phase heat transfer devices. They make use of the highly efficient heat transport process of evaporation and condensation to maximize the thermal conductance between a heat source and a heat sink. They are often referred to as thermal superconductors because they can transfer large amounts of heat over relatively large distances with small temperature differences between the heat source and heat sink. The amount of heat that can be transported by these devices is usually several orders of magnitude greater than pure conduction through a solid metal.

They are proven to be very effective, low cost and reliable heat transfer devices for applications in many thermal management and heat recovery systems. They are used in many applications including but not restricted to passive ground/road anti-freezing, baking ovens, heat exchangers in waste heat recovery applications, water heaters and solar energy systems and are showing some promise in high-performance electronics thermal management for situations which are orientation specific. They have highly used in nuclear reactor and in large electronic equipments also.

1.2. *Thermosyphon Geometry and Working Principle*

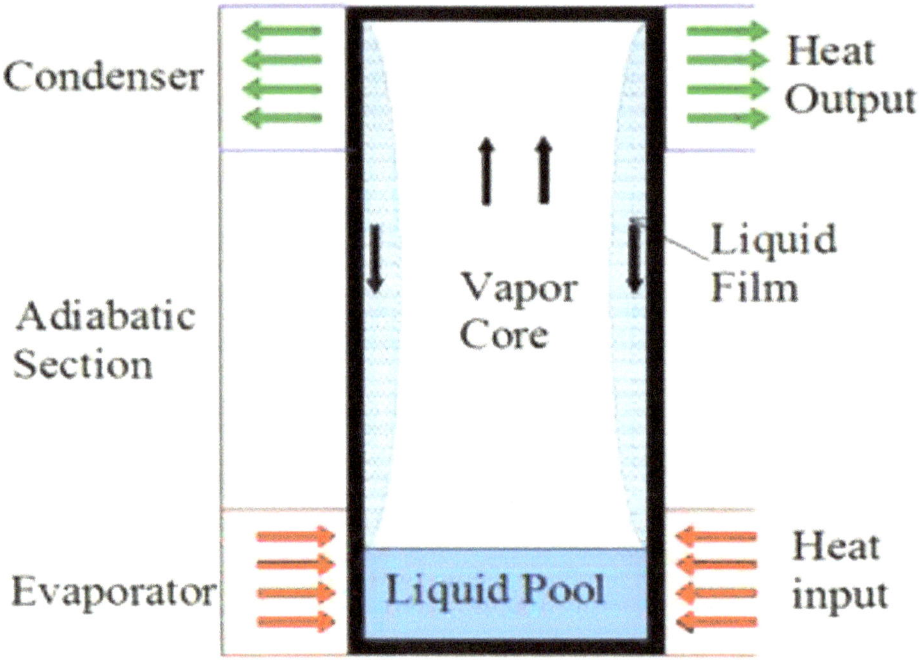

Figure 1. Two-phase closed thermosyphon working principle

A cross section of a closed two-phase thermosyphon is illustrated in Figure 1. The thermosyphon consists of an evacuated sealed tube that contains a small amount of liquid. The heat applied at the evaporator section is conducted across the pipe wall causing the liquid in the thermosyphon to boil in the liquid pool region and evaporate and/or boil in the film region. In this way the working fluid absorbs the applied heat load converting it to latent heat.

The vapour in the evaporator zone is at a higher pressure than in the condenser section causing the vapour to flow upward. In the cooler condenser region the vapour condenses and thus releasing the latent heat that was absorbed in the evaporator section. The heat then conducts across thin liquid film and exits the thermosyphon through the tube wall and into the external environment. Within the tube, the flow circuit is completed by the liquid being forced by gravity back to the evaporator section in the form of a thin liquid film. As the thermosyphon relies on gravity to pump the liquid back to the evaporator section, it cannot operate at inclinations close to the horizontal position.

1.3. *Experimental Setup*

It consists of an enclosed evacuated copper tube having evaporator section at base and condenser section at the top. 8-10 thermocouples are attached on the copper tube at similar distances. Temperature indicator displays the temperature. Coil heaters or band heaters are attached to the evaporator section for heat supply and it is controlled by controlling the voltage and current. Condenser section is surrounded by concentric cylinder through with coolant flows. Flow of coolant is measured by rotameter and controlled by a valve. For initial evacuation of tube arrangement is made to attach vacuumed pump at the top and also pressure gauge is attached to measure the pressure inside the tube.

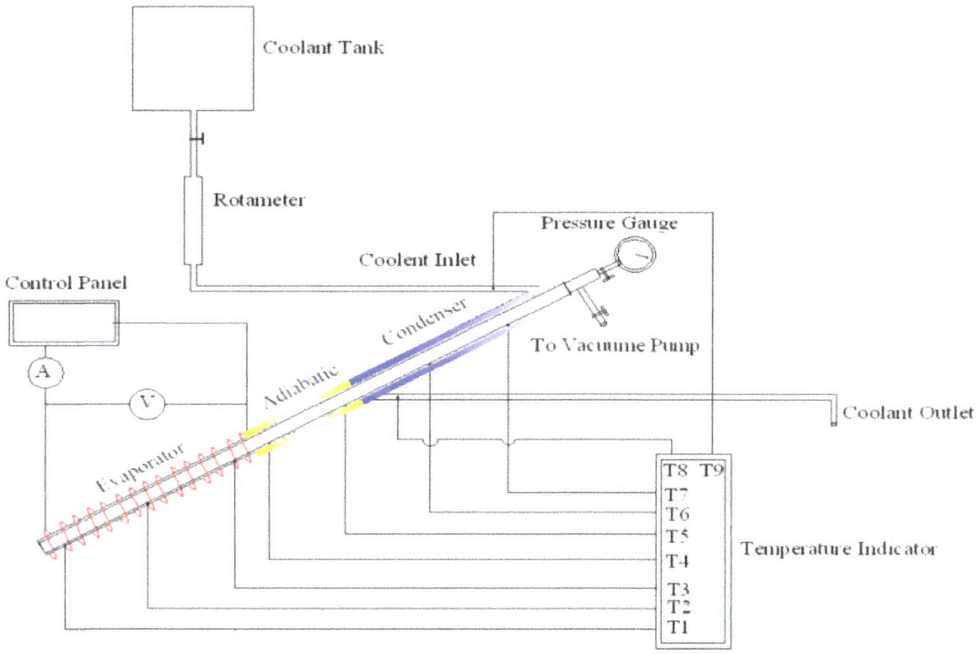

Figure 2. Experimental Setup

Following table shows the general configuration of experimental setup which may vary according to researchers requirement.

Table 1: Experimental Setup Description

Tube Material		Copper
Diameter (mm)	Internal	26
	External	32
Dimensions (mm)	Total	1000
	Evaporator	300
	Condenser	450
	Adiabatic	250
Aspect Ratio		11.53
Filling Fluid		Methanol-Ethanol Binary mixture
Inclination Angle (°)		50° to 90°
Heat Input (W)		Variable
Coolant Flow Rate (Kg/s)		Variable

2. Literature Survey

Many investigations were carried out in order to analyse and to enhance the thermal performance of thermosyphon. These are categorized into mainly three approaches.

2.1. Use of Efficient Working Fluid

In the first sort of investigation, researchers studied various working fluids as follows [1,2,3,4,5,6, 7 and 8]:

Z.Q.Long, P.Zang [1] investigated the the thermal performance of cryogenic thermosyphon charged with N2-Ar binary mixture. They have discussed heat transfer of the binary mixture in the thermosyphon theoretically by considering the mass transfer of the components. They built, an experimental setup for investigating the heat transfer performance of the cryogenic thermosyphon. They found that the N_2–Ar binary mixture can widen the operational temperature range of the cryogenic thermosyphon and it can work in the range of 64.0–150.0 K. The dry-out limit appears in the experiments for the cases with Ar fraction below 0.503. The heat transfer rate of the dry-out limit increases with the increase of Ar molar fraction until film boiling appears on the top of the condenser.

M. Karthikeyan, S. Vaidyanathan and B. Sivaraman [2] investigated the thermal performance of an inclined two phase closed thermosyphon with distilled water and aqueous solution of n-Butanol as a working fluid. They carried out the experiments for filling ratio of 60%. The thermosyphon was tested for various inclinations of 450, 600 and 900 to the horizontal. Flow rate of 0.08Kg/min, 0.1 Kg/min and 0.12 Kg/min and heat input of 40 W, 60 W and 80 W. The thermosyphon was of a copper material with inside and outside diameter of 17mm and 19mm respectively. The overall length of thermosyphon was 1000mm (400mm-evaporator length, 450mm-condenser length). They obtained the result that the thermosyphon charged with aqueous solution has the maximum thermal performance than compared to thermosyphon charged with distilled water.

H.Z. Abou-Ziyan, A. Helali, M. Fatouh and M.M. Abo El-Nasr [3] investigated the thermal performance of two phase closed thermosyphon under stationary and vibratory conditions with water and R134a as a working fluid. They carried out the experiments for filling ratio of range (40% to 80%). The thermosyphon was tested for various adiabatic lengths of (275,325 and 350mm), vibration frequency (0.0-4.33Hz) and input heat flux (160-2800 kW/m^2). They obtained the result that adiabatic length of 350mm and liquid filling ratio of 50% provide the highest heat flux.

M. R. Sarmasti Emami, S. H. Noie and M. Khoshnoodi [4] made an experimental study on the effect of aspect ratio and filling ratio on the thermal performance of inclined two-phase closed thermosyphon under normal operating conditions. They used distilled water as a working fluid. They carried out the experiments for filling ratio of range (20% to 60%) and aspect ratio of 15, 20 and 30 for an inclination angle of range (150to 900). The thermosyphon was of a copper material with inside and outside diameter of 14mm and 16mm respectively. The overall length of thermosyphon is 1000mm. They obtained the following results that the maximum thermal performance at inclination angle of 600 for all three aspect ratios and filling ratio of 45%.

K.S. Ong and Md. Haider-E-Alahi [5] investigated performance of an R134a filled thermosyphon. They carried out the experiments to study the effects of temperature difference between bath and condenser section, fill ratio and coolant mass flow rate. The thermosyphon was of a copper material with inside and outside diameter of 25.5mm and 28.2mm respectively. The overall length of thermosyphon was 780mm (300mm-evaporator length, 300mm-condenser length). They obtained the results that the heat flux transferred increased with increasing coolant mass flow rate, fill ratio and temperature difference between bath and condenser section.

Sameer Khandekar, Yogesh M. Joshi and Balkrishna Mehta [6] investigated the thermal performance of closed two-phase thermosyphon using water and various water based nanofluids (of Al_2O_3, CuO and laponite clay) as a working fluid. They observed that all these nanofluids show inferior performance than pure water.

Gabriela Humnic, Angel Huminic, Ion Morjan and Florian Dumitrache [7] performed an experiment to measure the temperature distribution and compare the heat transfer rate of thermosyphon with diluted nanofluid (with 0%, 2% and 5.3% concentration) in DI-water and DI-water. The thermosyphon was a copper tube with internal and external diameter of 13.6mm and 15 respectively. The overall of length of thermosyphon was 2000mm (evaporator length-850mm, condenser length-850mm, adiabatic section-300). They obtained the results that the addition of 5.3% (by volume) of iron oxide nanoparticles in water improved thermal performance of thermosyphon.

A.D.Patil, Dr. R.B. Yarasu [8] performed experimental and mathematical investigation on the thermal performance of a two phase closed thermosyphon charged with distilled water. The thermosyphon was a copper tube with internal and external diameter of 26 mm and 32 mm respectively. The overall of length of thermosyphon was 1000mm (evaporator length-300mm, condenser length-450mm, adiabatic section-250). Experiments were carried out on filling ratio range of 30% to 100%, inclination angle range of 40° to 90°. They obtained the results that the maximum thermal performance of thermosyphon occurred at filling ratio 45% and inclination angle 50°. Mathematical model shows an acceptable correlation between the investigated parameters.

2.2. *Mechanical and surface modification*

In the second sort of investigation researchers studied mechanical and surface modifications to enhance the thermal performance as follows [9, 10 and 11]:

P.G. Anjankar and Dr. R.B. Yarasu [9] investigated the effect of condenser length, coolant flow rate and heat load on the performance of two-phase closed thermosyphon. The thermosyphon was a closed copper tube of length 1000mm (evaporator length-300, condenser length- 450mm/400mm/350mm) and internal and external diameter of 26 and 32mm respectively. They obtained the results that thermal performance of a thermosyphon was higher at flow rate 0.0027kg/s and heat input 500W with a condenser length of 450mm.

H. Mirshahi and M. Rahimi [10] investigated experimentally the effect of heat loads, fill ratio and extra volume on the performance of a partial-vacuumed thermosyphon. They obtained the results that the change in heat flux, fill ratio and employing different extra volumes has a significant effect on the performance of thermosyphon.

Masoud Rahimi, Kayvan Asgary and Simin Jesri [11] studied the effect of the condenser and evaporator resurfacing on overall performance of thermosyphon. They obtained the result that by making the evaporator more hydrophilic and the condenser more hydrophobic the thermal performance of thermosyphon increases by 15.27% and thermal resistance decreases by 2.35 times compared with plane one.

2.3. *Mathematical and Computational Modelling*

Turning to third sort of investigation, researchers has done Numerical and computational modelling of heat transfer in thermosyphon [12 and 13]:

Asghar Alizadehdakhel, Masoud Rahimi and Ammar Abdulaziz Alsairafi [12] carried out experiments to investigate the effect of various heat loads and fill ratio on the performance of thermosyphon. They obtained the results that increasing the heat load up to certain limit increases the performance of thermosyphon further increase in heat load decreases the performance of thermosyphon. Also there is an optimum value of fill ratio for every energy input. Experimental results were compared with CFD modelling (FLUENT^{TM} version 6.2) and there was a good agreement observed between CFD and experimental results.

S. R. Raja Balayanan, V. Velmurugan, R. Sudhakaran and N. Shenbagavinayaga Moorhy [13] have been carried out experimental and theoretical research to investigate the thermal performance of water to air thermosyphon heat pipe heat exchanger. They selected independent controllable process parameters heat input, water temperature and air velocity to carry out experimental work and correlation was developed for effectiveness of heat pipe heat exchanger. They developed mathematical model using regression coefficient method which is helpful in analyzing the performance of heat pipe heat exchanger. Mathematical model shows an acceptable correlation between the investigated parameters.

3. Heat Transfer Limitations

Limitations of heat transfer are also important parameter affecting performance of heat pipe or thermosyphon. The maximum heat transfer rate of thermosyphon is limited due to the various parameters. Each working fluid has its own limiting points. These limiting parameters are as follow.

3.1. Flooding Limitation

Flooding limitation occurs due to the interaction between the counter current liquid and vapour flows occurring at the liquid-vapour interface in the thermosyphon. Liquid- vapour interface maximum occur in the evaporator section.

3.2. Boiling Limitation

Boiling limitation is due to the large liquid filling ratio and high radial heat fluxes in the evaporator section. Under this limitation, at the critical heat flux, vapour bubbles coalesce near the pipe wall prohibiting the contact of working liquid to wall surface, resulting in the rapid increase in evaporator wall temperature. This limitation affect the thermosyphon in greater extent as applying maximum heat load to evaporator, boiling of working fluid will occur to higher limit.

3.3. Dry-out Limitation

Dry-out limitation occurs due to the relatively small filling ratio. The condensate falls down along the wall and reaches the evaporator. The condensate starts evaporating and boiling by the input power and as it comes closer and closer to the bottom, the thickness of the condensate film is thinner. It eventually dries out, so the wall temperature rises from the bottom of the evaporator at the limitation.

4. Factors Affecting Thermal Performance of Thermosyphon

From the literature survey it is observed that following factors affects the thermal performance of thermosyphon.

1. Properties of working fluid
2. Filling ratio
3. Coolant flow rate
4. Coolant temperature
5. Heat load
6. Inside pressure of tube
7. Tube material properties and dimensions
8. Length of various sections (Evaporator section, adiabatic section and Condenser section).

5. Conclusion

Researchers have done experimental, mathematical and computational investigation to find out various factors affecting the thermal performance of thermosyphon and their effects. The following results are observed.

- Working fluid, filling ratio, tube material and dimensions, lengths (evaporator, condenser and adiabatic section), heat load, Coolant flow rate and temperature, operating pressure affects the thermal performance of thermosyphon.
- Copper (k = 386W/m-K) is having better thermal conductivity therefore during heat transfer it shows very small variation in temperature distribution of entire tube which is favourable condition for effective heat transfer. Also it is most economical metal to use as a tube material.
- For the effective heat transfer surface area of condenser section should be greater than the surface area of evaporator section. This condition can be achieved by varying diameter or length of sections.
- For lower temperature range, refrigerants show effective heat transfer performance. Considering the effect of global warming due to the refrigerants having high global worming potential (GWP), it is necessary to use and research new refrigerants having less GWP.
- Considering the flooding and dry-out limitations the filling ratio between the ranges of 45% to 65% show the best heat transfer performance.
- Evacuation of thermosyphon tube is compulsory to eliminate the inferior effects of non condensable gases. So considering the boiling point of working fluid and effect of non condensable gases, inside pressure of tube should be kept at appropriate level.
- Circulation of working fluid in the tube complete due to the gravity effect, so thermosyphon can't work at horizontal position. Heat transfer performance is superior between the angles of 50° to 90°.
- As per the necessity of heat transfer coolant temperature and coolant flow rate can be controlled and varied.

REFERENCES

[1] Z.Q.Long, P.Zang, 2013, "Heat transfer characteristics of thermosyphon with N_2–Ar binary mixture working fluid", *International journal of heat and mass transfer*, Vol. 63, pp 204–215.

[2] M. Karthikeyan, S. Vaidyanathan and B. Sivaraman, 2010, "Thermal performance of a two Phase closed thermosyphon using aqueous solution", *International Journal of Engineering Science and Technology*, Vol. 2(5), pp 913-918.

[3] H.Z. Abou-Ziyan, A. Helali, M. Fatouh and M.M. Abo El-Nasr, 2001, "Performance of stationary and vibrated thermosyphon working with water and R134a", *Applied thermal engineering*, Vol. 21, pp 813-830.

[4] M. R. Sarmasti Emami, S. H. Noie and M. Khoshnoodi, 2008, "Effect of aspect ratio and filling ratio on thermal performance of an inclined two-phase closed thermosyphon", *Iranian Journal of Science & Technology, Transaction B, Engineering*, Vol. 32, No. B1, pp 39-51.

[5] K.S. Ong and Md. Haider-E-Alahi, 2003, "Performance of an R-134a filled thermosyphon", *Applied thermal engineering*, Vol. 23, pp 2373-2381.

[6] S. Khandekar, Y. M. Joshi and B. Mehta, 2008, "Thermal performance of closed two-phase thermosyphon using Nanofluids", *International Journal of Thermal Science*, Vol. 47, pp 659-667.

[7] G. Humnic, A. Huminic, I. Morjan and F. Dumitrache, 2011, "Experimental study of the thermal performance of thermosyphon heat pipe using iron oxide nanoparticles", *International Communications in Heat and Mass Transfer*, Vol. 54 , pp 656–661.

[8] Patil A.D., Dr. R.B. Yarasu, 2013, "Experimental and Mathematical Investigation on the Thermal Performance of a Two Phase Closed Thermosyphon Charged with Distilled Water", *Proceedings of the National conference on Advances in Engineering, Technology and Applied Sciences, SPANDAN-13,* pp.726-736.

[9] P.G. Anjankar and Dr. R.B. Yarasu, 2012, "Experimental Analysis of Condenser Length Effect on the Performance of Thermosyphon", *International Journal of Emerging Technology and Advanced Engineering*, Vol. 2, Issue 3, pp 494-499.

[10] H. Mirshahi and M. Rahimi, 2009, "Experimental Study on the Effect of Heat Loads, Fill Ratio and Extra Volume on Performance of a Partial-Vacuumed Thermosyphon", *Iranian Journal of Chemical Engineering*, Vol. 6, No. 4 (autumn), IAChE.

[11] M. Rahimi, K. Asgary and S. Jesri, 2010, "Thermal characteristics of a resurfaced condenser and evaporator closed two phase thermosyphon", *International Communications in Heat and Mass Transfer*, Vol. 37, pp 703–710.

[12] A. Alizadehdakhel, M. Rahimi and A. A. Alsairafi, 2010, "CFD modelling of flow and heat transfer in a thermosyphon", *International Communications in Heat and Mass Transfer* ,Vol. 37, pp 312–318.

[13] S. R. Raja Balayanan, V. Velmurugan, R. Sudhakaran and N. S. Moorhy, 2011, "Optimization of Thermal Performance of Water to Air Thermosyphon Solar Heat Pipe Heat Exchanger using Response Surface Methodology", *European Journal of Scientific Research* ,Vol. 59, pp 451-459.

9

Experimental Analysis of Heat Transfer Enhancement through Porous Medium

Ashwin A. Kamble[1] , Sanjaykumar A. Borikar[2]

[1]P.G. Student, Department of Mechanical Engg, K.I.T.S, Ramtek, R.T.M. Nagpur University, Maharastra, India E-mail address: ashwinkamble31@gmail.com
[2] Asst. Professor, Department of Mechanical Engg, K.I.T.S, Ramtek, RTM Nagpur University, Maharastra, India , E-mail address: sanjaykumarborikar@gmail.com
Author Correspondence: Department of Mechanical Engg, K.I.T.S, Ramtek, Telephone: +919579940720 Email address: ashwinkamble31@gmail.com

Abstract

Heat transfer is the basic phenomenon in thermal application. It plays a vital role in heating and cooling of the material and system as well as it is very important phenomenon in many of the thermal applications like internal combustion engine for its cooling and lubrication, Heat Exchanger to transmit the heat energy from one system to another in refrigeration, air conditioning etc., Heat transfer enhancement motivate to identify new technique and the material through which it can be improved.
Enhancement in heat transfer depends upon the various dimensionless parameters along with porosity of the porous material and its aspect ratio. It has been proven that the porous material can enhance the heat transfer rate. During the experimentation by the researchers it is observed that the to evaluate the heat transfer rate from the system and the result were quiet impressive to extend the study for further research.

Keywords: heat, porous material, dimensionless number, porosity.

1. Introduction

Comparative analysis of heat transfer on the basis of temperature variation. Evaluated the performance testing of heat transfer rate without using porous material and with using porous material i.e., aluminum chips.

1.1 Heat Transfer

Heat transfer is defined as transfer of heat from one body to another due to temperature gradient. The convective heat transfer is one mode of heat transfer from hot surface to surrounding between two fluids. It can be increase by increasing the surface area, heat transfer coefficient or temperature gradient. Enhancement occurs due to passage of fluid through the open, interconnected void structure; exposing the fluid to large internal surface area.

Modes of Heat Transfer

Heat transfer is defined as transmission of energy from one region to another as a result of temperature gradient takes place by the following modes:

(i) Conduction *(ii)* Convection *(iii)* Radiation

1.2 POROSITY

A **porous medium** (or a **porous material**) is a material containing pores (voids). The skeletal portion of the material is often called the "matrix" or "frame". The pores are typically filled with a fluid (liquid or gas). Thus, it causes increase in heat transfer in the flow field. The skeletal material is usually a solid, but structures like foams are often also usefully analyzed using concept of porous media. Porous mediums have been identified as a means to reduce the thermal resistance at the gas/solid interface and enhance heat transfer rates. By using the porous medium as transport medium, fluid mixing and the heat exchange area between the solid matrix and fluid phase, flowing through the porous medium, increase and this leads to significant enhancement in heat transfer.

Porosity (ϕ) is defined as a ratio of pore volume (V_P) to bulk volume (V_B) of a reservoir rock.

$$\phi = \frac{V_P}{V_B}$$

$V_P = V_B - V_S$

Where is volume occupied by the solid grains of the rock. The pore spaces in reservoir rocks are most frequently the intergranular spaces between the sedimentary particles. The void space is microscopic in scale, rarely exceeding a few, or very few, tens of microns. The higher porosity, larger pore diameter foam layers produce the greatest enhancements in heat transfer. The porous medium with random porosities is used to enhance heat transfer and the random porosities Therefore, a porous medium with larger porosity and the proper bead diameter could provide more heat dissipation.

The porosity is one of the main parameters for studying porous media. Two major models are used to define the distribution of porosity in porous media for conveniently investigating the phenomena of thermal and flow fields. The one in which the porosity distributed in the porous medium is uniform is called the constant porosity model.

Variable porosity model : The one in which there is a variation in porosity and is assumed to be an exponential function of the distance from solid wall. Analysis shows that the porous material enhanced the heat transfer rate and also increases the efficiency over heat exchanger.

Examples of porous materials include: dry wall, wood, concrete, gunite, rubber, some plastic.

Porous media has a wide range of applications such as catalytic and inert packed bed reactors, enhancing drying efficiency such as food drying, filtering, geothermal energy management and harvesting, insulation, lubrication.

2. Specification Of Experimental Set Up

S.N.	Elements	Description
1	Aluminum block	Dimension= 120 *50*150 mm Thickness of Aluminum Sheet = 3 mm
2	Bakelite	Thickness of Bakelite Sheet = 4mm
3	Heater	Capacity of Heater =1000 W
4	Digital Thermometer	Temp. Range = - 50 °C to 70 °C.
5	Water pump	Capacity of water pump = 0.2 hp
6	Digital temp. indicator	Temp. Range = -200 ° C to 800 °C
7	Down Flow Radiator	No. tubes=10, length=30 cm, OD = 3mm, ID= 2mm, pitch =1.5 cm

The setup consists of basic elements as-

a) Aluminum rectangular model
b) Heater
c) Bakelite material (Insulator)
d) Water pump
e) Down Flow Radiator
f) Temperature indicator and Digital thermometer

2.1 Experimental Set Up

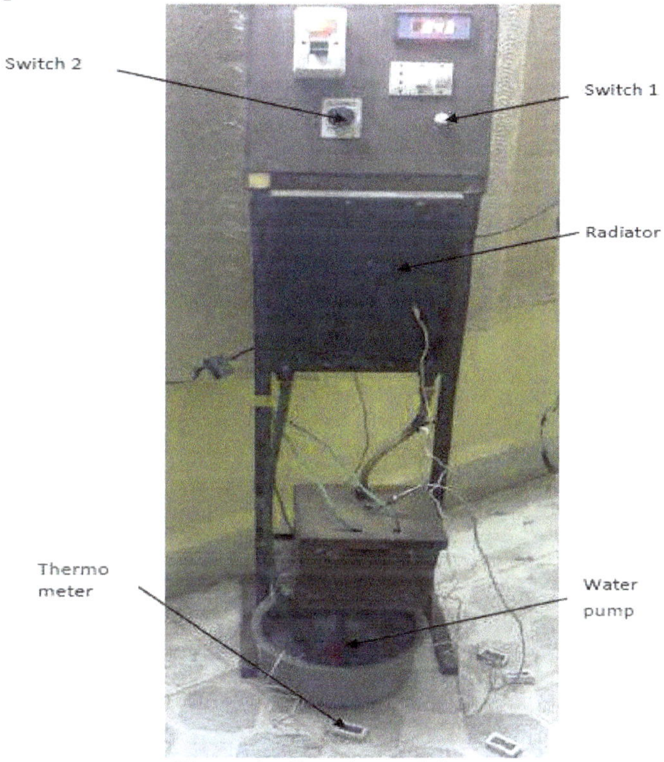

Figure 1: Experimental Setup

2.2 Experimental Procedure

The steps carried out for experimentation are as follows :

a) Fill up the tank with fresh water.
b) Make the proper connection and put the main Switch on.
c) Put knob 1 on which makes temperature indicator shows some reading
d) Put on switch 2 which makes the water circulating through the system right from tank to via. Aluminum block and radiator and return back to the tank again and also start the radiator fan.
e) Now adjust the heater by regulator 2 at some specific watt.
f) Now the water from the tank goes the heat making water warm and then circulate towards radiator here we note the temperature T_1, T_2, resp. T_1 is temperature at inlet and T_2 outlet of the aluminum block with the help of thermocouple at regular interval of time for 200 watt.
g) Here note surface temperature of aluminum block i.e., T_s.
h) Now warm water flow through radiator to dissipate the heat.
i) By flowing through no. of coils and radiator fan fitted back and the coil blows the air on tubes which makes dissipate the heat from water to temperature i.e, Ta and T_b resp.
   Ta is the temperature at the back of radiator fan.
   T_b is the temperature at the front of radiator.

j) Now the cold water recirculate to back to the tank.
k) Now same above procedure is repeat again by inserting porous material i.e., Aluminum chips is in specific proportion i.e., 50 gm. For the same coil watt.

3. Testing and Experimentation

3.3.1 Testing of experimental set up without porous material with input 200 W

In this table after every five minute of time interval the temperature like T_1, T_2, T_s, T_a T_b, T_{ro}, is note down. On the basis of that the heat gain by water ($Q = mC_p\Delta T$) heat transfer coefficient and Nusselt number is calculated.

Table 1: Testing of experimental set up without porous material input at 200 W

Time interval (min.)	Inlet temp (T_1) ℃	Outlet temp. (T_2) ℃	Surface temp. (T_s) ℃	Radiator temp. air in (T_a) ℃	Radiator temp. air out (T_b) ℃	Radiator water outlet temp. (T_{ro}) ℃	Heat gain by water $Q = mC_p\Delta T$ Watt	Heat transfer coefficient h (w/m²k)	Nusselt no. N_u	$Q' = hAs\Delta T$ Watt
5	26.8	27.2	43	27	27.8	25.4	29.74	46.61592	3.694451	53.00197
10	27	27.5	49	27	27.8	25.4	37.18	48.00307	3.780064	66.161
15	27.1	27.9	56	27	27.9	25.5	59.49	50.20056	3.899598	86.9714
20	27.3	28.3	63	27	27.9	25.6	79.57	50.95605	3.924479	108.3764
25	27.5	28.7	70	27	27.9	25.7	91.47	52.94101	4.046497	129.941
30	27.7	28.9	75	27	27.9	25.8	93.70	52.16201	3.958199	145.6216
35	27.8	29.2	81	27.1	27.9	25.9	107.09	49.93177	3.762391	164.4119
40	28	29.8	86	27.1	28	26	139. 06	54.33709	4.082153	179.174
45	28.4	30.4	90	27.2	28	26.1	150.22	54.69079	4.088047	191.4802
50	28.7	30.9	94	27.2	28	26.2	169.55	59.14945	4.405024	203.478
55	29.1	31.5	99	27.2	28	26.3	181.45	65.66323	4.872162	218.2281
60	29.4	32	105	27.2	28	26.5	198.56	67.72895	5.017327	236.707
65	29.6	32.8	110	27.2	28	26.7	244.67	69.85127	5.166211	251.9114
70	29.9	33.6	114	27.2	28	26.9	257.31	75.51279	5.576770	263.9613
75	30	34.1	119	27.2	28	27	309.37	78.65759	5.798844	287.9571

3.3.2 Testing of experimental set up without porous material with input 200 W

On the basis of temperature like T_1,T_2,T_s and area, perimeter of Aluminum block we calculated Re, P_r, Nusselt no. N_u.

Table 2: Testing of experimental set up without porous material input at 200 W

T_f	D_h	k	u	v	T_b	$Re = (u\ D_h)/v$	P_r	h_t	Nusselt no. N_u
47.825	0.0505	0.6372	0.00268	0.000000582	27.65	232.5429553	3.85	71.59645	5.853961073
53.2	0.0505	0.6413	0.00268	0.000000538	28.4	251.5613383	3.49	71.93642	5.851071128
59.025	0.0505	0.6501	0.00268	0.000000485	29.05	279.0515464	3.09	72.81103	5.824157757
65.325	0.0505	0.6557	0.00268	0.000000449	29.65	301.4253898	2.82	73.34829	5.840928328
71.55	0.0505	0.6607	0.00268	0.000000415	30.1	326.1204819	2.56	73.79389	5.837265529
76.325	0.0505	0.6655	0.00268	0.000000387	30.65	349.7157623	2.39	74.33725	5.838223434

82.525	0.0505	0.6702	0.00268	0.000000355	31.05	381.2394366	2.15	74.7373	5.834276275
85.775	0.0505	0.6722	0.00268	0.000000344	31.55	393.4302326	2.08	74.94992	5.823351096
91.575	0.0505	0.6756	0.00268	0.000000327	32.15	413.883792	1.94	75.20669	5.833273649
96.4	0.0505	0.6781	0.00268	0.00000031	32.8	436.5806452	1.84	75.488	5.837735755
101.375	0.0505	0.6806	0.00268	0.000000291	33.75	465.0859107	1.73	75.7767	5.825821239
105.725	0.0505	0.6817	0.00268	0.000000281	34.45	481.6370107	1.65	75.81888	5.820662211
110.6	0.0505	0.6828	0.00268	0.00000027	35.2	501.2592593	1.58	75.91903	5.821333211
115.125	0.0505	0.6838	0.00268	0.000000258	36.25	524.5736434	1.52	760741	5.821011821
120.1	0.0505	0.685	0.00268	0.000000247	37.2	547.9352227	1.44	76.16631	5.816523024

3.3.3 Testing of experimental set up with porous material with input 200 W at 50 gm chips

In this table after every five minute of time interval the temperature like T_1 , T_2 , T_s ,T_a T_b , T_{ro}, is note down. On the basis of that the heat gain by water ($Q = mC_p\Delta T$) heat transfer coefficient and Nusselt number is calculated.

Table 3: Testing of experimental set up with porous at 50 gm chips with material input at 200W

Time interval (min.)	Inlet temp. (T_1) ℃	Outlet temp. (T_2) ℃	Surface temp. (T_s) ℃	Radiator temp. air in (T_a) ℃	Radiator temp. air out (T_b) ℃	Radiator Water outlet temp. (T_{ro}) ℃	Heat gain by water $Q = mC_p\Delta T$ Watt	Heat transfer coefficient h (w/m²k)	Nusselt no. N_u	Q'= hAs (ts-tb)
5	27.2	28.2	44	27	28	26	60.17	87.89074	7.14041558	48.502
10	27.4	28.5	51	27	28	26.1	66.19	68.37104	5.50939416	69.028
15	27.9	29.6	57	27	28	26.1	102.3	86.21997	6.89814436	85.044
20	28.1	29.9	65	27	28	26.2	108.31	71.6336	5.68967708	109.11
25	28.5	30.4	72	27	28	26.3	114.33	63.97516	5.03780655	130.29
30	28.9	31.2	77	27	28	26.4	138.40	70.18611	5.50885732	143.94
35	29.3	32.1	81	27	28	26.6	168.47	79.74534	6.22817745	155.07
40	29.7	32.9	87	27	28	26.7	192.51	82.29033	6.39234226	172.28
45	30	33.7	93	27.1	28	26.9	222	86.4385	6.68680186	190.19
50	30.4	34.4	97	27.1	28	27	240.7	88.71443	6.83981507	201.46
55	30.9	35.2	103	27.1	28	27.2	258.7	88.05609	6.75707760	218.54
60	31.2	35.9	108	27.1	28	27.4	282.8	90.44101	6.91801135	233.57
65	31.6	36.7	113	27.1	28	27.5	306.9	92.67144	7.05655548	248.41
70	31.9	37.2	118	27.1	28	27.7	318.4	90.84425	6.89351533	263.52
75	32.3	38	122	27.1	28	28	343	94.03186	7.11508647	275.04

3.3.4 Testing of experimental set up with porous material with input 200 W at 50 gm chips

On the basis of temperature like T_1,T_2,T_s and area, perimeter of Aluminum block we calculated Re, P_r , Nusselt no. N_u.

Table 4: Testing of experimental set up with porous material at 50 gm chips with input at 200 Watt

T_f	D_h	k	u	v	T_b	P_r	Re = $(uD_h)/v$	h_t	N_u
34.35	0.0505	0.6216	0.00306	0.000000728	27.7	4.98	212.2664835	70.84788	5.75582005
39.475	0.0505	0.6267	0.00306	0.00000067	27.95	4.49	230.641791	71.30283	5.745640252

42.875	0.0505	0.6312	0.00306	0.000000647	28.75	4.24	238.8408037	71.67676	5.734594797
47	0.0505	0.6358	0.00306	0.000000592	29	3.86	261.0304054	72.16772	5.732100787
50.725	0.0505	0.6413	0.00306	0.000000554	29.45	3.68	278.9350181	72.90852	5.74127606
53.525	0.0505	0.6434	0.00306	0.000000538	30.05	3.49	287.2304833	72.99823	5.72957821
55.85	0.0505	0.6466	0.00306	0.000000513	30.7	3.35	301.2280702	73.40321	5.7328523
59.15	0.0505	0.6501	0.00306	0.000000485	31.3	3.09	318.6185567	73.64419	5.72070679
62.425	0.0505	0.6528	0.00306	0.00000046	31.85	2.98	335.9347826	74.05615	5.728914777
64.7	0.0505	0.655	0.00306	0.000000445	32.4	2.86	347.258427	74.25495	5.724999719
68.025	0.0505	0.6581	0.00306	0.000000433	33.05	2.69	356.8822171	74.38906	5.708322931
70.775	0.0505	0.6602	0.00306	0.000000417	33.55	2.62	370.5755396	74.69882	5.713859907
73.575	0.0505	0.6632	0.00306	0.000000398	34.15	2.49	388.2663317	75.01081	5.711770113
76.275	0.0505	0.6655	0.00306	0.000000387	34.55	2.39	399.3023256	75.18719	5.705413799
78.575	0.0505	0.6674	0.00306	0.000000374	35.15	2.31	413.1818182	75.40264	5.705474098

3.3.5 Testing of experimental set up without porous material with input at 400 Watt.

In this table after every five minute of time interval the temperature like T_1, T_2, T_s, T_a T_b, T_{ro}, is note down. On the basis of that the heat gain by water ($Q = mC_p\Delta T$) heat transfer coefficient and Nusselt number is calculated.

Table 5 : Testing of experimental set up without porous material with input at 400 Watt

Time interval (min.)	**Inlet temp. (T_1) ℃**	**Outlet temp. (T_2) ℃**	**Surface temp. (T_s) ℃**	**Radiator temp. air in (T_a) ℃**	**Radiator temp. air out (T_b) ℃**	**Radiator outlet temp. (T_{ro}) ℃**	**Heat gain by water $Q = mC_p\Delta T$ Watt**	**Heat transfer coefficient h (w/m²k)**	**Nusselt no. N_u**	**$Q' = hA_s (T_s - T_b)$**
5	27	27.5	54	26.8	27.6	25.3	37.18	33.09301	2.657756	81.7562
10	27.3	27.9	65	26.8	27.6	25.4	44.62	28.40591	2.261189	115.239
15	27.6	28.6	76	26.8	27.7	25.5	74.36	36.96192	2.906536	149.443
20	28	29.1	88	26.8	27.7	25.7	89.29	35.76034	2.782584	186.811
25	28.4	29.7	101	26.8	27.7	25.9	96.67	31.98981	2.464133	228.273
30	28.7	30.2	114	26.8	27.8	26	111.55	31.4128	2.401009	270.097
35	29	30.8	127	26.8	27.8	26	133.86	32.8233	2.485123	312.753
40	29.3	31.4	136	26.8	27.8	26.2	156.17	35.19483	2.651557	341.523
45	29.6	31.8	145	26.8	27.8	26.3	163.61	34.08116	2.557353	371.301
50	30	32.4	153	26.8	27.8	26.5	178.04	34.80335	2.600724	396.944
55	30.4	33	161	26.8	27.8	26.7	193.35	35.6038	2.651906	422.174
60	30.7	33.6	169	26.9	27.8	26.9	223.1	38.81553	2.880929	448.427
65	30.9	34.3	176	26.9	27.8	27.1	252.85	41.98213	3.111842	470.738
70	31.2	34.9	184	26.9	27.8	27.3	275.16	43.40132	3.212321	495.401
75	31.6	36	193	26.9	27.8	27.5	327.22	48.93814	3.617354	523.466

3.3.6 Testing of experimental set up with porous material with input at 400 Watt

On the basis of temperature like T_1,T_2,T_s and area, perimeter of Aluminum block we calculated Re, P_r, Nusselt no. N_u.

Table 6: Testing of experimental set up with porous material with input at 400 Watt

T_f	**D_h**	**K**	**u**	**T_b**	**v**	**$Re = (uD_h)/v$**	**P_r**	**h_t**	**N_u**
40.625	0.0505	0.6288	0.00378	27.25	0.000000657	290.5479	4.34	72.7692	5.844218473
46.3	0.0505	0.6344	0.00378	27.6	0.0000006	318.15	3.93	73.36332	5.839923741
52.05	0.0505	0.6422	0.00378	28.1	0.000000542	352.1956	3.56	74.28325	5.841333503

58.275	0.0505	0.649	0.00378	28.55	0.000000497	384.0845	3.14	74.81755	5.821704537
65.025	0.0505	0.6556	0.00378	29.05	0.000000449	425.1448	2.82	75.5397	5.818723317
71.725	0.0505	0.6607	0.00378	29.45	0.000000415	459.9759	2.58	76.06017	5.813589254
78.45	0.0505	0.667	0.00378	29.9	0.000000377	506.3395	2.31	76.68915	5.806299892
83.175	0.0505	0.6703	0.00378	30.35	0.000000353	540.7649	2.13	76.9664	5.798602696
87.85	0.0505	0.673	0.00378	30.7	0.000000333	573.2432	2.03	77.34484	5.803736114
92.1	0.0505	0.6758	0.00378	31.2	0.00000032	596.5313	1.93	77.59488	5.798374778
96.35	0.0505	0.678	0.00378	31.7	0.00000031	615.7742	1.84	77.73993	5.790363749
100.575	0.0505	0.6804	0.00378	32.15	0.000000293	651.5017	1.74	78.01862	5.790623384
104.3	0.0505	0.6813	0.00378	32.6	0.000000283	674.523	1.69	78.1594	5.793409191
108.525	0.0505	0.6823	0.00378	33.05	0.000000275	694.1455	1.61	78.14029	5.78350406
113.4	0.0505	0.6832	0.00378	33.8	0.000000263	725.8175	1.55	78.28821	5.786818929

3.3.7 Testing of experimental set up with porous material at 50 gm chips with input at 400 Watt

In this table after every five minute of time interval the temperature like T_1, T_2, T_s, T_a T_b, T_{ro}, is note down. On the basis of that the heat gain by water ($Q = mC_p\Delta T$) heat transfer coefficient and Nusselt number is calculated.

Table 7: Testing of experimental set up with porous material at 50 gm chips with input at 400 Watt

Time interval (min.)	Inlet temp (T_1) ℃	Outlet temp. (T_2) ℃	Surface temp. (T_s) ℃	Radiator temp. air in (T_a) ℃	Radiator temp. air out (T_b) ℃	Radiator water outlet temp. (T_{ro}) ℃	Heat gain by water $Q = mC_p\Delta T$ Watt	Heat transfer coefficient h (w/m^2k)	Nusselt no. N_u	$Q'=hA_s(t_s - t_b)$
5	26.8	27.8	55	27.1	27.9	25.1	66.19	56.89359	4.563415	83.0008
10	27	28.4	67	27.1	27.9	25.2	84.24	51.03599	4.05366	119.137
15	27.3	29.3	78	27.1	28	25.3	120.3	57.6315	4.525565	152.305
20	27.5	29.9	89	27.1	28	25.5	144.4	57.0165	4.429744	186.348
25	27.8	30.5	103	27.1	28	25.7	162.47	52.38095	4.029918	230.246
30	28	31.1	116	27.1	28	25.8	186.55	51.37845	3.914032	272.489
35	28.4	31.7	128	27.1	28	26	198	48.12951	3.639078	310.252
40	28.7	32.5	138	27.1	28	26.2	228	50.54536	3.803517	341.384
45	29	33.1	146	27.1	28	26.4	246	50.95383	3.855512	361.953
50	29.3	33.7	155	27.1	28	26.5	264	50.89647	3.800491	395.450
55	29.6	34.5	162	27.1	28	26.7	294	53.86687	4.00807	416.805
60	29.9	35.3	171	27.1	28	27	318	54.70685	4.059207	445.506
65	30.2	35.9	177	27.1	28	27.2	343	56.73266	4.203961	463.971
70	30.6	36.5	186	27.2	28	27.4	355	55.44363	4.10122	491.799
75	31.2	37.9	194	27.2	28	27.6	385	57.48929	4.247562	514.831

3.3.8 Testing of experimental set up with porous material at 50 gm chips with input at 400 Watt

On the basis of temperature like T_1,T_2,T_s and area, perimeter of Aluminum block we calculated Re, P_r, Nusselt no. N_u.

Table 7: Testing of experimental set up with porous material at 50 gm chips with input at 400 Watt

T_f	D_h	K	u	v	T_b	$Re = (uD_h)/v$	P_r	h_t	N_u
41.15	0.0505	0.6296	0.00306	0.000000671	27.3	230.2980626	4.29	71.34336	5.722426069

47.35	0.0505	0.6358	0.00306	0.000000591	27.7	261.4720812	3.86	72.17821	5.732934497
53.15	0.0505	0.6431	0.00306	0.000000538	28.3	287.2304833	3.49	72.96419	5.72957821
58.85	0.0505	0.65	0.00306	0.000000497	28.7	310.9255533	3.14	73.57978	5.716583114
66.075	0.0505	0.6564	0.00306	0.000000445	29.15	347.258427	2.78	74.23243	5.711056603
72.775	0.0505	0.6629	0.00306	0.0000004	29.55	386.325	2.53	75.04729	5.717134374
79.025	0.0505	0.6679	0.00306	0.00000037	30.05	417.6486486	2.27	75.41563	5.702184896
84.3	0.0505	0.6711	0.00306	0.000000349	30.6	442.7793696	2.11	75.68164	5.695012215
88.525	0.0505	0.6674	0.00306	3.462E-07	31.05	446.3604853	2	74.97128	5.672833978
93.25	0.0505	0.6763	0.00306	0.000000319	31.5	484.4200627	1.92	76.23872	5.692821834
97.025	0.0505	0.6787	0.00306	0.000000309	32.05	500.0970874	1.82	76.36735	5.682262188
101.8	0.0505	0.6806	0.00306	0.000000291	32.6	531.0309278	1.73	76.64228	5.686798816
105.025	0.0505	0.6815	0.00306	0.000000281	33.05	549.9288256	1.67	76.74146	5.686638207
109.775	0.0505	0.6827	0.00306	0.000000272	33.55	568.125	1.6	76.80891	5.6816317
114.275	0.0505	0.6835	0.00306	0.000000261	34.55	592.0689655	1.53	76.87614	5.679948665

4. Result And Discussion

4.4.1 The Graph on the variation in Nusselt number with time input at 200 W

The following graph is plotted between Nusselt number verses time at 200 Watt. The graph shows the comparison of Nusselt number between with and without porous material. On the basis of graph at 60 gm chips Nusselt number increases as compared to the without porous media and also at 50 gm chips.

It also concluded that for constant heat supply Nusselt number is increased for 60 gm chips

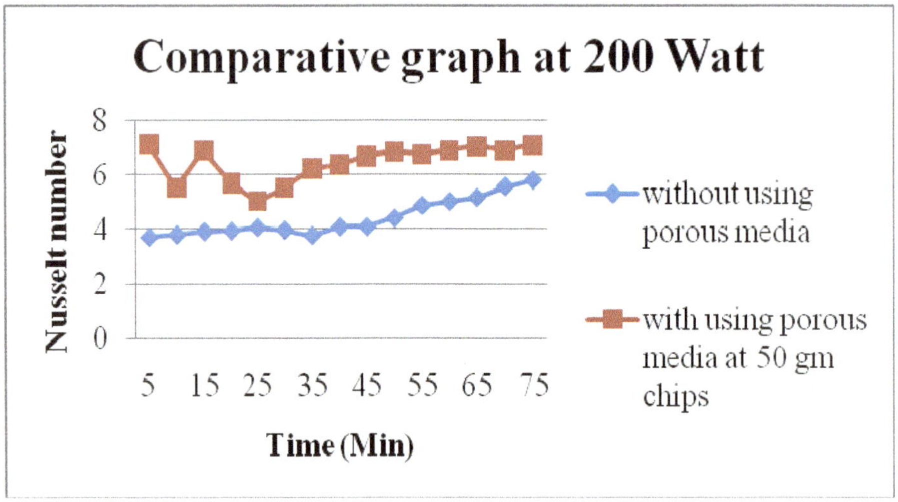

Graph 1: The variation in Nusselt number with time input at 200 W

4.4.2 The graph on the variation in Heat transfer with time input at 200 W

The following graph is plotted between heat transfer rate verses time at 200 Watt. The graph shows the comparison of heat transfer rate between with and without porous material. From the graph it has been concluded that the heat transfer rate is increased for 60 gm chips of porous media it increased by 68 % .It is also observed that as the porosity decreases the heat transfer rate is increases.

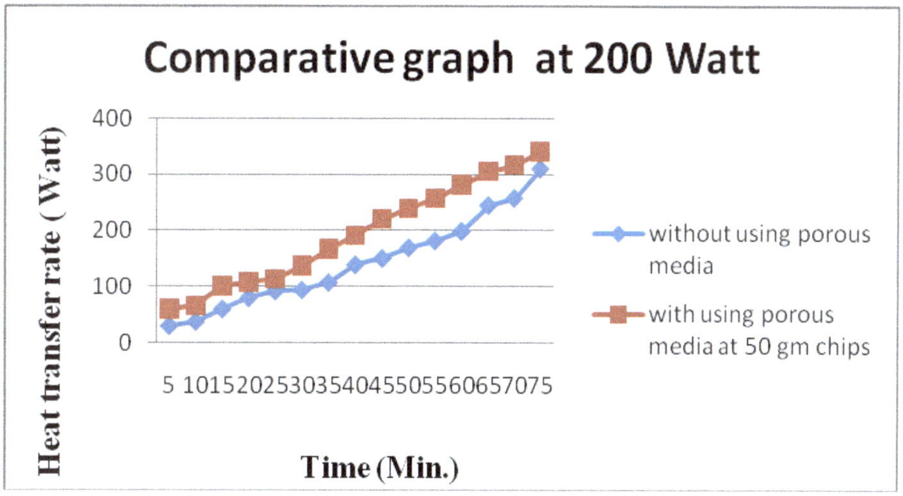

Graph 2: The variation in Heat transfer with time input at 200 W

4.4.3 Graph on the variation in Nusselt number with Time input at 400 W

The following graph is plotted between Nusselt number verses time at 400 Watt. The graph shows the comparison of Nusselt number between with and without porous material. On the basis of graph at 60 gm chips Nusselt number increases as compared to the without porous media and also at 50 gm chips.It also concluded that for constant heat supply Nusselt number is increased for 60 gm chips.

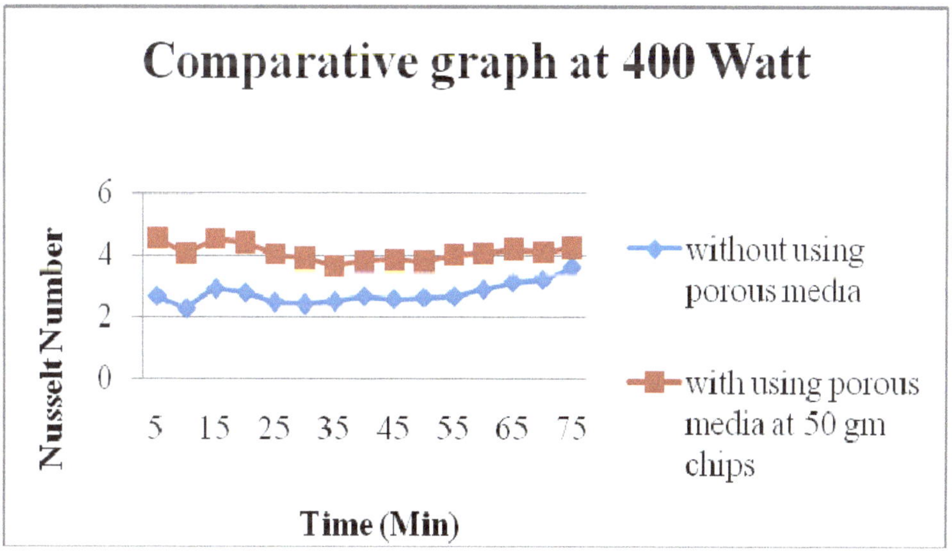

Graph 3: The variation in Nusselt number with Time input at 400 W

4.4.4 Graph on the variation in Heat Transfer rate with Time input at 400 W

The following graph is plotted between heat transfer rate verses time at 400 Watt. The graph shows the comparison of heat transfer rate between with and without porous material. From the graph it has been concluded that the heat transfer rate is increased for 60 gm chips of porous media it increased by 71% .
It is also observed that as the porosity decreases the heat transfer rate is increases.

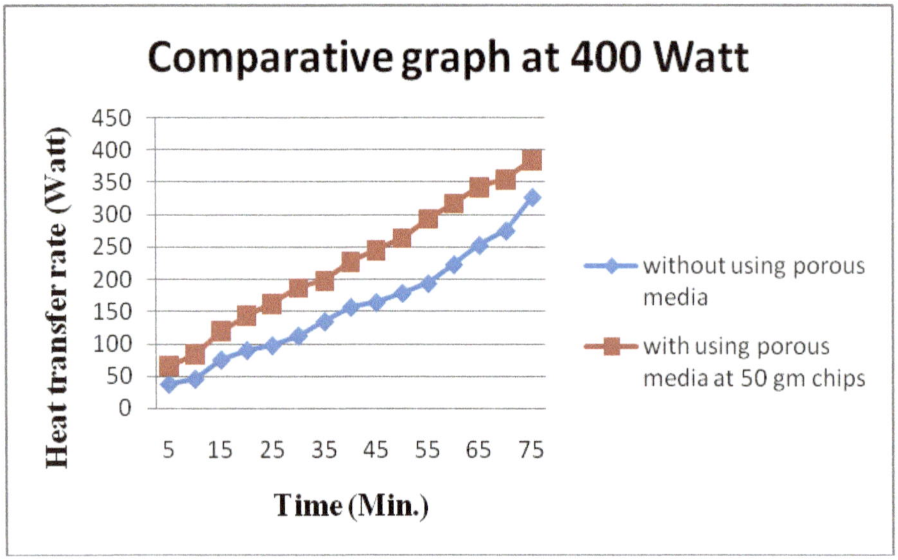

Graph 4: The variation in Heat Transfer rate with Time input at 400 W

5. CONCLUSIONS

On the basis of experimentation carried out on different temperature and different condition. It has been concluded that as the quantity of porous media increases the heat transfer rate will get increased. Overall conclusion is as follows.

i. Heat transfer rate at 200 W without using porous media is 309 Watt.
ii. Heat transfer rate at 400 W without using porous media is 327.22 Watt.
iii. Heat transfer rate at 200 W for 50 gm chips is 343 Watt.
iv. Heat transfer rate at 400 W for 50 gm chips is 385 Watt.

From the experimentation the heat transfer rate is more by using porous material as compare to the 50 gm chips of porous material and without porous material.

Heat transfer rate also govern by the maximum quantity of chips.

REFERENCES

[1] Wirtz Richard.A, (1997) "A semi-Empirical model for porous media heat exchanger design", Proceedings American Society of Mechanical Engineers National Heat Transfer Conference, Baltimore, MD. Pp1-8.

[2] Togashi Hisae, Yuki Kazuhisa, Hashizume Hidetoshi, "Heat transfer enhancement technique with copper fiber porous media" Dept. of Quantum Sci. and Eng., Tohoku Univ., Sendai, Miyagi

[3] Boomsma K, Poulikakos D. Zwick F, (2003) "Metal foams as compact high performance heat exchangers" Mechanics of Materials 35, p.p. 1161-1176.

[4] Baez Elsa, Nicolas Alfredo, (2007), "Natural convection fluid flow and heat transfer in porous media", Journal of Mechanics of Materials and Structures Vol. 2, No. 8, P.P. 1571-1584.

[5] Yang C, Liu W, Nakayama A., (2009) , "Forced convective heat transfer enhancement in a tube with its core partially filled with a porous medium" The Open Transport Phenomena Journal, School of Energy and Power Engineering, Huazhong University of Science and Technology, Wuhan, Japan. p.p. 1-6

[6] Al-Sumaly Gazy .F, Sheridan J, Thompson Mark C.,(2012), "Analysis of forced convection heat transfer from a circular cylinder embedded in a porous medium", International Journal of Thermal Sciences 51, p.p 121-131.

[7] Zehforoosh A, Hossainpour S, Tahery A. A.,(2010), "Numerical investigation of forced convection heat transfer for laminar flow in various parallel porous channels" , International Journal of Innovation, Management and Technology, Vol. 1, No. 3, p.p. 2010-0248.
[8] Fu Wu-shung, Wang Ke-nan, Ke Wen-wang, (2001), "Heat transfer of porous medium with random porosity model in a laminar channel flow", Journal of the Chinese Institute of Engineering, Vol.24,No.4,. pp.431-438 .
[9] Waheed M.A, Odewole G.A, Alagbe S.O, (2011), "Mixed convective heat transfer in rectangular enclosures filled with porous media", ARPN Journal of Engineering and pplied Sciences, Asian Research Publishing Network (ARPN)." VOL. 6, NO. 8 . p.p. 1819-6608
[10] Keyhany M.H, Karimi G, Nazari M., (2006), "Heat transfer enhancement in channel partially filled with porous material", Proceeding of the 2nd WSEARS Int. Conference on Applied and Theoretical Mechanics, Venice, Italy.. p.p. 269-274.

A Brief Author Biography

Ashwin A. Kamble[1] – Presently pursuing M Tech in Heat power engineering from RTMNU Nagpur University Nagpur. Maharashtra, India.

Sanjay Kumar Borikar[2] – A Presently working as a Assistant Professor in KITS Ramtek, RTMNU, Nagpur University Nagpur, Maharashtra, India.

10

AN EXPERIMENTAL STUDY ON PROPERTIES OF FLY ASH BRICKS

Er. Rinku Kumar[1], Er. Naveen Hooda[2]

[1]*Er. Rinku Kumar (M-Tech student at RPIIT Technical Campus, Karnal), E-mail- rinkukapoor307@gmail.com*
[2]*Er. Naveen Hooda (Assistant Professor at RPIIT Technical Campus, Karnal), hoodanaveen723@gmail.com*

Abstract

In fast growing today's world development of new building materials and Processing & utilization of industrial waste is being given the top priority in the program of building research at a very high rate. This is important for achieving maximum disposal of wastes and conservation of scarce resources and materials.
The bricks produced were about 29% lighter than clay bricks. The compressive strength of bricks manufactured from fly ash possessed higher than 20 MPa. This generally exceeds the best of load carrying clay bricks available by more than 25% and is several times better than acceptable commercially available common clay bricks. Other important characteristics of the fly ash bricks have been evaluated. The values of these characteristics for fly ash bricks are excellent and have exceeded those pertaining to clay bricks. Moreover, fly ash bricks have been produced with a naturally occurring reddish colour similar to that of normal clay bricks. The new bricks and process have been patented. This dissertation presents the results of testing and the advantages gained by this type of fly ash bricks over conventional clay bricks.
In the present study, the effects of fly ash on the properties of bricks are studied and the behaviour of fly ash bricks is compared with conventional burnt clay bricks.
The various properties of fly ash bricks with different materials were tested. The properties studied water absorption, hardness, efflorescence, soundness, shape and size, crushing strength and basic compressive strength of the prism using different mortar mixes normally 1: 3, 1: 4 and 1: 5 cement-sand mortars.
In general bricks are made by top fertile agricultural soil but by using fly ash, 28 percent of top fertile agricultural soil is saved. Use of fly ash in brick making also is beneficial in diverse ways. As compared to conventional clay bricks fly ash bricks are stronger, more durable and yet more economical. Also, the process of fly ash brick manufacturing results in lesser pollution. Being less permeable as compared to clay bricks dampness related issues are far lesser in case of fly ash bricks than their clayey counterparts.

Keywords: Conventional Bricks, Fertile agriculture soil, crushing strength of bricks, efflorescence, fly ash

1. Introduction

Burnt clay bricks are being used extensively almost throughout India and are perhaps the most important building construction material. But the unlimited use of clay is harmful to society as all the bricks kilns in India depend on good quality clay available from agricultural fields and presuming a weight of 3 kg. per brick. The total clay taken out from the agricultural fields per day was over 300 million tonnes for 10,000 Crore bricks. Moreover, clay bricks available in certain regions are poor in quality and costly which have forced engineers to look for better material capable of reducing the cost of construction. At present, India has production capabilities of over 10,000 Crore bricks through around 45,000 local kilns (bhattas), in the unorganized sector. So the use of industrial waste products such as fly ash, for making bricks is ecologically and economically advantageous since apart from saving precious top agricultural soil, it meets the social objective of disposing industrial waste i.e. fly ash which otherwise is a pollutant and a nuisance.

The ever increasing volume of fly ash quantities in the world has not been remotely matched by its utilization. Australia is an example where such utilization has been minimal. The most important and popular use of fly ash in Australia has been in the partial replacement of portland cement. The use of fly ash in concrete is to the extent of maximum of 25% replacement of Portland cement. This conservatism can be understood in the context of concrete where the ash is mixed raw, and the effects of high volume replacement are still subject to research. It is however not quite justifiable that the brick industry should take similar conservative attitude. Environmental concerns have been raised in some parts of the world where coal is the main power generating resource and where bricks are also the main building material. Such concerns have resulted in legislation to oblige the brick industry to incorporate at least 25% by weight of fly ash and or bottom or pond ash in the brick making mixture if the industry is within 50 km from a coal power generation plant. Some successful ventures have been reported where fly ash was incorporated in the mixture at the rate of 20% to 50%. Nevertheless, there is only little evidence that incorporation of fly ash in the brick mixture has exceeded the 30% by volume, even when the legislation was obeyed. Reasons behind such reluctance are not clear. A most probable reason is the fear of change in many small factories and the ingrained conservatism in the attitude of stake holders of the large producers. Added to this is the fact, that with an existing clay brick factory, the incorporation of fly ash is a potential addition of cost. The possible incompatibility of the ash with the clay and shale during the various processes of production including the crucial one of firing may be a legitimate difficulty.

At high temperatures beyond 1000 °C, the temperature and length of time of firing become very sensitive to the type of ash and of course to the clay and shale if in the same mixture. This would be the case as long as the factory still uses the ash as partial replacement to the main clay and shale ingredients. The situation may become completely different when the ash is the only ingredient of the bricks mixture. Compatibility is no more an issue in such a case. So far, few attempts at manufacturing bricks from more than 80 % fly ash have been made.

The engineers now believe that fly ash on its own can be an excellent raw material for brick making. This has now been proven and a patent is taken for the manufacture of bricks from fly ash.

The response of the ash to firing temperature at 1000 °C and beyond can be accurately controlled even in small factories. The potential savings with this approach are many. Savings in production and transportation costs and producing bricks of superior qualities to those of standard clay bricks are in addition to the environmental solution that such venture may bring about.

2. Objective and Scope

The objective of the present paper is to study the effect of fly ash bricks on the performance and the properties of bricks with the view to study the comparison between clay bricks and a fly ash brick because fly ash is enriched with silica, is the main constituent for conventional building material. From the experiment, it is further desired to compare the strength of fly ash brick by that of the conventional clay brick. The salient properties of bricks like crushing strength, water absorption, shape and size, soundness, hardness and efflorescence are to be determined.

Bricks will always be the necessity of building materials and in future soil for the making of bricks would not be available in plenty as the use of land is rapidly increasing, at that time fly ash will fulfil the necessity of clay bricks as these bricks are made by the industrial wastes. Moreover the building structures are getting heavier with time, the use of fly ash bricks would be more efficient as its strength is more than that of clay bricks.

3. Experimental Programme

In the present study, fly ash bricks are subjected to the following tests to find out its suitability for the construction work:

3.1. Absorption Test:

This test is carried out to determine the amount of water absorbed by the brick. When immersed in water for a period of 24 hours it should not, in any case, exceed 20% of weight of dry brick. This test is carried out for all the samples of fly ash bricks and clay bricks.

1.Normal Bricks:				
S.NO	Dry Weight (kg)	Wet weight (kg)	Water absorption (gm)	Water absorption (gm)
1.	3.13	3.50	370	10.57
2.	3.21	3.64	430	11.81
3.	2.78	3.21	430	13.40
				11.93

2. **Fly Ash Bricks**:				
S.NO	Dry Weight (kg)	Wet weight (kg)	Water absorption (gm)	Water absorption (gm)
1.	3.17	3.47	300	8.64
2.	2.98	3.30	320	9.70
3.	3.00	3.37	320	10.97
Average				9.77

3.2. Hardness Test:

This test is carried out to see that the brick is sufficiently hard or not. We can judge hardness of the brick by making impression on the surface of the brick with the help of a finger nail. This test is carried out for all samples of fly ash bricks and clay bricks.

Normal Bricks	*Fly Ash Bricks*
No impression after scratching with the help of a finger nail.	No impression after scratching on the surface of brick.

3.3. Efflorescence Test:

This test is conducted for finding out the presence of soluble salts in a brick when it is immersed in water for 24 hours and taken out and allowed to dry in shade. Absence of grey or white deposits on its surface indicates absence of soluble salts. If the white deposits cover about 10%surface, the efflorescence is said to be slight and it is considered as moderate, when the white deposits cover about 50% of surface. If grey or white deposits are found on more than 50% of surface, the efflorescence becomes heavy and it is treated as serious, when such deposits are converted into powdery mass. This test is carried out forth fly ash bricks and clay bricks.

Normal Bricks	*Fly Ash Bricks*
Slight to moderate	The grey deposit is less than 10 %.

3.4. Soundness Test:

This sound is carried out to find out that a clear ringing sound is produced or not when the two bricks are struck with each other without breaking any of the two bricks. If the two bricks are not broken after striking with each other and a clear ringing sound is produced then it means that the bricks are sufficiently sound. The procedure of this test is self explanatory.

Normal Bricks	*Fly Ash Bricks*
Good	A clear ringing sound produced.

3.5. Shape and Size Test:

This test is done to examine the structure of the brick when the brick is broken. It is seen that the structure of the brick is homogeneous, compact and free from any defects such as holes, lumps etc. or not. Mainly the defects such as holes, lumps should not be there.

3.6. Crushing Strength Test:

This is the main test conducted to test the suitability of the brick for construction work. This test is executed with the help of compression testing machine.

A brick is placed in a compression testing machine. It is pressed till it breaks. Then the compression strength of the brick is recorded from meter of the compression testing machine. A brick after undergoing compression test, this test is carried out for both fly ash bricks and clay bricks.

Normal Bricks:					
S.	Length	Breadth	Depth(mm)	Load	Crushing Strength
No.	(mm)	(mm)		KN	(N/mm^2)
1.	228	110	72	220	8.70
2.	221	110	70	160	6.58
3.	220	105	69	160	6.93
4.	218	104	70	220	9.70
5.	227	104	74	200	8.47

6.	226	103	71	190	8.47

Fly Ash Bricks:					
S.	Length	Breadth	Depth(mm)	Load	Crushing Strength
No.	(mm)	(mm)		KN	(N/mm^2)
1.	228	114	75	387	14.89
2.	227	114	76	522	21.33
3.	228	113	74	390	15.13
4.	227	114	74	613	23.68
5.	227	114	74	453	17.50
6.	226	113	73	520	20.36
		Average		18.81	

		Average		8.14	

3.7. Prism Test:

Mortar	Prism Size (mm)		Correction	Failure	Basic Compressive	Basic Compressive Stress of Class

Type	(widthxthickx hight)	h/t	Factor	Load (KN)	Stress of Fly Ash Bricks MPa	A Bricks As Per IS: 1905-1987 MPa
1:3	600×230×685	3	0.86	290	1.8	1.31
1:4	600×230×610	2.65	0.82	264	1.56	1.19
1:5	600×230×610	2.65	0.82	240	1.42	1.13

4. Analysis of results

4.1. Absorption Test:

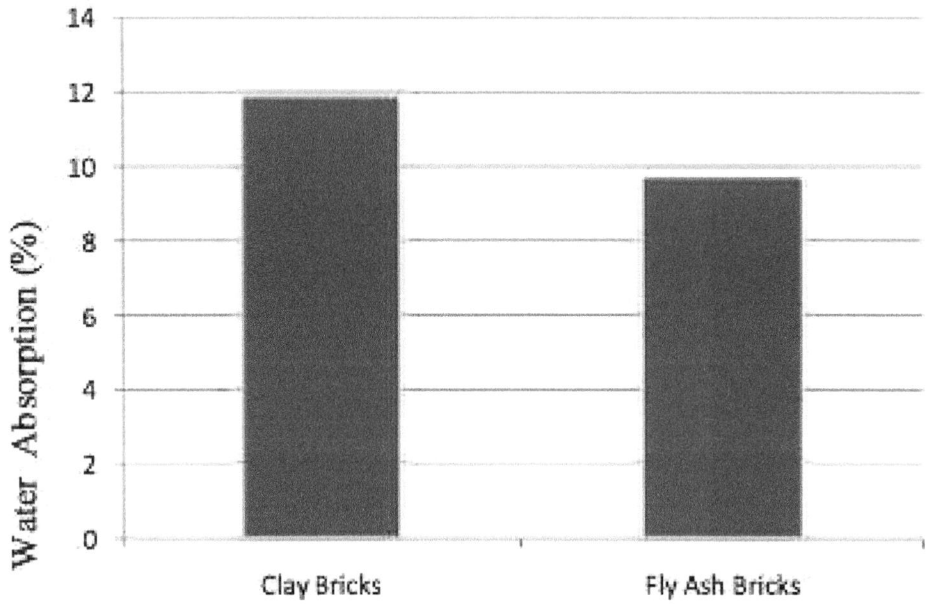

Figure: 4.1. Water Absorption (%)

The average absorbed moisture content of clay bricks is found to be 11.93% and for fly ash bricks are found to be 9.77%. Thus there is net 18.10% decrease in moisture absorbed for fly ash bricks as a part to clay bricks.

4.2. Hardness Test:

The hardness test for clay bricks and fly ash bricks was conducted, test brick was taken and scratch was made on bricks surface with the help of finger nail and found no impression after scratching in both the cases.

4.3. Efflorescence Test:

The Efflorescence test for clay bricks and fly ash bricks was conducted and the results were compared in which Grey or white deposits are slight to moderate in normal bricks and less than 10% on the surface area in fly ash bricks.

4.4. Soundness Test:

The Soundness test for clay bricks and fly ash bricks was conducted and the results were compared in which two bricks are struck with each other, It was found that a normal brick shows good results when struck with each other but fly ash bricks show clear ringing sound.

4.5. Shape and Size Test:

The Shape and Size test is done for clay and fly ash bricks to examine the structure of a brick when the brick is broken and it was found both types of bricks are free from any defects such as holes, lumps etc. but fly ash bricks are compact and homogeneous.

4.6. Crushing Strength Test:

Crushing strength test for clay bricks and fly ash bricks was conducted and the result are compared in Table: 4.2

Table: 4.2

Type of Bricks	Average Crushing Strength (N/mm^2)	% Increase Average Crushing Strength
Class A Clay Bricks	8.14	---
Fly Ash Bricks	18.81	56.72

The crushing strength of clay bricks is found to be 8.14 N/mm^2 and for fly ash bricks is found to be 18.81 N/mm^2. Thus there is net 56.72% increase is crushing strength for fly ash bricks as a part to clay bricks.

4.7. Prism Test:

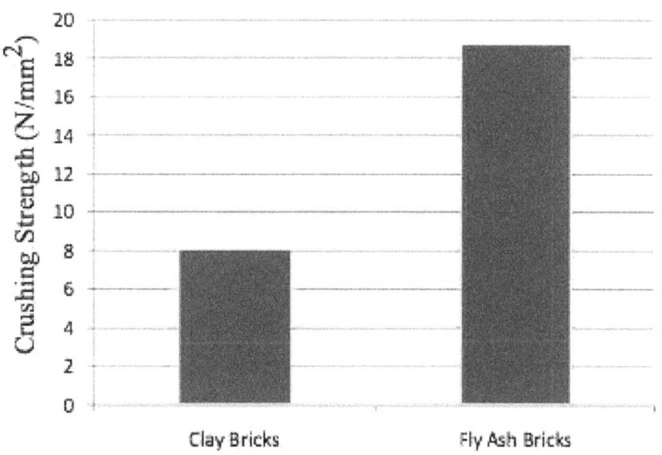

Figure: 4.2. Crushing Strength (N/mm^2)

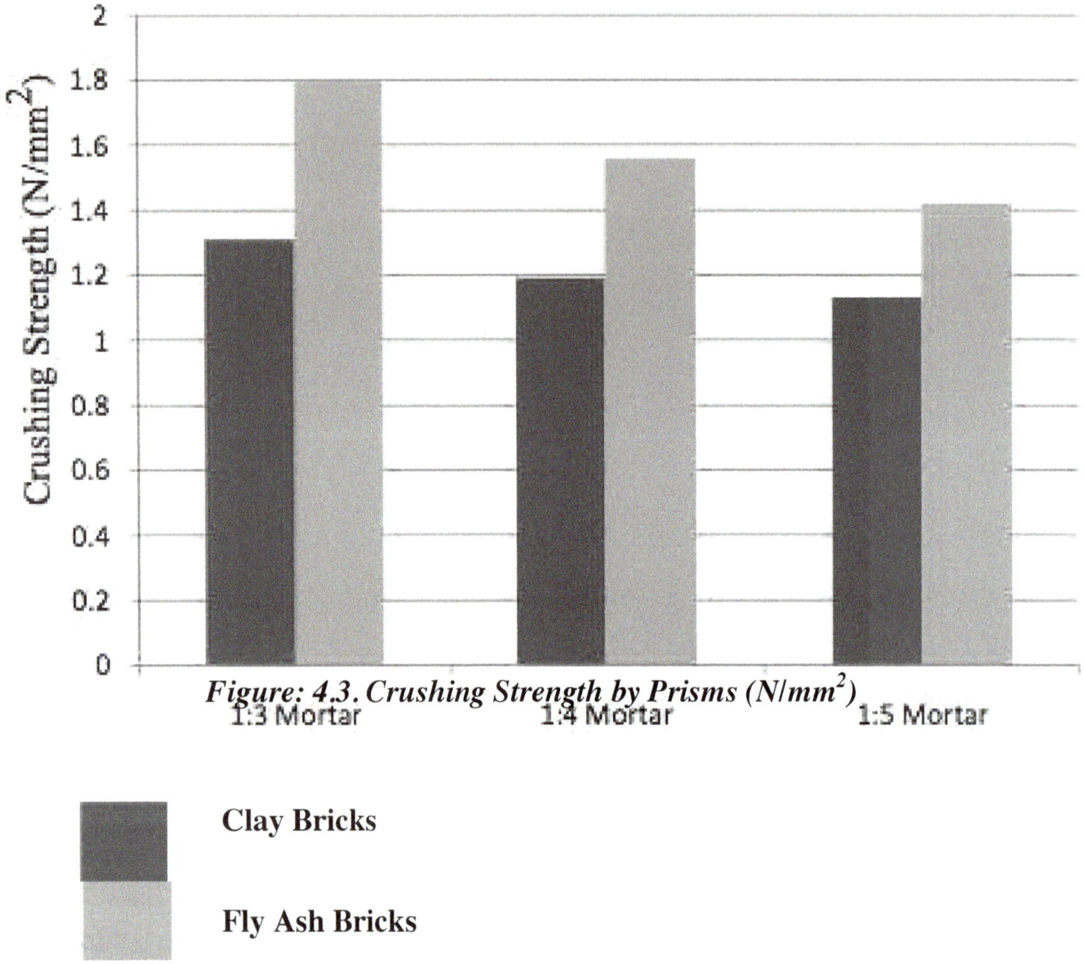

Figure: 4.3. Crushing Strength by Prisms (N/mm^2)

Table: 4.4 Cracks in 1: 5 Cement Sand Mortar Prism

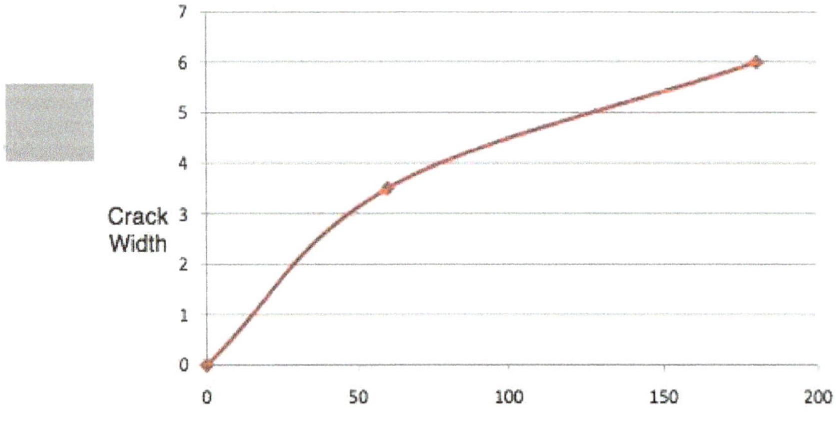

Load

In prism testing of 1:5 cement sand mortar maximum width of crack appeared to be 3.5 mm and average width of crack was 1 mm on applying load of 60 KN. On again applying load crack were measured with the maximum width of 6 mm and average crack width of 2 mm by applying load of 180 KN.

Table: 4.5. Cracks in 1:4 Cement Sand Mortar Prism

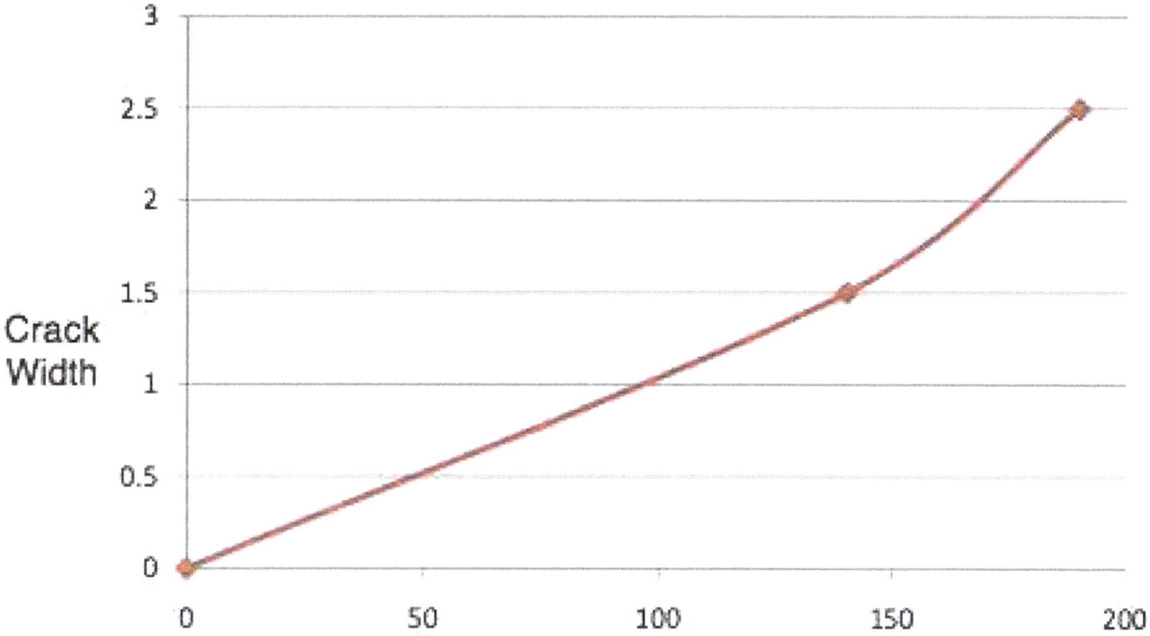

Load

In prism testing of 1:4 cement sand mortar maximum width of crack appeared to be 1.5 mm and average width of crack was 1 mm on applying load of 140 KN. On again applying load crack were measured with the maximum width of 2.5 mm and average crack width of 1.5 mm by applying load of 190 KN.

Prism of 1:4 cement sand mortar failed on applying load of 264 KN.

Table: 4.6. Cracks in 1:3 Cement Sand Mortar Prism

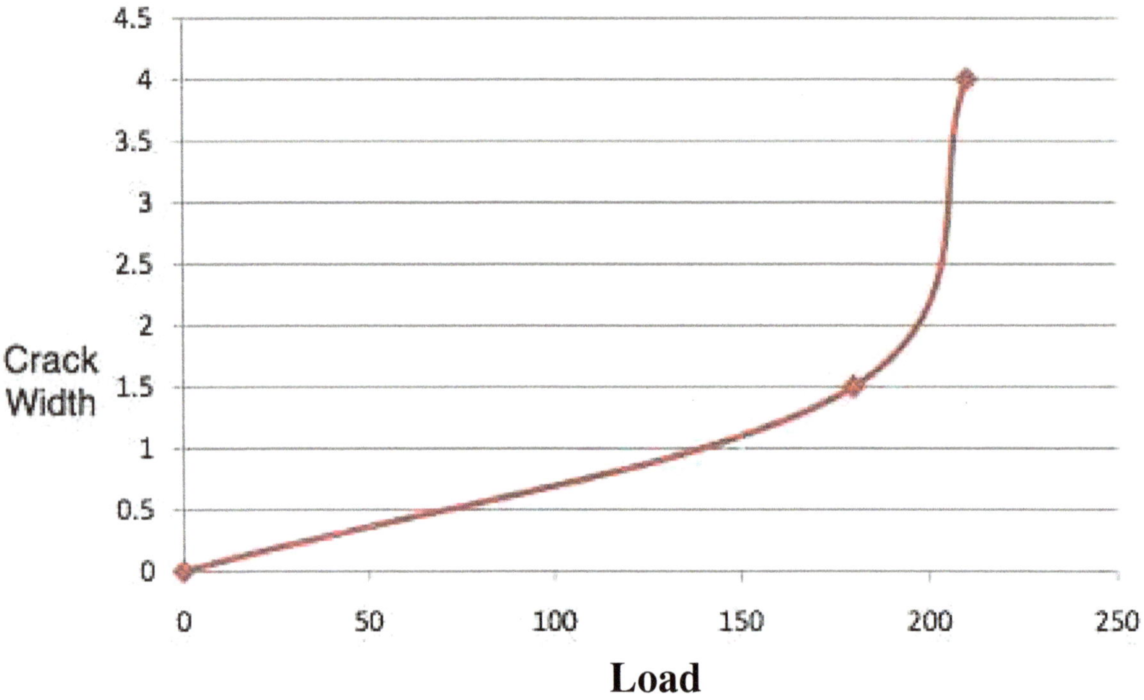

In prism testing of 1:3 cement sand mortar maximum width of crack appeared to be 3.5 mm and average width of crack was 1.5 mm on applying load of 180 KN. On again applying load crack were measured with the maximum width of 6 mm and average crack width of 2 mm by applying load of 210 KN.

Prism of 1:3 cement sand mortar failed on applying load of 290 KN.

5. Conclusion

1. Fly Ash Bricks were found to be sufficiently hard as scratching by the finger nail on the surface left no impression on it as compared to normal bricks.
2. The Efflorescence of all bricks tested were found to be slight as white or grey deposits were less than 10% on surface of the bricks which is almost same as that in the normal bricks.
3. A ringing sound in the Fly ash Bricks was observed to be far better than that in normal bricks.
4. Structure of the bricks was found to be compact, homogeneous and free from any defects like holes, lumps etc as compared to normal bricks.
5. The average absorbed moisture content of clay bricks is found to be 11.93% and for fly ash bricks are found to be 9.77%. Thus there is net 18.10% decrease in moisture absorbed for fly ash bricks as a part to clay bricks.
6. The crushing strength of clay bricks is found to be 8.14 N/mm^2 and for fly ash bricks is found to be 18.81 N/mm^2. Thus there is net 56.72% increase is crushing strength for fly ash bricks as a part to clay

bricks.

7. The crushing strength by prism of clay bricks is found to be 1.31 N/mm^2 and for fly ash bricks is found to be 1.8 N/mm^2. Thus there is net 27.22% increase is crushing strength by prism for fly ash bricks as compared to clay bricks.
8. The crushing strength by prism of clay bricks is found to be 1.19 N/mm^2 and for fly ash bricks is found to be 1.56 N/mm^2. Thus there is net 23.71% increase is crushing strength by prism for fly ash bricks as compared to clay bricks.
9. The crushing strength by prism of clay bricks is found to be 1.21 N/mm^2 and for fly ash bricks is found to be 1.59 N/mm^2. Thus there is net 23.90% increase is crushing strength by prism for fly ash bricks as compared to clay bricks.

6. Future Extension

Instead of Fly Ash, investigations can be done on waste material such as saw dust, lime sludge, rice husk etc. Thermal conductivity can also be investigated for such type of brick. Work can be extended by using Fly Ash. Fly ash bricks can be used partially with clay so that economically fly ash bricks can be better used.

Fly ash bricks can be used partially with rice husk so that bonding with mortar will be stronger. Industrial wastes should be better used in brick making for strengthening purpose and also profitable for cleaning society.

7. References

1. Dass. Mohan, "New Substitutes of Bricks", Civil Engineering and Construction, June 1992, PP22.

2. Gupta P. C. and Ray S. C., "Commercialisation of Fly ash", The Indian Concrete Journal vol- 167, now. 1993, PP 554-560.

3. IS: 1077-1992 Specification for common Burnt Clay Building Bricks (Fifth revision).

4. IS: 1905-1987 Code of Practice for Structural use of Unreinforced Masonry (Third revision)

5. IS: 2117-1975 Guide for Manufacture of Hand- Made Common Burnt Clay Building Bricks (Second Revision).

6. Lal. A. K., "Alternate Building Material for low cost housing, "Civil Engineering and Construction, Aug 1991, PP22.

7. S.C. Rangwala, Engineering Materials, (Fifteenth Edition), 1991, PP 72 - 112.

8. Dr. Durgesh C Rail, "Review of design Codes for Masonry Buildings", Department of Civil Engineering, Indian Institute of Technology, Kanpur.

9. Sengupta. J, "Availability of Fly ash and its Application in Construction Industry", NBO Journal, Vol. XXXIX, April 1984, PP 17-22. K S Nanjunda Rao, "Structural Masonry: Properties and Behavior",Department of Civil Engineering, Indian Institute of Science, Banglore Kayali, O. (2005) High Performance Bricks from Fly Ash. Proc. World of Coal Ash, Lexington, Kentucky, USA..

10.Sarangapani, G., Venkatarama Reddy, B. V. and Jagadish, K. S., Structural characteristics of bricks, mortar and masonry. J. Struct. Eng. (India), 2002, 29,

Author Biography

Er. Rinku Kumar **–** pursuing M-Tech in structural engineering from RPIIT Technical Campus, Bastara, Karnal

Er. Naveen Hooda (Structure Engineer) - B-Tech in Civil Engineering from KUK, M-Tech in Structural Engineering from Indo Global College of Engineering, Mohali, Presently working as an assistant professor in RPIIT Technical Campus, Bastara, Karnal.

11

Composite Fiber-Resin Lamina, Compared by Finite Element Analysis and Analytical Solution

Mr.Zeeshan Khan[1], Prof. R. J. Patil[2]

[1]*Dr. D.Y.Patil Institute of Engg. & Tech. Ambi, Talegaon Dabhade, Pune University*
Zeeshan_khann@rediffmail.com
[2]*Dr. D.Y.Patil Institute of Engg. & Tech. Ambi, Talegaon Dabhade, Pune University*
Rupesh1002001@yahoo.com

Abstract

The objective of this work is to compare ligament system with the composite fibre-resin lamina which is not exactly but matches nearer to the ligament structure. The skeletal ligaments are short bands of tough fibrous connective tissue which are same as composite material. This paper presents 3D finite element analysis performed for a composite Fiber-Resin lamina and compare this result with the orthotropic stress -strain relationship formulation for validation of the result. Nonlinear static analysis performed for composite material to evaluate stress-strain relationship for a composite material.

Keywords: composite lamina, FEA, Ligament.

1. Introduction

Ligaments are composed of closely packed collagen fibre bundles oriented in a parallel fashion to provide for stability of joints in the musculoskeletal system. ligaments can be compared with the composite material because of the similar structural system. Composites are made from two or more constituent materials with significantly different chemical and material properties that when combined, produce a material which characteristics different from the individual components. The current paper attempt to combine the two approaches and the result for stress-strain relationship is compared.

Analytical solutions are developed based on proposed solution by (P. Boresi and Richard J.Schmidt).The fibre-resin lamina stress-strain relationship case from Advanced Mechanics of Materials. FEM analysis was carried out by simulating the composite model using ANSYS 14.5 the stress strain relation obtained by applying shear force on one face of model along the axis of fibre.

2. ANALYTICAL SOLUTION.

Advance Mechanics of Material (P. Boresi and Richard J.Schmidt) proposed the solution for unidirectional model of a lamina of a section of an airplane wing composed of fibres and resin. The volume fraction is consider for the determination of stress strain relations of the lamina

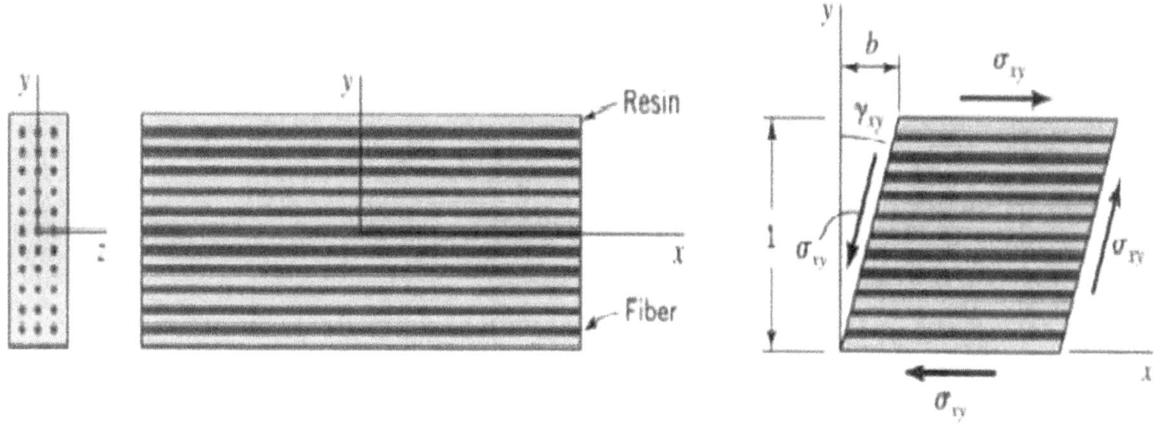

Fig.1 Fiber-Resin lamina, fiber: resin volume fraction = f, resin volume fraction=1-f

The modulus of elasticity and Poisson's ratio of Fibre and Resin be denoted as EF, μF and ER since the lamina is thin, the effective state of stress in lamina is approximately one of plane stress in the x-y plane of the laminate. stress strain relations for the fibres and resins are

$$\epsilon_{xxR} = \frac{1}{E_R}(\sigma_{xxR} - \mu_R \sigma_{yyR})$$

$$\epsilon_{yyR} = \frac{1}{E_F}(\sigma_{yyR} - \mu_R \sigma_{xxR})$$

$$\epsilon_{xxF} = \frac{1}{E_F}(\sigma_{xxF} - \mu_F \sigma_{yyF})$$

$$\epsilon_{yyF} = \frac{1}{E_F}(\sigma_{yyF} - \mu_F \sigma_{xxF}) \quad (1)$$

Where $(\sigma_{xxF}, \sigma_{yyF})$, $(\sigma_{xxR}, \sigma_{yyR})$, $(\epsilon_{xxF}, \epsilon_{yyF})$ and$(\epsilon_{xxR}, \epsilon_{yyR})$ denotes stress and strain components in fiber(F) and resin(R) respectively.

Since the fibers and resin are bonded, the effective lamina strain ϵ_{xx} is same as that in the fibers and in the resin; x direction

$$\epsilon_{xx} = \epsilon_{xxF} = \epsilon_{xxR} \quad (2)$$

In y- direction, the effective lamina strain ϵ_{yy} is proportional to the amount of fiber per unit length in y direction,

$$\epsilon_{yy} = f\epsilon_{yyF} + (1-f)\epsilon_{yyR} \quad (3)$$

Also, by equilibrium of the lamina in x direction, the effective lamina stress σ_{xx} is

$$\sigma_{yy} = f\sigma_{xxF} + (1-f)\sigma_{xxR} \quad (4)$$

In the y direction, the effective lamina stress σ_{yy} is same as in the fibers and in resin

$$\sigma_{yy} = \sigma_{yyF} = \sigma_{yyR} \quad (5)$$

Solving equations.(a) through (e) for ϵ_{xx} and ϵ_{yy} in terms of σ_{xx} and σ_{yy}

$$\epsilon_{xx} = \frac{1}{E}(\sigma_{xx} - \mu\sigma_{yy})$$

$$\epsilon_{yy} = \frac{1}{E}(\beta\sigma_{yy} - \mu\sigma_{xx}) \quad (6)$$

Where,

$$\mathrm{E} = f\mathrm{E_F} + (1-f)\mathrm{E_R}$$

$$\mu = f\mu_F + (1-f)\mu_R$$

$$\beta = f(1-f)\left[\left(1-\mu_R^2\right)\frac{E_F}{E_R} + \left(1-\mu_F^2\right)\frac{E_R}{E_{RF}} + 2\mu_F\mu_R + \frac{1-f}{f} + \frac{f}{1-f}\right] \quad (7)$$

To determine the shear stress-strain relation, apply a shear stress σ_{xy} to a rectangular element of the lamina and calculated angle change γ_{xy} of the rectangle. by figure the relative displacement b of the top of the element is

$$\mathrm{b} = f\gamma_F + (1-f)\gamma_R \quad (8)$$

Where γ_F and γ_R are the angle changes attributed to the fiber resin

$$\gamma_F = \frac{\sigma_{xy}}{G_F} \quad , \gamma_R = \frac{\sigma_{xy}}{G_R} \quad (9)$$

And G_F and G_R are the shear modules of elasticity for fiber and resin respectively.

hence ,the change γ_{xy} in angle of element(the shear strain is)

$$\gamma_{xy} = 2\epsilon_{xy} = \frac{b}{1} = \left[\frac{fG_R+(1-f)G_F}{G_F G_R}\right]\sigma_{xy} \quad (10)$$

by equation (j) shear stress relation is

$$\sigma_{xy} = G\,\gamma_{xy} = 2G\epsilon_{xy} \quad (11)$$

Where,

$$G = \frac{G_F G_R}{f G_R + (1-f) G_F} \tag{12}$$

By equation (6),(7),(11),(12) we obtained the stress-strain relations of lamina

$$\sigma_{xx} = c_{11}\epsilon_{xx} + c_{12}\epsilon_{yy}$$

$$\sigma_{yy} = c_{12}\epsilon_{xx} + c_{22}\epsilon_{yy}$$

$$\sigma_{xy} = c_{33}\gamma_{xy} \tag{13}$$

Where,

$$C_{11} = \frac{\beta E}{\beta-\mu^2}, \quad C_{22} = \frac{E}{\beta-\mu^2}, \quad C_{12} = \frac{\mu E}{\beta-\mu^2}, \quad C_{33} = G \tag{14}$$

3. FINITE ELEMENT ANALYSIS

Basically finite element method considers a structure is constructed from simple elements which are connected at their nodes and fulfills equilibrium and compatibility conditions. Based on this definition the composite fiber-resin model of size 20mm×50mm×20mm is drawn with fiber of size 3mm diameter and 20mm in length inside the resin. The finite element analysis was carried out using ANSYS 14.5 software. There are two option which can be used: ANSYS Parametric Design Language (APDL) and Graphics User Interface (GUI) in this paper (GUI) is used.

Properties of composite materials:-

Glass fiber:-
EF=72.4GPa, GF=27.8GPa, poissons ratio=0.30
Epoxy resin:-
EF=3.50GPa, GF=1.35GPa, poissons ratio=0.30
The volume fraction of fiber is f=0.70

FEM analysis is done with 77475 nodes to describe 20mm×50mm×20mm block and meshing is kept automatic to generate the result. Static nonlinear analysis is done on the model. a shear force of value 10N is applied on the upper face of the block keeping lower face fixed to get shear effect.

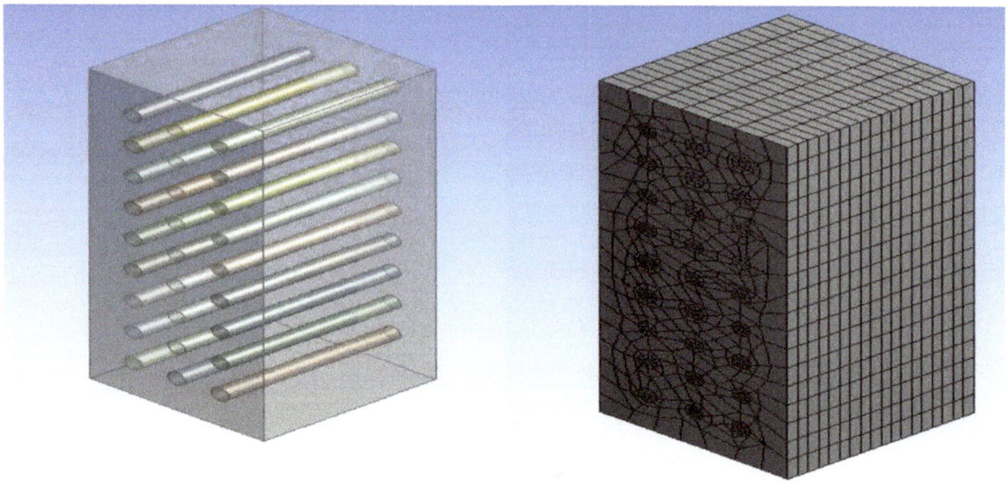

(a) Composite Fiber-resin model. inside tubes showing fibers and outside block showing resin.

(b) FEM mesh generated model.

The mesh configuration for the composite model is shown in Fig.(b) the stress strain relation for the fiber and Resin is found by solving the problem with above boundary conditions and the following results are showing the shear elastic strain and shear stress for the model.

4. RESULTS AND DISCUSSION

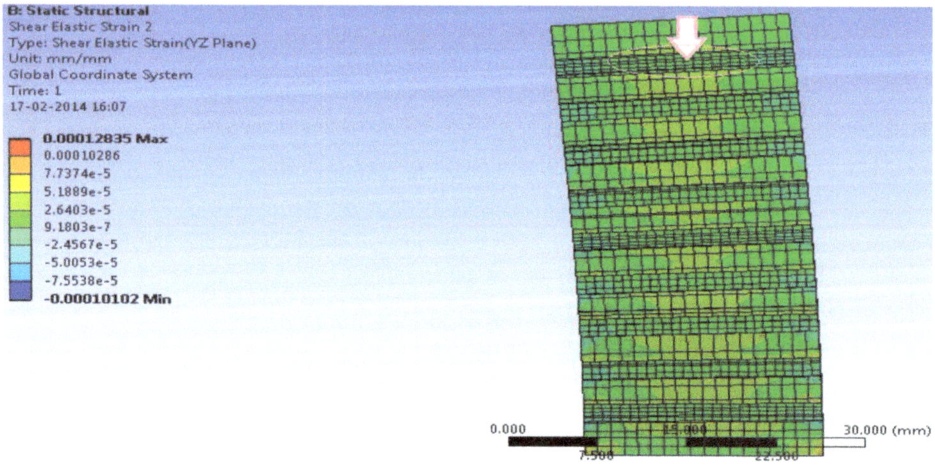

Shear elastic strain of composite model at y-z plane

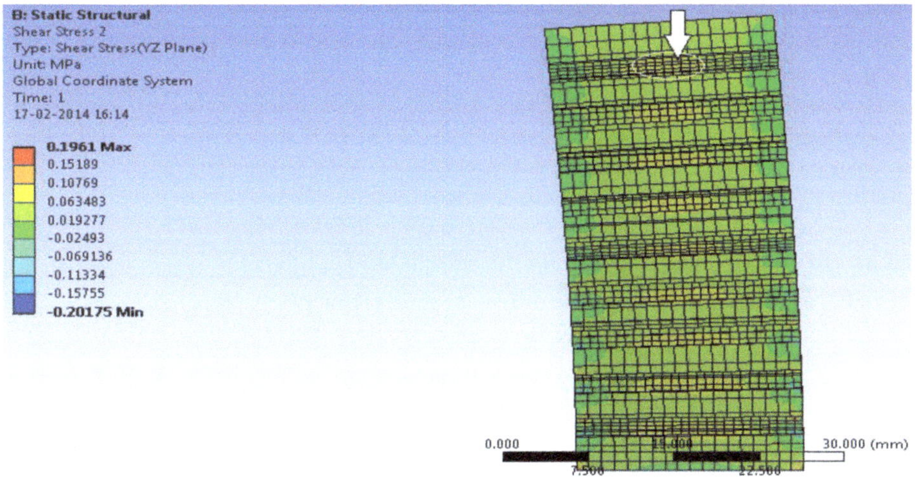

Shear stress of composite model at y-z plane

For the stress-strain relation the results are presented in Fig.4 and 5 for y-z plane. On this plane the shear stress and shear strain are calculated on the fiber of the model. the above model shows the circular portion at which both results are taken. the green zone of fiber showing the shear strain value and same portion is observed in yellow zone for shear stress. Evaluated value is used to validate the analytical solution, the values for analytical solution are calculated by putting the above table data in the equation (13).results are as

E= 0.7×72.4 + 0.3×3.50 =51.73 GPa
μ= 0.7×0.3 + 0.3×0.3 = 0.3

$$\beta = (0.7 \times 0.3)\left[(1-0.009)\frac{72.4}{3.50} + (1-0.009)\frac{3.50}{72.4} + (2 \times 0.3 \times 0.3) + \frac{0.3}{0.7} + \frac{0.7}{0.3}\right]$$

$$= 4.9325$$

$$G = \frac{27.8 \times 1.35}{(0.7 \times 1.35) + (0.3 \times 27.8)}$$

$$= 4.0420$$

C_{11}=52.6914
C_{12}=3.2047
C_{22}=10.6824
C_{33}=4.0420

The stress-strain relation for shear stress is given from equation(13) is given below. this equation is used to validate the FEA result by putting the value of FEA strain into the given equation.

$\sigma_{xy} = C_{33}\gamma_{xy}$

$\sigma_{xy} = 4.0420\text{e}3 \times \gamma_{xy}$

$\sigma_{xy} = 4.0420\text{e}3 \times 2.6403\text{e}^{-5}$

$= 0.108\text{Mpa}$

Comparison will be made with the analytical solution obtained in the equation (13). It shows that good agreement is achieved between FEA and analytical results. The difference is less than 5%.

5. CONCLUSION

From this analysis it can be conclude that a good agreement is found between the analytical solution and FEA results for shear stress of the fiber-resin lamina and the remaining results for stresses will also match with the FEA results. Finite element analysis can predict results for the composite lamina and in future it will be used to find out the Ligament analysis by referring this result. the lamina is independent from both dimension as well as force.

REFERENCES

[1] P. Boresi, Richard J. Schmidt. "Advanced Mechanics Of Materials" 2009

[2] Ever J. Barbero. "Finite Element Analysis of Composite Materials" 2011

[3] Jeffrey A Wiess, John C. Gardiner, Benjamin J Ellis, Trevo J. Lujan, Nikhil S. Pathak Three dimensional finite element modeling of ligament technical aspects, medical engineering and physics27(2005);845-861

[4] Ligament tissue engineering: an evolutionary materials science approach. Biomaterials 26(2005)7530-7536

[5] Savi L-Y Woo, Steven D Abramowitch, Biomechanics of knee ligaments: injury, healing, and repair; journal of mechanics 39(2006)1-20

[6] Wrinkling of sandwich column: comparison between finite element analysis and finite element solution. composite structures 53(2001)447-482.

12

Friction Stir Welding of Dissimilar Aluminium alloys AA1100 to AA6101-T6

Vinayak D. Yadav[1], S. G. Bhatwadekar[2]

[1]*PG Student ME CAD/CAM/CAE, vinayakyadav.h@gmail.com*
[2]*Associate Professor in Production Engineering, sgbhatwadekar@rediffmail.com*

Abstract

Friction stir welding (FSW) is primarily used for aluminium. FSW is used to replace rivets joints in the aeronautical industry. AA1100 and AA6101 are the most common aluminium alloys used for electrical busbar applications. In present study FSW of dissimilar aluminium alloys AA1100 and AA6101-T6 of 5mm thick plates was carried out. The FSW trial was carried on vertical machining center and cylindrical profile tool of H13 material was used. Tensile testing of joint is done using computerized UTM.Tensile strength of dissimilar joint is less than the stronger base metal alloy (AA6101-T6) but very nearer to the weaker material (AA1100).

Keywords: Friction Stir Welding, Aluminium Alloys, Tensile Strength.

1. Introduction

Friction stir welding (FSW) is the solid state welding process was invented at The Welding Institute (TWI) of the United Kingdom in 1991. Due to the rotational and translational motion of tool shoulder, plunged pin and forging pressure exerted by tool shoulder on the metal pieces joint is formed without fusion and filler material.

AA 1100 Aluminium is 1000 series aluminium alloy. It is pure aluminium with a minimum 99% aluminium content. AA 1100 is widely used in the chemical and food processing industries also used for electrical bus bars and bus bar supports. It is comparatively soft, low strength between all the aluminium alloys. AA 6101 Aluminium is 6000 series aluminium alloy. In 6000 series aluminium is alloyed with magnesium and silicon. Aluminum 6101 alloy has good mechanical strength and high electrical conductivity.Therefore, it is used for electrical bus bar conductor where it requires minimum loss of electrical conductivity and good mechanical properties.T6 is temper designation for solution heat treated and artificially aged to maximum mechanical property levels.

S. Nansaarng et al. [5] studied dissimilar friction stir welded joint of AA1100 and AA6063-T6 aluminium alloy. They investigated the effect of tool rotational speed and welding speed on hardness, ultimate tensile strength, elongation and yield strength using factorial design. They found the maximum of hardness as 82.08 HV, ultimate tensile strength as 100.49 MPa and yield strength as 66.373 MPa.

N. T. Kumbhar et al. [6] studied friction stir welding of two aluminium alloys AA6061 and AA5052 at various combinations of tool rotation speeds and tool traverse speeds. From The microstructural studies, they suggested that there was no rigorous mixing of both materials in the nugget. They found that tensile properties of the FSW AA5052-AA6061 specimens were better than the properties of the softest of the similar friction-stir-welded systems (i.e., FSW AA6061).

Ranjith R. et al. [7] carried out friction stir welding of dissimilar aluminium alloys AA2014 T651 and AA6063 T651. They showed that better interlocking and bonding of materials occurs at 40 tilt angle.They found the tensile strength is better when the tool is offset towards AA2014 side because of complete fusion of harder material.

S. Ravikumar et al. [8] investigated the effects of rotational and welding speeds with the pin profiles on micro hardness distribution and tensile property of the dissimilar Aluminium AA6061-T651 and AA7075-T651 joints. They found that good mixing of both nthe materials joined was obtained at lower welding and higher rotational speed.

There is no work reported on dissimilar Friction Stir Welding of Aluminium alloys AA1100 to AA6101-T6 at lower tool traverse speed 3mm/min. In this study experiment is conducted at 3 mm/min tool travers speed and 1500 rpm tool rotational speed.

2. Experimental Details

In this work frictional stir welded dissimilar Aluminium alloys AA1100 and AA6101-T6 specimens are compared for tensile properties. In this study FSW specimens are prepared at 3mm/min feed rate and at 1500 rpm spindle speed. The plate size of aluminium alloys are same and having 100 mm length, 50 mm width and 5 mm thickness. H13 material is used to manufacture the tools. [9] Tool has pin diameter of 6 millimeter size. Tool dimensions: Shoulder Diameter- 18mm, Pin Diameter- 6mm.

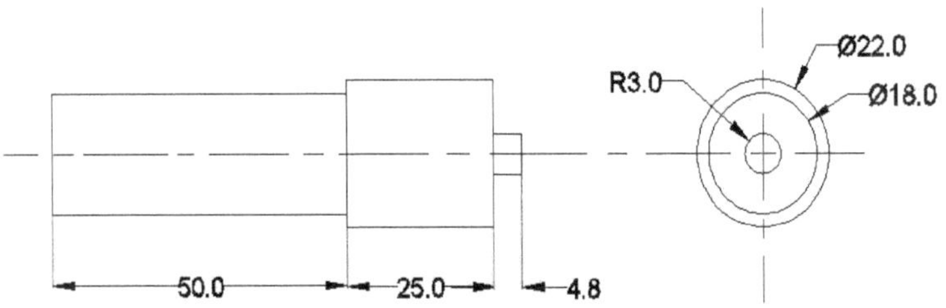

Figure 1: Design of straight cylindrical pin

Following are materials and parameters used for experiment:
Material: AA1100 Aluminium, AA6101-T6 Aluminum
Sheet Thickness: 5mm
Tool: Cylindrical
Spindle Speed: 1500 rpm
Welding Speed: 3mm/min
The FSW process was carried on vertical milling machining centre.
As per test report composition of alloys is as shown in Table 1 and Table 2

Table 1: Composition of AA1100

Element	Al	Si + Fe	Cu	Mg	Mn	Zn	Ti	Ca	Pb	Zr
AA1100	99	0.68	0.2	0	0.001	0.02	0.02	0.002	0.002	0.003

Table 3: Composition of AA6101-T6

Element	Al	Si	Fe	Cu	Mg	Mn	Zn	Ti	Ca	Pb	Zr	Ni
AA6101-T6	Balance	0.5	0.2	0.2	0.5	0	0.025	0.02	0.002	0.002	0.001	0.01

Workpiece Specification: Plate Length=100mm, Plate Width=50mm,Plate Thickness=5mm
Raw material was cut into the specified dimensions, then machined at the sides in order to make them flat to ensure accurate face-to-face contact at weld joint as shown in Figure2.

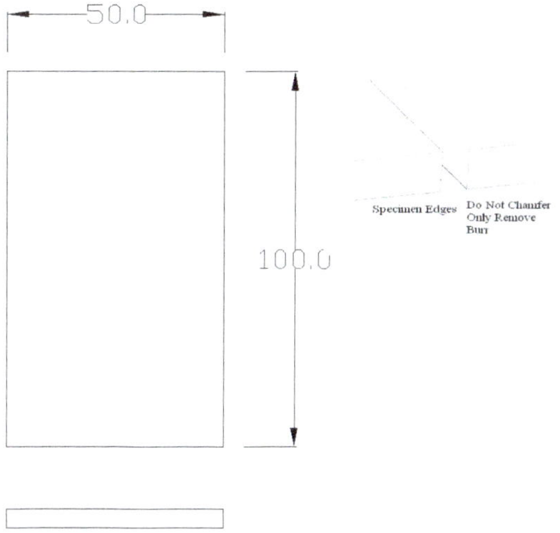

Figure 2: Workpiece Dimensions

At advancing side Aluminium AA6101 –T6 alloy and at retreating side Aluminium AA1100 alloy was kept in fixture. The joint is successfully obtained.The joint is formed as shown in Figure3

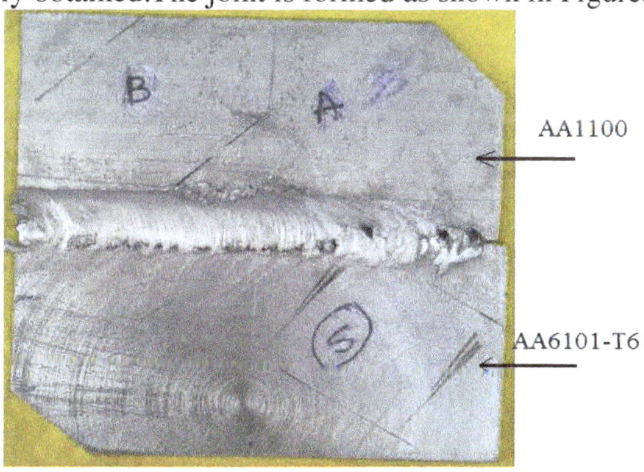

Figure 3: Photograph of Dissimilar joint Aluminium AA1100 and Aluminium AA6101-T6

3.Testing and Analysis

1.Visual Inspection

From Figure3 following are the observations:

a) Dissimilar joint of Aluminium AA1100 to AA6101-T6 found satisfactory in visual inspection.

b) Tunnel like defects are present at the surface of joint.

2.Tensile Test

Tensile testing specimen was made as per the drawing shown in figure4. Care was taken during this stage to align the centre of the weld with the centre of the tensile specimen. Tensile testing was carried out on an UTM.

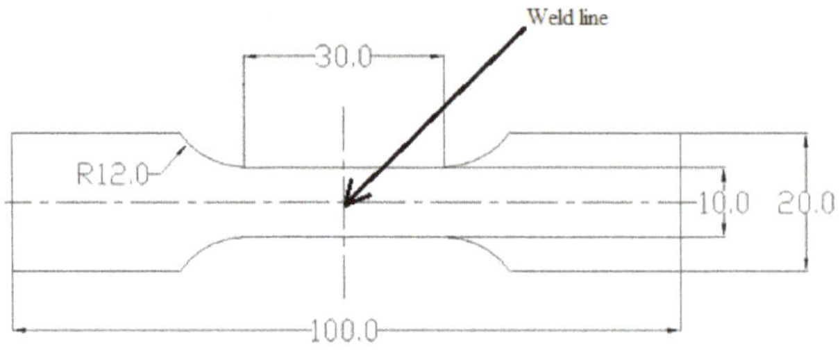

Figure 4: Tensile test specimen [4]

Tensile test results

Following graph shows result of tensile test of the joint.

Figure 5: Graph of Stress vs Strain of Dissimilar Friction Stir Welding Aluminium AA1100 to AA6101-T6

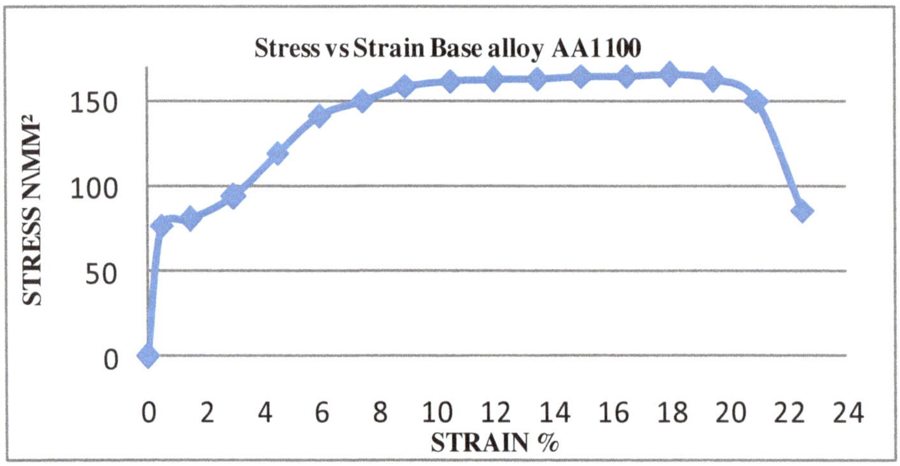

Figure 6: Graph of Stress Vs Strain of Base Alloy AA1100

Figure 7: Graph of Stress Vs Strain of Base Alloy AA6101-T6

Table 3: Comparison of Tensile Strength

Material	Tensile Strength N/mm^2
Dissimilar Joint	153.33
AA6101 –T6 Aluminum	284.4
AA1100 Aluminum	165.60

As seen from figure and table:

1. For Dissimilar Friction Stir Welding Aluminium AA1100 to AA6101-T6 joint Tensile strength for FSW is less than that for both base metals.

4. Conclusion

1)Dissimilar Friction Stir Welding Aluminium AA1100 to AA6101-T6 joint is formed
2)Tunnel like defects are present at the surface of joint.
3)Tensile strength of dissimilar joint is less than stronger base metal alloy (AA6101-T6) but very nearer to weaker material (AA1100) shows that tensile properties of aluminum alloy retained after friction stir welding.
4)Extensive experimentation is required to study effect of parameters on properties of dissimilar Aluminium AA1100 to AA6101-T6 friction stir welded specimen.

References

1.P L Threadgill, A J Leonard, H R Shercliff, "Friction stir welding of aluminum alloys", *International Materials Reviews*, Volume.54. Issue.2. March 2009. Pages 49-93

2.D. Trimble, J. Monaghan, G.E. O'Donnell Department of Mechanical and Manufacturing Engineering, Trinity College, Dublin, Ireland "Force generation during friction stir welding of AA2024-T3". CIRP Annals - *Manufacturing Technology*, Vol. 61, issue 1, 2012, Pages 9–12

3.C Chen & R Kovacevic "Thermomechanical modeling and force analysis of friction stir welding by the finite element method" *Journal of Mechanical Engineering Science* 2004, Vol. 218, Issue No. 5, pages 509-519

4.L.V. Kamble, S.N. Soman, P.K. Brahmankar, "Effect of Tool Design and Process Variables on Mechanical Properties and Microstructure of AA6101-T6 Alloy Welded by Friction Stir Welding", *IOSR Journal of Mechanical and Civil Engineering* 2012, Volume 1, Issue 334, pages 30-35

5.S. Nansaarng, K. Chaivanich, " Study of influence of parameters of dissimilar materials joining on friction stir welding process by design of experimental", *Proceedings of the 5th IASME/WSEAS Int. Conference on Heat Transfer, Thermal Engineering and Environment,* Athens, Greece, pages 129-136

6.N. T. Kumbhar, K. Bhanumurthy, "Friction StirWelding of Al 5052 with Al 6061 Alloys", *Journal of Metallurgy*, Hindawi Publishing Corporation2012, Volume 2012, pages 1-7

7.Ranjith R., Senthil Kumar B., "Joining of dissimilar aluminium alloys AA2014 T651 and AA6063 T651 by friction stir welding process", *WSEAS Transactions on Applied and Theoretical Mechanics* 2014, Volume 9, pages 179-186

8.S. Ravikumar, V. Seshagiri Rao, R.V. Pranesh. "Effect of Process Parameters on Mechanical Properties of Friction Stir Welded Dissimilar Materials between AA6061-T651 and AA7075-T651 Alloys", *International Journal of Advanced Mechanical Engineering* 2014, Volume 4, Issue 1, pages 101-114

9.Rajiv S. Mishra, Murray W. Mahoney "Friction Stir Welding and Processing" ASM International 2007

A Brief Author Biography

Mr. Vinayak D. Yadav– PG Student ME CAD/CAM/CAE KIT's college of engineering Kolhapur. Maharashtra

Prof. S. G. Bhatwadekar– Head of Department of Production Engineering, KIT's College of Engg. Kolhapur

EXPLORATION OF EFFECT OF COPPER FILLER MATERIAL ON THERMAL PROPERTIES OF PALMYRA FIBER REINFORCED COMPOSITE

P.PADMAVATHI[1], C.J.RAO[2]

[1.] M.Tech Student (Thermal Engineering), Aditya Institute of Technology and Management, Tekkali.
E mail: padma.vathi89@gmail.com
[2.]Professor &HOD. Mechanical Engg. Dept. Aditya Institute of Technology and Management, Tekkali.

Abstract

Recently, thermoplastic and thermoset polymers are combined with natural fibers to produce the composites, which possess better strength and good resistance to fracture. Due to an excellent property profile, these composites find wide applications in packaging, building and civil engineering fields. The present work aims to elucidate the optimization of thermal properties such as thermal conductivity, specific heat.Thermal degradation measurement by Thermo-Gravimetric Analysis (TGA) and Thermal Diffusivity of Palmyra reinforced polymer composites with chemical treatment and addition of copper powder materials. The composite specimens were prepared with different weight percentages of palmyra fiber and copper powder in polymer matrix.

Keywords: Palmyra fiber, natural composite, Chemical treatment, Copper powder, Thermal properties.

1. Introduction

Material characteristics play an important role in manufacturing and design engineering. Knowledge of the response of the work material during manufacturing is essential for adopting more efficient, effective and economical processing methods. Proper understanding of the response of the work material under different situations is possible only if the characteristics of a material are known. Composites with natural fibres are gaining increasing attention for a variety of applications. Natural fibres such as jute, coir, sisal, and Palmyra belong to this category [1]. Glass, carbon, boron and Kevlar fibres are being used as reinforcing materials in Fibre Reinforced Plastics [FRP] which have been widely accepted as materials for structural and non-structural applications. Synthetic fibres are not eco-friendly. Hence attention has been focused on the utilization of natural fibres for the production of fibre-reinforced materials. One of the natural fibre-polymer composites are investigated by Paramasivan and Abdulkalam, A.P.J. [2] using sisal fibres and epoxy matrix. The fabrication process attempted by them includes winding and lamination. It is found that the fabrication of these composites is fairly easy and cost of production is quite low. Winding of cylinders with longitudinal or helical and hoop reinforcements is successfully carried out. Tensile strength of the sisal-epoxy composites is found to be 250-300 MPa, which is nearly half the strength of fibre glass –epoxy composites for the same volume fraction. However due to low density of sisal fibre, however, the specific strength of sisal composites is comparable with that of glass composites. The unidirectional modulus of sisal-epoxy composites is found to be about 8.5GPa. This study indicates the feasibility of developing composites incorporating one of the abundantly available natural fibres, to be used in the field of consumer goods, low-cost housing and civil structures. However, no attempt is made in this work on the performance of such composites to assess the effect of exposure to weather or the performance of the components in actual use. There are also reports [2] of sisal-polyester composites using electron probe microanalysis for evaluating filler dispersions in polyester composites. Chopped sisal fibre polyester composites are prepared by the press moulding technique. Mechanical properties of the composites are evaluated through accelerated weathering tests conducted pertaining to

ASTM D-520 specification. It is found that the specific modulus of the composite is 1.90 compared with 2.71 for glass fibre reinforced plastics, while the specific strength is of the same order as that of polyester and 30% less than glass fiber reinforced plastics. Accelerated testing revealed little change in initial modulus, reductions of 5% in ultimate tensile strength, 16% in flexural strength and 5.4% in water absorption. Lakkad.S.C. et al [6] compared the values of ultimate tensile and compressive strength and Young's modulus of elasticity of bamboo specimens with those of mild steel and glass reinforced plastics. But they have not specified the specic-name of the bamboo specimens tested. There are more than 500 species of bamboo available in India and each has different mechanical properties. Jindal.U.C. [7] studied the mechanical properties of bamboo specimens made from Dendrocalamus Stricutus specie of bamboos. The specific ultimate tensile strength of bamboo specimens is nearly six times the specific ultimate tensile strength of the mild steel. Bamboo specimens have maximum stiffness along the fibers and minimum transverse to the fibers. The variation of the modular ratio (i.e. the ratio of Young's modulus along the fibers and transverse to the fibers) of bamboo specimens under different fiber orientation is similar to the variation of modular ratio of other fibre-reinforced composites. Fiber incorporated plastics have been very popular due to their flexibility, their lightness and the ease of fabrication of complicated shapes with economic savings in contrast to fiber reinforced metals/alloys. In addition, these composites can be easily substituted for conventional materials in several areas such as the building industry, transportation and consumer goods. Some of these attempts made in recent times for the utilization of natural fibres through composite material technology have indicated their potential as substitute for the conventional materials such as wood and glass fibre reinforced plastics (GFRP) in many applications. There are, however, a number of limitations, including cost factor and their performance over a long time duration, which need further research. Extensive literature is available on the production and mechanical behavior of composites obtained by reinforcing epoxy with fibre of glass, boron, carbon silicon carbide etc. Many researchers in the past have developed composites with natural fibres such as sisal henequen, jute, banana, cotton, etc. [3, 4, 5, 8], but the work on the Palm Fibre Reinforced Plastic Composites [PFRP] is not available in the literature.

2. Preparation of Samples

Steps in preparation of polymer fiber reinforced composite samples using Hand Lay technique.

2.1 Extraction of Palmyra Fibers from Palm Tree

Palmyra fiber is available in the form of bract on a Palmyra tree. First dried bracts are collected from Palm trees. Then the fibers inside the bract are segregated. They are soaked in water for 24 hours. Using a knife, the black layer on top of the fiber is scrapped off. These fibers are dried in sun for 2 days to remove the moisture. If necessary they should also be put in an oven for 2 hours at 70°C to ensure that all the moisture is completely removed.

2.2 Chemical Treatment of Fibers

In this work only chemically treated fibers are used. Cleaned fibers are kept in 4% (NaOH) solution for 4 hours for chemical treatment. The fibers are then removed and cleaned in water. They are kept in sunlight for 24 hours to remove moisture entirely. The fibres are then cut as per the required dimensions as per ASTM test sample specifications.

2.3 Mold Preparation

Thermal conductivity test samples should be of circular shape with 50 mm diameter and 10mm thickness. To prepare the mold, a 50 mm diameter hole is cut in a 10 mm thick rubber sheet. This cut rubber sheet is affixed to a cleaned tile with mansion hygienic wax. The tile is cleaned thoroughly with shellac NC thinner solution.

During preparation of the circular sample, it is observed that the fiber concentration is much higher at the center than at the periphery. To get a nearly homogenous fiber distribution in the sample, in this research a large rectangular sample of 165 mm x 55 mm x 10 mm is prepared, out of which the required circular samples are cutout.

Fig.1: Chemically Treated Palmyra Fibers

Fig.2: Mold for Sample Preparation

2.4 Determining Quantities of Fiber and copper

The maximum amount of fiber or copper powder that can fit in the mold is taken as 100%. To fit in maximum amount of fiber, fibers are spread layer upon layer very closely and ramming is done after each layeı mold is found to be 8 gm, which is taken as 100% (FB100). The weight of maximum copper powder quantity iı found to be 32gm, which is taken as 100% (CU100).

2.5 Polyester Bonding Material

Polyester resin is durable, comparatively inexpensive, has superior corrosion resistance, has good ran 4413, which is a general purpose polyester resin is used as matrix material.

Property	Value
Resin	ECMALON 4413
Colour	pale yellow
Viscosity	500-600 CPS (Brokfield Viscometer)
Specific Gravity	1.13 grams/c.c.
Temperature	25^0 C
Acid Number (mg KOH/g)	22
Monomer content	35%

Table.1: Properties of Polyester Resin

2.7 Catalyst and Accelerator

Curing or cross-linking of polyester is achieved by adding a catalyst (initiator) plus an accelerator (promoter) at room temperature. The function of catalyst is to speed up a chemical reaction by providing an alternate reaction pathway with lower activation energy. The function of accelerator is to alter chemical bonds and speed up the chemical process. In this work, cobalt accelerator along with Methyl Ethyl Ketone Peroxide (MEKP) catalyst is used.

Optimum quantity of catalyst and accelerator must be used. If more quantity is used, the specimen cures faster but will be of lesser strength and poor appearance. If lesser quantity is used, then the sample takes very long time (more than 8 hours) to cure. In this work approximately 2ml catalyst and 2 ml accelerator is used, which gave a curing time of around 4 hours.

2.8 Hand Lay Technique

2.8.1 Clean the mold with shellac NC thinner solution. Apply a thin coating of poly-vinyl alcohol on the interior tile surface and along the edges of the rubber sheet. Dry it for a day.

2.8.2 Fill the mold with required mass of fibers by spreading them as homogenously as possible.

2.8.3 Take the required mass of copper powder in a measuring jar.

2.8.4 Pour small amounts of liquid polyester in the copper powder jar and stir it thoroughly.

2.8.5 Add catalyst to this paste using a syringe and stir it fast.

2.8.6 Add accelerator to this mix and stir it fast. Extreme caution should be taken in ensuring that the catalyst and accelerator does not get into direct contact with each other. Else they both react chemically extremely rapidly with issuing out fire.

2.8.7 Immediately apply this paste on top of the fibers which are filled in the mold, otherwise it would solidify rapidly in the measuring jar itself.

2.8.8 To ensure that no air bubbles are trapped inside, take a transparency sheet and cover it over the mold immediately by using rolling operation.

2.8.9 Place a tile on top covering the entire mold and its contents. Place sufficient weight (roughly 50 kg) on top of the mold and leave it undisturbed in a closed room for 1 day until the composite cures.

Fig.2: 100Wt%Fiber composite

Fig.3: Copper powder

3. Thermal Testing Results

Thermal Tests conducted are:

1) Thermal Conductivity measurement
2) Specific Heat Capacity measurement
3) Thermal Diffusivity is computed.
4) Thermal Degradation measurement by Thermo-Gravimetric Analysis (TGA)

To investigate the effect of aluminum powder filler material on the thermal properties of the chemically treated palmyra fiber composites, 5 samples are prepared.

Sample Name	Polyester gm	Fiber Gm	Al gm	Relative CU %
CU0	13	8	0	0%
CU25	10.5	6	3.75	25%
CU50	7	4	7.5	50%
CU75	4.5	2	11.25	75%
CU100	26	0	15	100%

Table.2: Sample's Composition by Weight

Sample	Weight	Volume	Density

Name	(kg)	(m^3)	(kg/m^3)
AL0	0.020	2.50E-05	1277.5
AL25	0.022	1.78E-05	1844.4
AL50	0.023	2.50E-05	1197.63
AL75	0.028	2.14E-05	1210.95
AL100	0.031	1.79E-05	1732.6

Table.3: Sample's Density

3.1 Thermal Conductivity

Thermal conductivity of the samples is measured at 50°C using guarded heat flow test method as per ASTM E1530 specifications. Unitherm Model 2022 manufactured by ANTER Corp., Pittsburgh, PA is used for this test

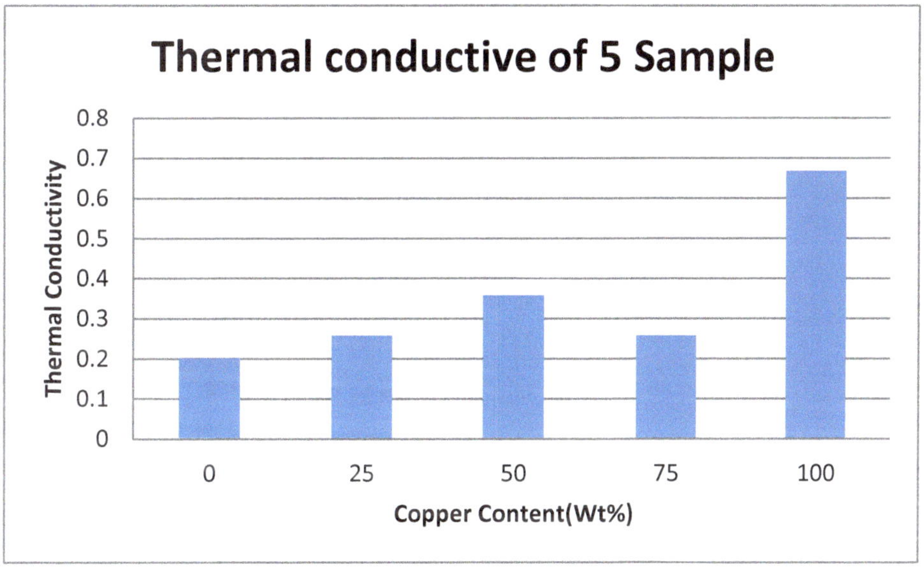

Fig.4: Thermal conductivity of specimens at 55°C.

3.2 Specific Heat Capacity

Differential Scanning Calorimeter (DSC) technique using Double Furnace setup is used for measuring specific heat capacity. DSC is a thermo- analytical technique in which the difference in the amount of heat required to increase the temperature of a sample and reference is measured as a function of temperature. Compared to Single Furnace setup, Double Furnace DSC gives more accurate readings over larger temperature range, with more rapid response time as it measures the heat flow change of the sample directly. Test equipment used is Netzsch Simultaneous Thermal Analyzer STA 449F5 Jupiter.

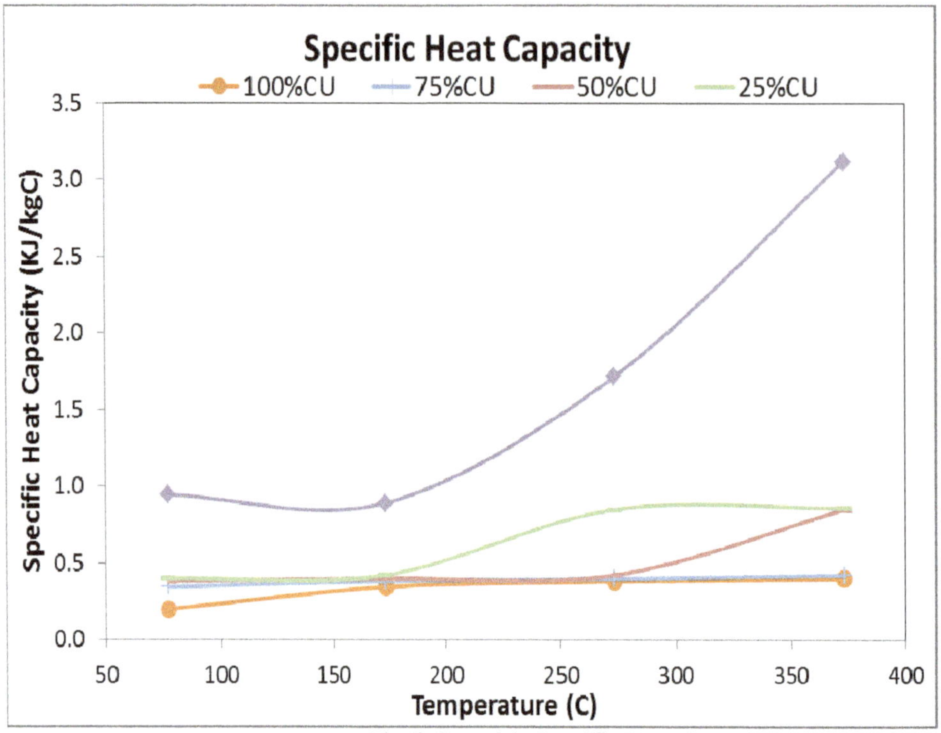

Fig.5: Sample's Specific Heat Capacity

3.3 Thermal Diffusivity

Although thermal diffusivity can be experimentally measured by flash method, in this work it is calculated after knowing thermal conductivity, specific heat capacity and density.

$$\alpha = \lambda / (\rho C_p)$$

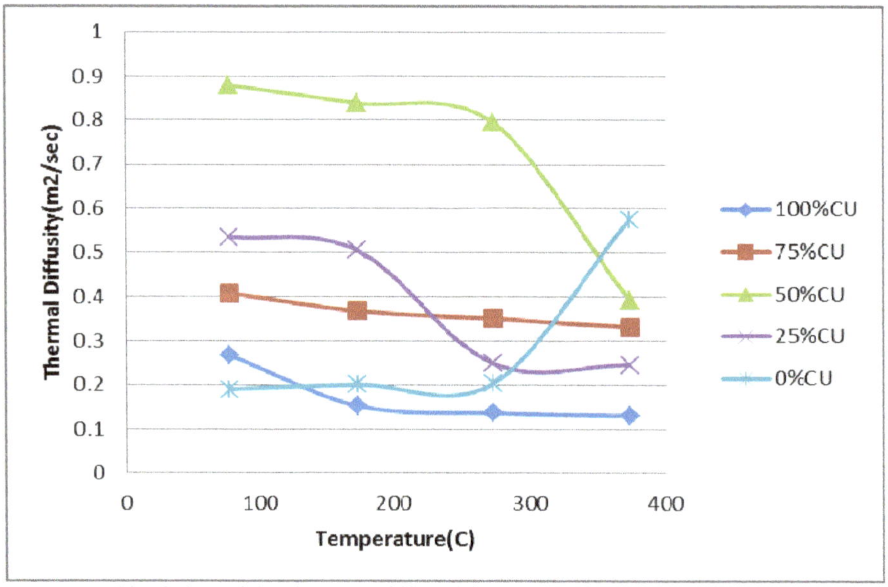

Fig.6: Sample's Thermal Diffusivities at Different Temperatures

3.4 Thermal Degradation by TGA

Thermo-Gravimetric Analysis (TGA) is a technique in which the mass of a substance is monitored as a function of temperature or time as the sample specimen is subjected to a controlled temperature program in a controlled atmosphere. TGA measures a sample's weight as it is heated or cooled in a furnace. The loss in weight over specific temperature ranges provides an indication of the composition of the sample, including volatiles and inert filler, as well as indications of thermal stability.

All the measurements were performed as per ASTM E1131 standard using high resolution Perkin Elmer TGA7 Thermo-gravimetric Analyzer. The samples weighing between 10 and 20mg are placed in a platinum pan and tests are performed within the temperature range of 20–700°C at a heating rate of 5°C/min under nitrogen atmosphere at flow rate of 50 ml/min. TG and DTG curves were analyzed to study the high temperature degradation behavior.

The TGA curves (Fig.7) of specimens provide three distinct temperature regions, wherein the samples experience major weight loss. A small weight loss was observed during Phase-1 attributed to the evaporation of moisture. Actual degradation happens in second region attributed to the thermal degradation of hemicelluloses, cellulose and lignin together with polymeric matrix and thereafter the rate of decomposition was slow.

As can be seen from curves, the thermal decomposition onset temperature is highest for CU100% with the corresponding least amount of weight loss %. Thus 100% Copper powder composite has the greatest thermal stability or the least thermal degradation. Among the remaining composites.

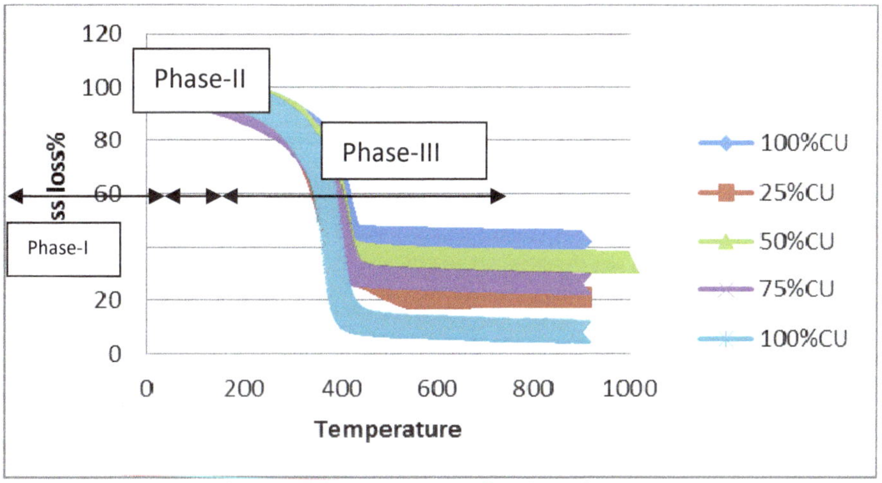

Fig .7: Sample's Thermo-Gravimetric Curve

4. CONCLUSIONS

(i) Thermal conductivity of the composite increases with increase in the copper filler content. But it gives more thermal conductivity at 50Wt% of copper compering than 25and 75Wt% of copper.

(ii) The response of specific heat capacity for a 75Wt% copper composite for a given temperate is varying considerable But other % copper does not showing any remarkable variation.

(iii) Thermal diffusivity of the composite decreases with increase in temperature except for 0Wt%copper.

(iv) 100% copper filler composite has the highest thermal degradation onset temperature and also the least amount of weight loss. Thus it has the highest thermal stability.

References

[1] ***A.R.Prasad*** *"Bamboo reinforced plastic composites"* .XIII National conference on Indian society of mechanical engineering ISME2003, (December 2003).

[2] ***Parmasivam, T and Abdulkalam, A.P.J.***, "On the study of natural fiber composites" Fiber Science and Technology I (1974) pp. 8588.

[3] ***Satyanarayana, K. G., Sukumaran, K., Kulkarni, A. G., Pillai, S. G. K. and Rohatgi, P. K.***, *"Fabrication and properties of natural fiber reinforced polyester composites. Composites"*, 17 (October 1986) pp 329-33.

[4] ***Madhava, M. R., Suha Raman, Pavithran, C., Prasad, S.V. and Rohatgi, P. K.***, *"Ultrasonic velocity and attenuation in coir-polyester composites"*. Paper presented at the 2nd National Symposium on Ultrasonics, New Delhi. Ultrasonics Society of India, New Delhi, (February 1982).

[5] ***Fardons Mubark and Han Augustin***, *"Composite hard board from pith and depithed bagasse-filled plastics". Research & Industry (India),* 29 (February 1954). pp. 108-113.

[6] ***Sefain, M. Z., Fedl, M. A. and Rakha, M.***, *"Influence of pressing time in fiber board making". Research &Industry (India),* 29 (April 1984) pp269- 272.

[7] ***Pakotiprapha, B., Pama, R. P. and Lee, S. I.***, "Study of Bamboo Pulp and Fibre Cement Composites," *Int. J. Hous. Sci. Appl.*, Vol. 3, No.3, (1979) pp. 167-190.

[8] ***Jindal, U. C.***, *"Development and testing of bamboo fibre reinforced plastic composites"* J. of comp. mat. Vol.20.Jan1986. pp.19-29.

[9] ***V.V.S.Prasad, KNS Suman, DN Rao, NRMR Bhargava*** "Testing of Palm Fibre as Reinforcement Material In Polyester Composites"

14

Simulation of Different Geometries and Flow Parameters of the Micro channel Heat Sinks

Neeraj kumar[1], Dilbagh Singh[2], Harry Garg[3], Nirmal Kant Singh[4]

[1,4]National institute of Technology (NIT) Kurukshetra, India, 136119

[2]NILLM UNIVERSITY, KAITHAL, 136027

[3]Central Scientific Instruments Organisation (CSIR-CSIO), Chandigarh, India, 160030

Corresponding author email id: neeraj07326@gmail.com

Abstract

Micro channel Heat Sinks are used for heat transfer for miniature electronic circuits. The high performance of the micro channel depends upon the number of parameters like geometrical & flow parameters. There is wide variation in the performance of the micro channel heat sinks as the geometry of these channels changes. The performance is measured in terms of the thermal resistance of the channels. A three- dimensional numerical model of the interrupted micro channel heat sink is presented to study the effect of pumping power and thermal resistance and heat transfer characteristics due to various dimensions of micro channel. Numerical investigation was conducted to predict pumping power and thermal resistance in a rectangular cross section micro channel heat sink by numerical analysis of different geometries. The pumping power and thermal resistance was analysed with different cooling liquid used as water and ethylene glycol while micro channel substrate used was made of copper. The fluid flow and heat transfer mechanism of the newly proposed micro channel heat sink are analysed.

Keywords: Simulation; Heat sinks; microelectronic; electronic circuits.

1. Introduction

The developments in the micro electro mechanical devices naturally require a critical design consideration for efficientt removal of internally generated heat in microelectronic components.

Significant amount of research has been made to develop innovative cooling techniques that have the potential to deliver high-heat flux rates for microelectronic applications. A number of studies have investigated the thermal design optimization of micro-channel heat sinks to determine the geometric dimensions that give optimum performance. Over the past few decades, there is an increasing interest in the development of heat sink processes for heat dissipation and many design methodologies regarding how to optimize heat sinks have been proposed For optimal design of plate- and pin-fin. Most of the

research in the area of micro channel heat exchangers has focused on solving the problem of electronics cooling, where heat generated by the microelectronic components is removed by a coolant flowing through channels located as close as possible to the heat source. Hong and Cheng [1] presented a numerical study on laminar forced convection of water in offset strip-fin micro channels network heat sinks for microelectronic cooling. The heat transfer and fluid flow characteristics in offset strip-fin micro channel, the heat transfer enhancement mechanism and the effects of geometric size of strip fin on the heat sink performance were analysed. Recently Kim & Kim [2] have proposed a closed-form total thermal resistance correlation for the thermal design optimization of micro-channels. In addition to providing excellent heat transfer, the high conductivity typical in this class of fluids offers the potential of efficient, compact pumping. Of various such metals reviewed in literature liquid gallium has been found to be most suitable candidate.[3]The Pentium 4 processor in 2004 which produced heat fluxes of 30 W/cm^2 is now replaced by semi-conductor chips producing more than a 100 W/cm^2. The need to remove such high levels of heat fluxes in cooling systems poses a great challenge to thermal engineers who require new ways to enhance heat transfer rates. The performance, reliability and life expectancy of the electronic equipment hence depends on managing the operating temperature effectively. Micro-channels fabricated in metals with good conducting properties (e.g. copper) subjected to high heat fluxes from electronic chips exemplifies as a practical and economically feasible way of energy removal. The very first people to conceive the idea of high heat flux removal from micro-channels were Tuckerman and Pease [4].The idea involved using a substrate with several fins and micro-channels in parallel conducting heat efficiently from the substrate to the coolant. Water was used as a fluid under laminar conditions flowing through micro channels embedded in a silicon wafer. The experiment employed heat flux of up to 7.9 $MW/m^2$and a maximum temperature difference of

71°C between the substrate and the inlet water. A large pressure drop200kPa was recorded with plain micro-channels and 380kPa with pin fin-enhanced micro-channels. Given the deviation in past published results related to heat transfer of flow in micro-channels, there are no generally recognized correlations predicting their heat transfer characteristics. Though there has been a general agreement in published results that the heat transfer coefficient increases with increasing heat flux, there is hardly any numerical work done to obtain valid results for the application of high heat fluxes equal to or more than 100 W/cm^2on micro-channels employing water flows. A numerical investigation has been carried out in the present study using the FLUENT. Since the pioneering work of Tuckerman and Pease in 1981, many studies have been conducted on micro-channel heat sinks as summarized by Phillips [5] and more recently, by Morini. The need for cooling in high power dissipation (100 W/cm^2) systems in several scientific and commercial applications such as microelectronics requires something beyond the conventional cooling solutions. A number of studies have investigated the thermal design optimization of micro-channel heat sinks to determine the geometric dimensions that give optimum performance. For the heat transfer study purpose, the channel walls were assumed to behave as fins. The experimental study of Miner and Ghoshal[6] showed that heat transfer coefficients of the order 10$W/cm^2/K$ are achievable using different liquid metals as fluid. Besides analytical models, several numerical methods have been used to study the thermal performance of micro-channels heat sinks.[7]Gao et al. 2002 also studied the transition Reynolds number in micro channels by investigating how it varies with the channel height

In this paper micro channel heat sink is presented to study the effect of pumping power and thermal resistance and heat transfer characteristics due to various dimensions of micro channel. As the fluid is passing through the different section of the micro channel the distribution of the fluid in the passage is disturb the flow condition of the fluid that affect the velocity and thermal boundary layer of the flow. As flow is reached fully developed there is no change in the velocity of the fluid layer. The thermal and velocity boundary layer are playing a significant role in the fluid flow in micro channel. The different shapes of micro channel are used to dissipate the large amount of heat from the system or electronic circuit. As a practical cooling fluid, the liquid metal must satisfy the following requests: non-poisonous, non-caustic material, low viscosity, high thermal conductivity and heat capacity. Most studies in this approach employed the classical fin theory which models the solid walls separating micro channels as thin fins. The heat transfer process is simplified as one- dimensional, constant convection heat transfer coefficient and uniform fluid temperature. However, the nature of the heat transfer process in MCHS is conjugated heat conduction in the solid wall and convection to the cooling fluid. Using a nano fluid as the heat transfer working fluid has gained much attention in recent years. Xuan and Roetzel (2000) proposed two theoretical models to predict the heat transfer characteristics of nano fluid flow in a tube [8-9]. Li and Xuan (2002), Xuan and Li (2003). [10] and Pak and Cho (1998) experimentally measured the convection heat transfer and pressure drop for nano fluid tube flows. Their results indicated that the heat transfer coefficient was greatly enhanced and depended on the flow Reynolds number, particle Peclet number, particle size and shape, and particle volume fraction. These studies also indicated that the presence of nano particles did not cause an extra pressure drop in the flow. Recently, Yang et al. (2005) carried out an experimental study attempting to construct a heat transfer correlation among the parameters that affected heat transfer [11].

2.0 ANALYSIS PROCEDURE

The combination of the thermal resistance models and the optimisation algorithm served as useful tool in the design of the micro channel heat sink. The thermal resistance in the heat sink arises from three sources: conduction resistance in the heat sink, including the fin effects; convection resistance between the micro channel surfaces & the coolant & the resistance due to the temperature arise of the cooling fluid.

$$R_{th}= R_{cond}+R_{conv}+R_{cap} \quad (1)$$

The total thermal resistance is calculated:

$$R_{th}=T_{max}-T_{min}/Q \quad (2)$$

According to theoretical approach Pumping is defined as the product of flow rate and the pressure drop mathematically is defined as

$$\Omega = v.\Delta p \quad (3)$$

The analysis is based on the following assumptions:

- Steady state flow.
- Incompressible fluid.
- Laminar flow.
- Constant properties of both fluids and solid.

Effects of viscous dissipation are negligible.

3.0 COMPUTATIONAL DOMAIN

Fig. 1 shows computational domain of single micro channel heat sink. Various parameters considered in this study are shown in figure. t_{sb} and t_s represents thickness of solid substrate width from heated surface (base) and channel thickness wall respectively shown.

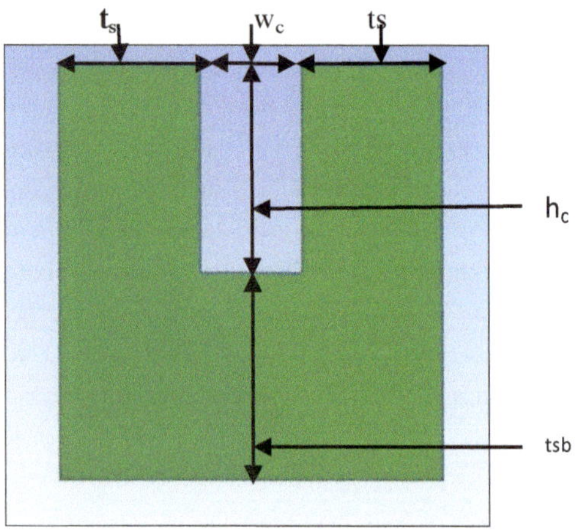

Fig 1 Computational Domain

t_{sb} define the width of solid region it is also referred to as wall thickness. w_c denotes channel width while h_c represents channel depth and their ratio being called as aspect ratio. Substrate materials are considered as copper (cu) whose properties along with the fluid of liquid water (H_2O) and ethylene glycols are given in Table.

3.1 EXPERIMENTAL DESIGN AND PROCEDURE:

The process parameters & their values are given in Table. A rectangular Copper micro channel of size 2.5 mm ×2.1mm ×1mm was used for experimental work. Different Geometrical & flow parameters are chosen as these directly affect the response parameters. The single channel geometry is chosen & all the corresponding flow parameters are selected with respect to the chosen geometry. The shape & size of the channel varies depending upon the selection of different geometrical parameters. Thermal resistance and pumping power for various dimensions of different geometries of the micro channel of heat sink are chosen. The Reynolds number chosen is in the laminar range of 50-200 & the simulation has been done using FLUENT for the optimum analysis. The Table 1 shows the input parameters of the micro channel height (h_c), width (w_c), substrate thickness (t_s) and fluid flow with different Reynolds number. Table 1 shows the input parameters of different values.. Table 2 and Table 3 show the response for the water and the ethylene glycol with different resistance and the pumping power. Though there are lot of parameters but to ease the computation three different geometrical parameters and one flow parameter is chosen. The geometrical parameters are to be chosen as the height of the micro channel ($h_{c)}$, width of micro channel ($w_{c)}$ and the thickness of substrate (t_s) There is variation in geometrical parameters as height of channel, width of channel, thickness of substrate. The flow parameter used as Reynold^s No. & are chosen for both water and ethylene glycol. The flow is varying with different velocities for each geometry parameter. The analysis is done on basis of the simulation of the micro channel heat sink. The geometry of the micro channel heat sink is considered to be as parametric to optimise the different parameters of the micro channel heat sink.

Table1: Input Parameters

S.NO.	R_e	h_c	w_c	t_s
1	150	1	0.5	0.7
2	200	1.5	0.3	0.3
3	150	1	0.5	0.5
4	150	1	0.5	0.5
5	100	1.5	0.7	0.7
6	100	1.5	0.3	0.3
7	100	1.5	0.7	0.3
8	150	1	0.7	0.5
9	50	1	0.5	0.5
10	100	0.5	0.7	0.3
11	100	0.5	0.7	0.7
12	150	2	0.5	0.5
13	200	1.5	0.7	0.3
14	200	0.5	0.7	0.7
15	150	1	0.1	0.5
16	200	1	0.5	0.5
17	200	0.5	0.3	0.3
18	100	0.5	0.3	0.3
19	200	1.5	0.7	0.7
20	200	0.5	0.7	0.3
21	150	1	0.5	0.5
22	150	1	0.5	0.5
23	150	1	0.5	0.5
24	100	1.5	0.3	0.7
25	200	0.5	0.3	0.7
26	150	1	0.5	0.1
27	150	0.5	0.5	0.5
28	150	1	0.5	0.5
29	150	1	0.5	0.5
30	200	1.5	0.3	0.7
31	100	0.5	0.3	0.7

Table 2: Response for Water

S.No	R_{cond}	R_{conv}	R_{th}	pp
1.	0.09	0.001189	3.49E-05	8.31E-06
2.	0.056	7.82E-05	2.21E-05	1.35E-05
3.	0.078	0.000677	3.04E-05	8.31E-06
4.	0.078	0.000677	3.04E-05	8.31E-06
5.	0.062	0.000445	2.42E-05	9.92E-06
6.	0.056	7.82E-05	2.21E-05	1.35E-05
7.	0.04	0.000129	1.85E-05	9.92E-06
8.	0.073	0.000733	2.87E-05	7.89E-06
9.	0.078	0.000677	3.04E-05	8.31E-06
10.	0.099	0.001448	3.85E-05	6.97E-06
11.	0.131	0.005227	5.11E-05	6.97E-06
12.	0.044	0.000101	1.73E-05	1.40E-05
13.	0.047	0.000129	1.85E-05	9.92E-06
14.	0.131	0.005227	5.11E-05	6.97E-06
15.	0.245	0.000733	9.53E-05	1.26E-05
16.	0.064	0.000309	2.50E-05	8.31E-06
17.	0.145	0.001558	5.65E-05	5.95E-06
18.	0.214	0.006855	8.30E-05	5.95E-06
19.	0.047	0.000129	1.85E-05	9.92E-06
20.	0.117	0.003041	4.55E-05	6.97E-06
21.	0.078	0.000677	3.04E-05	8.31E-06
22.	0.078	0.000677	3.04E-05	8.31E-06
23.	0.090	0.001189	3.49E-05	8.31E-06

24.	0.088	0.000337	3.42E-05	1.34E-05
25.	0.214247	0.006855	8.30E-05	5.95E-06
26.	0.048093	8.47E-05	1.86E-05	8.31E-06
27.	0.13339	0.003481	5.17E-05	6.25E-06
28.	0.078483	0.000677	3.04E-05	8.31E-06
29.	0.078483	0.000677	3.04E-05	8.31E-06
30.	0.088228	0.000337	3.42E-05	1.34E-05
31.	0.214247	0.006855	8.30E-05	5.95E-06

The table above contains the different values of the geometrical & flow parameters & the corresponding responses are derived with respect to the same. The reason for choosing no of variables is to increase he computation space & get the optimum from the different domains.

Table 3: Responsefor Ethylene Glycol

S.NO.	R_{cond}	R_{conv}	R_{th}	pp
1	0.091599	0.070052	3.54E-05	0.030848
2	0.034408	0.007625	1.33E-05	0.080351
3	0.084543	0.043574	3.27E-05	0.03085
4	0.084543	0.043574	3.27E-05	0.03085
5	0.219251	0.047669	8.49E-05	0.001565
6	0.060331	0.008424	2.33E-05	0.020018
7	0.193292	0.017924	7.48E-05	0.001565
8	0.139687	0.049372	5.41E-05	0.011066
9	0.16801	0.066221	6.50E-05	0.003375
10	0.083724	0.100822	3.24E-05	0.036895
11	0.097351	0.264032	3.77E-05	0.036895
12	0.139708	0.010355	5.41E-05	0.004496
13	0.142359	0.012432	5.51E-05	0.006332
14	0.056837	0.219604	2.20E-05	0.151047
15	0.013915	0.108535	5.38E-06	1.88354
16	0.068424	0.040147	2.65E-05	0.055274
17	0.010576	0.129858	4.09E-06	1.70424
18	0.020452	0.133829	7.91E-06	0.417331
19	0.167811	0.032939	6.49E-05	0.006332
20	0.048868	0.084042	1.89E-05	0.151047
21	0.084543	0.043574	3.27E-05	0.03085
22	0.084543	0.043574	3.27E-05	0.03085

23	0.084543	0.043574	3.27E-05	0.03085
24	0.082323	0.030127	3.19E-05	0.020018
25	0.012853	0.463331	4.97E-06	1.70424
26	0.059953	0.009444	2.32E-05	0.03085
27	0.282088	0.250094	1.09E-04	0.000471
28	0.084543	0.043574	3.27E-05	0.03085
29	0.084543	0.043574	3.27E-05	0.03085
30	0.047238	0.027218	1.83E-05	0.080351
31	0.02484	0.477563	9.61E-06	0.417328

Table 2 shows the response for the water which shows different value of the thermal resistance, conduction, convection and the pumping power with variable geometry parameters. The different plots for the same are also shown in the subsequent work. The reason to chosen different thermal resistance is to examine the most critical thermal resistance of the total thermal resistance. Table 3 shows the response for the ethylene glycol which shows the different value of thermal resistance, conduction, convection and the pumping power with variable geometry parameters.

4.0 RESULTS AND DISCUSSION

Results are discuss with the help of various graphs and plots as shown in Fig. the velocity and temperature contour are shown in Fig. for the coolant water and ethylene glycol .the temperature contour for the liquid water is shown in Fig. 2 and for the ethylene glycol is shown in Fig.3.Similarly for the velocity contour as shown in Fig. 4 and 5 for water and ethylene glycol.

There are the various plots are shown in Fig. for different geometry parameters for optimize the result of micro channel heat sink. Table 1 shows the input parameters of different geometries of micro channel heat sink and the Table 2 and 3 shows the results for the thermal resistance and the pumping power.

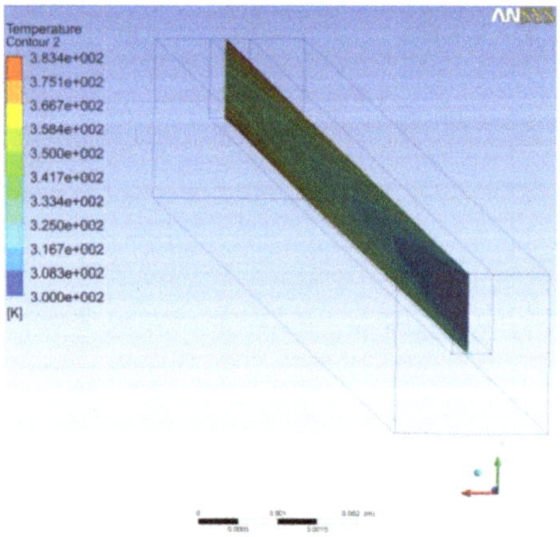

Fig.2 Temp Contour of water

Temperature contour for water is shown in figure 2. The thermal boundary layers are shown in the figure the temperature range is varying as shown in figure.

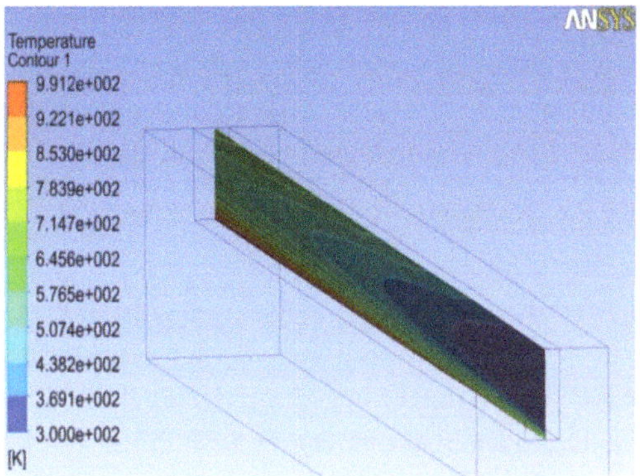

Fig.3 Temp contour of ethylene glycol

Temperature contour of ethylene glycol is shown in figure the thermal boundary layers are shown in figure.

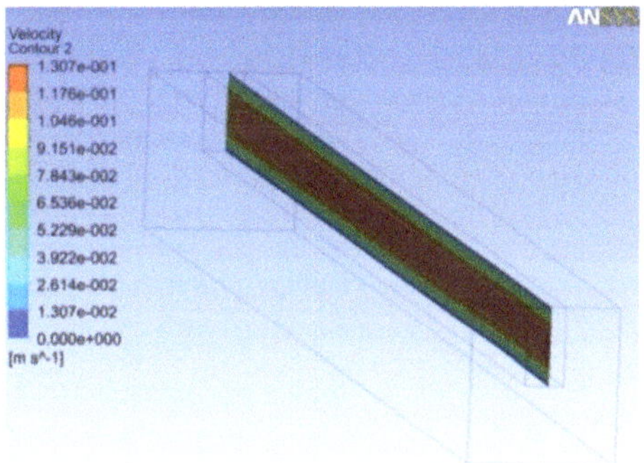

Fig. 4 velocity contour for ethylene

Velocity contour for ethylene glycol is shown on figure 4 velocity boundary layer is shown in figure.

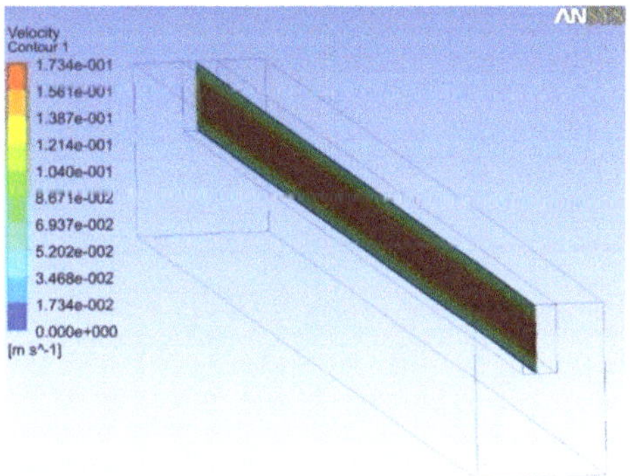

Fig. 5 velocity contour for water

The velocity and temprature contour is shown Fig.from the temprature contour for theethylene glycol the temprature is high so more coolant is required to remove the heat from the system . in the case of water the heat removed the system is easily as compared to ethylene glycol the max. heat is dissipate from the system.the maximum temprature reached the value is low as compared to ethylene glycol.

4.1 Plots for liquid Water:

The different plots for water considering different response parameters are shown & plotted below. plots are plotted between the Reynolds number and the heat transfer coefficient and temprtaure difference that shows the variation in the temprtaure of the micro channel and the amount of heat is transfered through the channel. as the optimization for the best geometrical parameter of the micro channel that can be prescribed by the various results are obtained from the simulation . the variation in heat transfer coeffiecient and the temprature difference are shown in the plots.

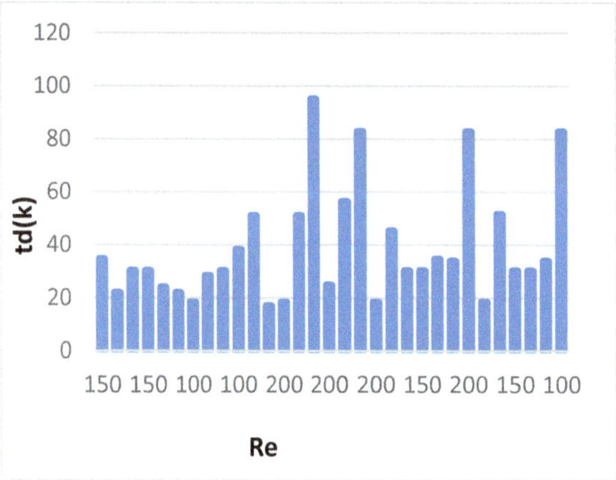

Fig.6 Temp Plot vs Reynold's No

From the fig. 6 the maximum temprature difference is reached a value 95 k at the maximum velocity as there is also varaition of geometry parameters.

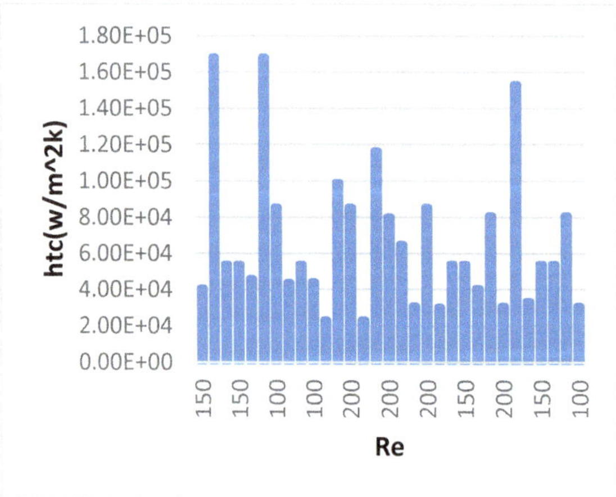

Fig.7 Heat Transfer Coefficient vs Reynold's No

From the above Figure 7 which shows the value of heat transfer cofficient reached the maximum value at velocity reache is an average value. maximum value of ΔT higher th heat treansfer cofficient . The maximum value of heat transfer cofficient is 1.70e+05w/m^^{2}k. It was noticed that heat transfer coeffciient varies a lot with change in different Reynold's No.

From the Fig. 8 the maximum value of velocity &the maximum value of $R_{th.}$ is 45m^^{2}k/w at the Re number from 100 to 200.

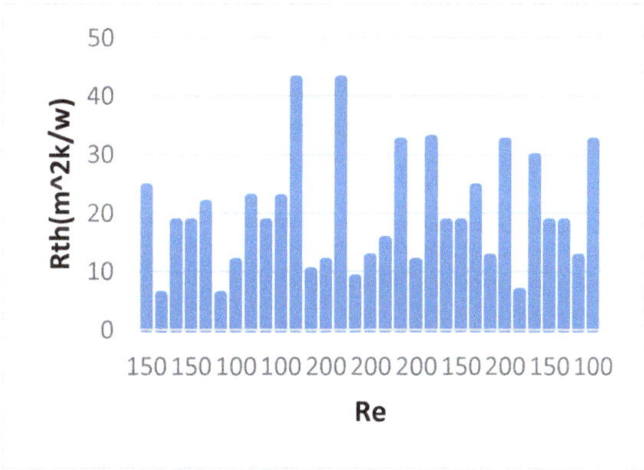

Fig.8:Thermal Resistance vs Reynold's No

4.2 Plots for liquid Ethylene Glycol:

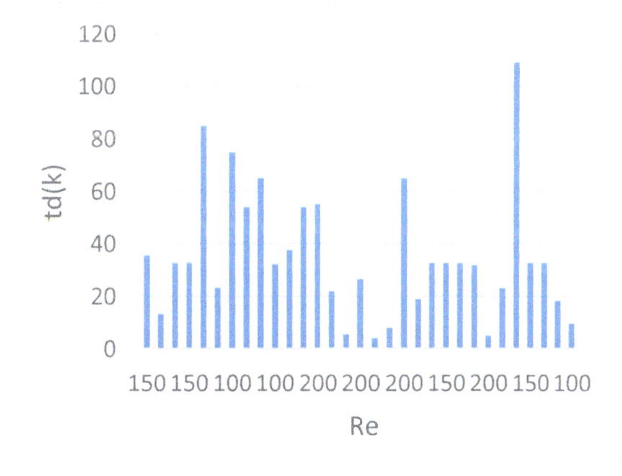

Fig.9Temp. difference vs Reynold's No.

The maximum value of temprature difference in the case of ethylenne glycol is reached a value 5k to 110k at the higher rate of flow the value of temprature difference is maximum.

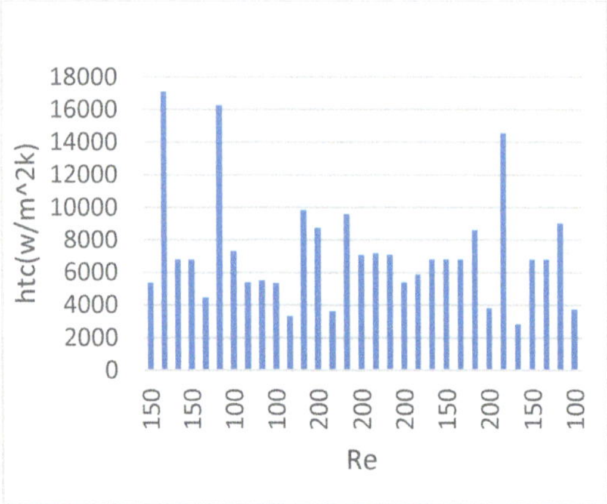

Fig.10 Heat Transfer Coefficient vs Reynold's No.

Fig. 10shows the value of heat transfer cofficient reached the maximum at the minimum value of temprature difference higher the heat treansfer cofficient . the maximum value of heat transfer cofficient is 16000w/m^^{2}k.

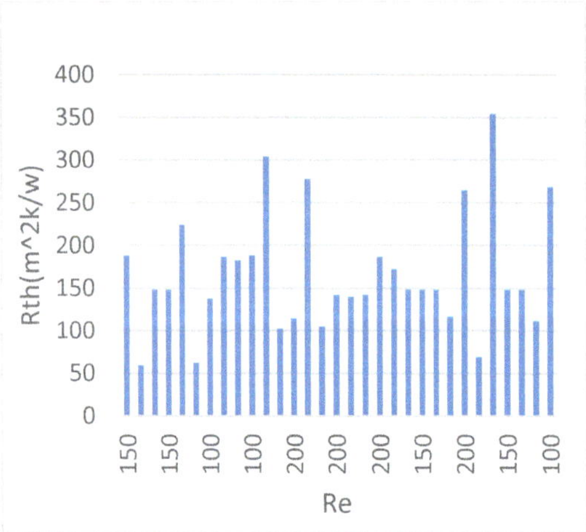

Fig.11Thermal Resistance vs Reynold's No.

maximum value of velocity the maximum value of $R_{th.}$ the value of thermal resistnace is reached at 355m^^{2}k/w at the Re number from50 to 200.

5.0 CONCLUSION

In the present study, a numerical investigation has been carried out to determine the heat transfer characteristics and the pumping power micro-channel employing water and ethylene glycol flows in the laminar regime. The performance is measured in terms of the thermal resistance of the micro channels heat sink. As it is clear from the graphs the water is more suitable coolant as compared to the ethylene glycol. The value of heat transfer coefficient for water is more suitable as compared to the ethylene glycol. Following conclusions are drawn

1. The thermal resistance & PP both are considerably affected and the variation is wide.

2. It was also noticed that the variation in both the parameters is very small. The reason for this can be concluded that the flow effects are being studied in micro scale. It has given a significant direction in optimizing the scale of the micro channel heat sink.

3. Both the responses have significant effect on the performance of micro channel heat sink. The Design and manufacturing constraints can be searched for the whole space.

Acknowledgement:

The research work is supported by Department of Science and Technology, India vide grant no (DST/TSG/FF/2011/187-C) & Authors are thankful to Director CSIR-CSIO & pay their gratitude for supporting the research work.

NOMENCLATURE	
t_s	thickness of channel (mm)
w_c	width of channel (mm)
h_c	channel height (mm)
t_{sb}	thickness of solid substrate width (mm)
R_e	Reynolds number
pp (Ω)	pumping power (W)
t_d	temperature difference (k)
R_{cond}	resistance due to conduction (k/w)
R_{conv}	resistance due to convection(k/w)
R_{th}	total thermal resistance (m^2k/w)
v	flow rate (m^3/s)
Δp	pressure drop
h_{tc}	Heat transfer coefficient (w/m^2k)

REFERENCES:

[1] F. Hong, P. Cheng, Three dimensional numerical analyses and optimization of offset strip-fin microchannel heat sinks, Int. Commun. Heat Mass Transf. 36(2009) 651e656.

[2] Kim, S. J., and Kim, D., Forced Convection in Microstructures for Electronic Equipment Cooling, *ASME J. Heat Transfer*, vol. 121, 1999 pp. 639–645,.

[3] Intel Pentium processor data sheet. [Online].Available:ftp://download.intel.com/designPentium4/datashts/303128.pdf.

[4] D.B. Tuckerman, R.F.W. Pease, High-performance heat sinking for VLSI, IEEE Electron Device Letters 2 (1981) 126–129.

[5] G.L. Morini, Single-phase convective heat transfer in micro-channels: overview of experimental results, Int. J. Thermal Sci. 43 (2004) 631–651.

[6] A. Miner and U. Ghoshal, 2004. "Cooling of high-power-density microdevices using liquid metal coolants", Applied Physics Letters 85 p. 506-508.

[7] Gao P., Person S. Le, and Favre-Marinet M., "Scale Effects on Hydrodynamics and Heat Transfer in Two-Dimensional Mini and Micro channels", International Journal of Thermal Sciences, vol.41, (2002) pp. 1017–1027.

[8-9]. Li, Q., Xuan, Y., 2002. Convective heat transfer and flow characteristics of Cu–water nano fluid. Science in China E 45, 408–416.

[10]. Pak, B.C., Cho, Y.L., 1998. Hydrodynamic and heat transfer study of dispersed fluids with submicron metallic oxide particles. Experimental Heat Transfer 11, 151–170.

[11]. Yang, Y., Zhang, Z., Grulke, E.A., Anderson,W.B.,2005.Heat transfer properties of nano particle in fluid dispersions (nano fluids) in laminar

A detailed review and Prediction of Ductile Fracture Behavior of Materials by Using ANN and DOE Techniques

P.YUVANARASIMMAN[1]
[1]ASSISTANT PROFESSOR / MECHANICAL, DHIRAJLAL GANDHI COLLEGE OF TECHNOLOGY, SALEM, TAMILNADU, pynmech@gmail.com

Corresponding Author: P.Yuvanarasimman, Assistant Professor / Department of Mechanical Engineering, Dhirajlal Gandhi College of Technology, Salem.

Abstract

Nowadays sheet metal applications are very wide in the automotive as well as aerospace industries, at the same time failures are also occurs occasionally during the applications which leads in losses in the resources. Earlier researchers were studied the failure analysis through various Failure Mode and Effective Analysis (FMEA) technique and tried to predict the fracture before failure occurs. This work attempts to predict the ductile fracture criterion for the sheet metal to avoid the failure by means of considering various controllable and uncontrollable parameters. With the Design of Experiments and the Artificial Neural Network techniques a ductile fracture initiation criteria modeled and the model will be validated through Finite element simulation for the verification of the quality. If comparison results are within the limit condition, the model can be utilized for the prediction of ductile fracture before it happens in the real application.

Keywords: Ductile Fracture, ANN, Doe.

1. Introduction

Fracture mechanics is the field of mechanics concerned with the study of the propagation of cracks in materials. It uses methods of analytical solid mechanics to calculate the driving force on a crack and those of experimental solid mechanics to characterize the material's resistance to fracture. In modern materials science, fracture mechanics is an important tool in improving the mechanical performance of mechanical components. It applies the physics of stress and strain, in particularly the theories of elasticity and plasticity, to the microscopic crystallographic defects found in real materials in order to predict the macroscopic mechanical failure of bodies. Fractograpy is widely used with fracture mechanics to understand the causes of failures and also verify the theoretical failure predictions with real life failures. The prediction of crack growth is at the heart of the damage tolerance discipline.

1.1 Deformation

Depending on the type of material, size and geometry of the object, and the forces applied, various types of deformation may result. The image to the right shows the engineering stress vs. strain diagram for a typical ductile material such as steel. Different deformation modes may occur under different conditions, as can be depicted using a deformation mechanism map.

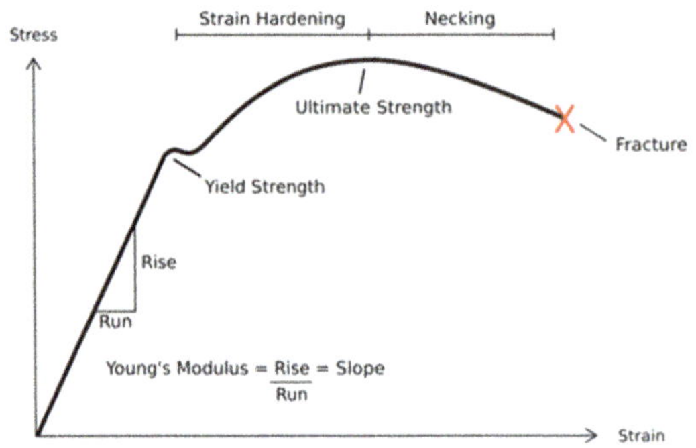

Figure 1: Typical stress vs strain diagram with the various stages of deformation

1.2 Elastic deformation

This type of deformation is reversible. Once the forces are no longer applied, the object returns to its original shape. Elastomers and shape memory metals such as Nitinol exhibit large elastic deformation ranges, as does rubber. However elasticity is nonlinear in these materials. Normal metals, ceramics and most crystals show linear elasticity and a smaller elastic range.

Linear elastic deformation is governed by Hooke's law, which states:

$$\sigma = E\varepsilon$$

Where σ is the applied stress, E is a material constant called Young's modulus, and ε is the resulting strain. This relationship only applies in the elastic range and indicates that the slope of the stress vs. strain curve can be used to find Young's modulus. Engineers often use this calculation in tensile tests. The elastic range ends when the material reaches its yield strength. At this point plastic deformation begins. Note that not all elastic materials undergo linear elastic deformation; some, such as concrete, gray cast iron, and many polymers, respond nonlinearly. For these materials Hooke's law is inapplicable.

1.3 Ductile Fracture

One of the most important and key concepts in the entire field of Materials Science and Engineering is fracture. In its simplest form, fracture can be described as a single body being separated into pieces by an imposed stress. For engineering materials there are only two possible modes of fracture, ductile and brittle.

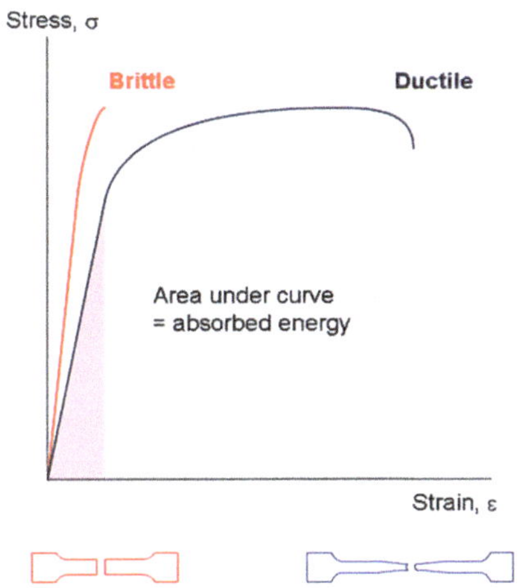

Figure 2: Brittle vs Ductile Stress Strain Behavior

In general, the main difference between brittle and ductile fracture can be attributed to the amount of plastic deformation that the material undergoes before fracture occurs. Ductile materials demonstrate large amounts of plastic deformation while brittle materials show little or no plastic deformation before fracture. Crack initiation and propagation are essential to fracture. The manner through which the crack propagates through the material gives great insight into the mode of fracture. In ductile materials (ductile fracture), the crack moves slowly and is accompanied by a large amount of plastic deformation. The crack will usually not extend unless an increased stress is applied. On the other hand, in dealing with brittle fracture, cracks spread very rapidly with little or no plastic deformation. The cracks that propagate in a brittle material will continue to grow and increase in magnitude once they are initiated. Another important mannerism of crack propagation is the way in which the advancing crack travels through the material.

1.4 Crack Growth

Ductile fracture occurs mainly due to the enlargement of voids due to the external loading conditions which are named as a crack growth. The Fig.- shows the steps incurred in the ductile fracture specimen at the tensile loading condition. Crack will have the enlargement with 900 to the stress formed in the material and the maximum shear stress will be at450 inclined positions.

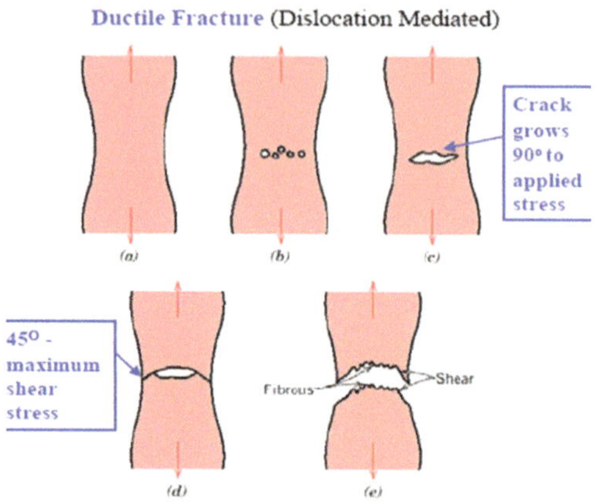

(a) Necking, (b) Cavity Formation,
(c) Cavity coalescence to form a crack,
(d) Crack propagation, (e) Fracture

1.5 Sheet Metal Blanking

Two common forms of sheetmetal fracture are brittle and ductile. Fractures in glass, rocks and ice have the characteristics of brittle fracture. However, brittle fracture in sheetmetal forming is uncommon. A rare example would be a stamping that cracks when dropped on the floor because the chemistry, processing, microstructure and amount of cold work all interact just right to produce a brittle condition.Instead, the more common brittle fractures in sheetmetal happen for specific metal chemistries when subjected to high impact loading at very low (-40 deg.) in-service temperatures. Unfortunately, too many statements are heard in press shops that deformation work hardens the steel so much that it becomes brittle and fails. Others will explain that high-strength grades or full hard tempers of sheet must be formed only once because the material already is so hard after the first hit that any added deformation makes it brittle and unable to withstand a second hit. The usual mode of stamping fracture is ductile fracture. The cross-sections through the sheet thickness in Fig. 1.8 illustrate the difference between brittle and ductile fracture. The brittle fracture (A) has no or very little localized plastic deformation surrounding the fracture. The fracture surface often occurs at a 45-deg. angle through the thickness of the sheet. In contrast, the ductile fracture (B) has significant deformation and thickness reduction prior to the onset of fracture. One can visualize the ductile flow of material before the sheet actually tears. The resulting fracture surface has a cup and cone topography.

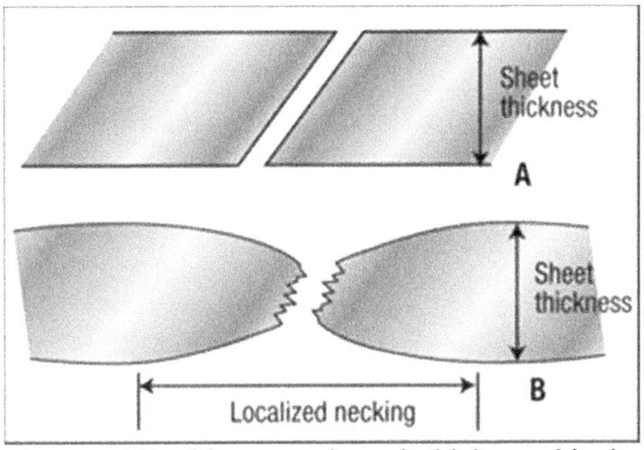

Figure4: Fracture profiles of brittle fracture (A) without any through-thickness thinning, and ductile fracture (B) showing extensive localized thinning.

The amount of deformation or strain that a material can withstand before ductile fracture is very difficult, if not impossible, to predict. Microstructure, grain size, inclusions, stress state, constraints, forming speed and many other factors control the onset of a ductile fracture.

In the press shops, the termination of useful deformation in most stampings is not the unpredictable ductile fracture. A local neck is the failure mechanism that terminates global stamping deformation. Local necking is defined as a narrow line of highly localized thinning with deformation across the neck but no deformation along the line of the thinning.

The forming mode changes to a rigid sheet above and below the neck that move apart as the local neck thins and widens as total deformation force decreases. In a normal tensile-test sample, the local neck is angled about 55 deg. from the major loading axis. This is the angle along which the resultant strain is zero. As specimen width increases, the angle increases until the local neck eventually occurs perpendicular to the major strain direction. As the local neck develops, a high rate of straining occurs within the neck that eventually leads to ductile fracture.

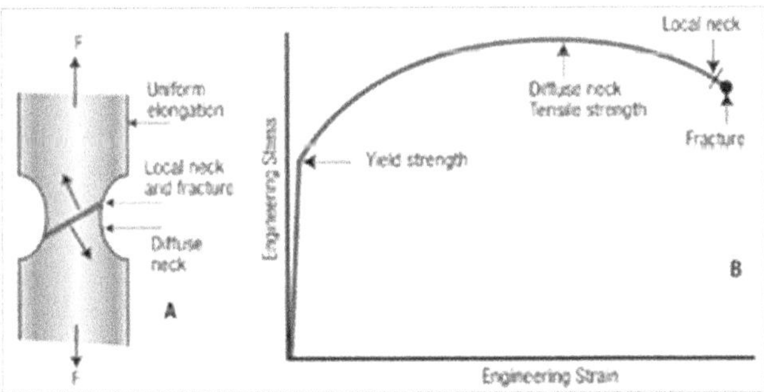

Figure 5: Schematic of a tensile-test sample (A) showing an angular local neck at the onset of fracture, and a stress-strain curve (B) with the local neck occurring just prior to the onset of specimen fracture

The criteria for a local neck defined above for a tensile test remains the same when forming stampings in the press shop. However, the direction of the maximum strain, as well as widely varying gradients of strain that change throughout the stamping with the stroke of the punch, make the theoretical prediction of the onset of local necking almost impossible. However, metal forming studies over the last three decades have collected sufficient data to generate experimental curves that predict the maximum allowable stretchability called forming-limit curves or forming-limit diagrams, these important curves and their application will be the topic of the December column.

So far the discussion has focused on excessive stretching of sheetmetal. Other types of failures occur when sheetmetal is formed compressively. For small amounts of compressive deformation of thick sheets, the compressive direction becomes smaller as sheet thickness (and sometimes sheet width) becomes larger, according to the constancy of volume rule. When the amount of compression becomes too large for a given sheet thickness, the sheetmetal simply forms buckles as the least-energy mode of deformation.

1.7 Applications of Fracture Mechanics

The design process for a component consists of choosing the appropriate geometry, the necessary material strength as per the loading conditions (either cyclic or constant loading), the temperature of usage and structural analysis (Testing and FEM analysis), so that it does not fail under load. The methodologies followed in design criteria traditionally pick up the conventional materials based on standard data and as per the loading conditions proportioning the geometry of the components on basis of analysis. The material strength is chosen keeping in mind the factor of safety, i.e. the ultimate stress (where it fails) is much higher than maximum stress in the component.

Fracture mechanics follows one of two design principles, either fail-safe or safe-life. In fail safe mode, even if a component fails, the entire structure is not at risk (failure of redundant members). According to the safe life principle throughout the life, no component of the structure may fail. Fracture mechanics estimated the maximum crack that a material can withstand before it fails through analysis taking into consideration the overall dimensions of the structure, the stress value where crack initiation takes place, notch toughness value (ability of a material to absorb energy in the presence of a crack for crack propagation), the behavior of materials under the action of stresses by finding out the stress intensity factor (K), fatigue crack growth and stress corrosion crack growth.

Major applications of fracture mechanics design are material selection, effect of defects, failure analysis and control/monitoring of components. Fracture analysis includes the usage of mathematical models such as linear elastic fracture mechanics (LEFM), crack opening displacement (COD) and J-integral approaches by using finite element analysis (FEM). The relationship used for estimating stress intensity factor is

$$K = c\sigma\sqrt{a}$$

where K is the critical fracture toughness value, c a constant that depends on crack and specimen dimensions, σ the applied stress, and a the flaw size. The above relation is very general and as per the shape of the crack, relations available in standard data books or course books are to be used, any general crack can be approximated to standard shapes used in writing the relations.

For a given material the value of K is dependent on stresses acting and flaw size. Flaw size decreases as the stress increases. Thus a design engineer can dictate the life of a component by choosing appropriate values of K, a and σ. Even there are other parameters that estimate the life of a component like working temperature, loading rate (fatigue), residual stress and stress concentration. The higher the K value, the higher is the resistance to crack growth, and the material can resist higher stresses.

Designers try to decrease the defects in the component arising in casting or manufacturing processes by following good fabrication processes and inspection, and estimate notch-toughness values of materials using methods like charpy V-notch impact test, or drop weight tests. In many investigations it was proved that the material failed at a very much lower than the critical stress intensity factor because of defects in the material or micro cracks. Analysis proved that for any component there are two phases for crack development, i.e. crack initiation and second phase crack growth until failure. Of the two, the first phase covers a larger percentage of fatigue life, and under very large high cycle loading conditions second phase is instantaneous.

The factor $(K/\sigma)^2$ is used for estimating design of component because it estimates crack size, more the value better the resistance to the forces(Stress). But how large this factor has to be is decided by considering type of the structure, frequency of inspection, access to inspection, design life of the structure, consequences of failure, probability of over load, methods of fabrication, required quality, material cost in addition to the results obtained by fracture mechanics analysis.

2. Literature Survey

A.M. Goijaerts, et al. (1) described that in critical applications, the development of a blanking process becomes trial-and-error due to the empirical blanking knowledge. A validated FEM-model of the process is present but a proper ductile fracture model is missing. For that development of ductile fracture model was carried out. From the experimental blanking process they determined the critical value and the characterized the process. The main goal of the research was to predict the product shape of a blanked product. An FEM-model, validated on the deformations in the blanking process, existed but the problem of ductile fracture initiation had not been solved yet. The category of local ductile fracture criteria was chosen for this application. For the characterization of such a model two approaches are discussed. To verify these approaches an experimental setup was built and results are presented for the punch displacement at ductile fracture initiation for five different clearances in the blanking process. David Hunt, (2), Finite Element Assisted Prediction of Ductile Fracture in Sheet Bulging of Magnesium Alloys, M.Eng Thesis.

The Finite Element Model of a sheet bulging process was built and validated with results obtained from physical testing. The FEA model uses Oyane's ductile fracture criterion to predict whether fracture has occurred in the material and also to predict the location of fracture if it occurs. This validated FEA model implements a failure range where the failure is predicted over a range of draw depths, and sensitivity analysis provides a confidence level in this range by varying some of the material properties and examining the effects on the prediction of fracture. Emad Al-Momani, et al. (3) metal blanking is a widely used process in high volume production of sheet metal components. The main objective was to present the development of a model to predict the shape of the cut side. The model investigates the effect of potential parameters influencing the blanking process and their interactions. This helped in choosing the process leading parameters for two identical products manufactured from two different materials blanked with a reasonable quality on the same mold. Finite Element Method (FEM) and Design of Experiments (DOE) approaches were used in order to achieve the intended model objectives. The combination of both techniques was proposed to result in a reduction of the experimental cost and effort in addition to getting a higher level of verification. It can be stated that the Finite Element Method coupled with Design of Experiments approach was provided a good contribution towards the optimization of sheet metal blanking process. A.M. Goijaerts, et al. (4) studied and focused on the evaluation of ductile fracture methodologies, which are needed to predict product shapes in the blanking process. Earlier two approaches were elaborated using local ductile fracture models. The First strategy incorporates the characterisation of a ductile fracture model in a blanking experiment. The second methodology was more favorable for industry. In that approach, instead of a complex and elaborate blanking experiment, a tensile test is used to characterize a newly proposed criterion, which was shown to predict accurately the ductile fracture for different loading conditions. Ahmad Rafsanjani, et al.(5) proposed a methodology to predict the ductile damage in the sheet metal blanking process using a coupled thermo mechanical finite-element method. A constitutive material model combined with the ductile fracture criteria was used. The effect of material softening due to the heat generated during plastic work in a specimen was considered in blanking simulations. The sheet metal blanking process was simulated using DEFORM2D, a commercial finite-element code. The effect of material softening due to heat generation because of plastic deformation was introduced to the simulation using a coupled thermo mechanical finite element method. The effect of punch speed and punch–die clearance on blanking quality was carefully examined. To verify the validity of the proposed model, several blanking simulations performed and the results compared with those obtained from an experimental study. I.M. Gunter, et al.(6) added some spinal additions>2.5 vol.%, the additional crack initiation sites associated with the brittle particles embrittle the molybdenum and hinder the ductilizing grain size effect, resulting in a decrease in ductility beyond 2.5 vol.% spinel. Thus, it would appear that controlling the molybdenum grain size, rather than adding spinel particles, is a more effective means of controlling ductility of molybdenum when fracture occurs inter granularly. Molybdenum-base materials exhibit excellent potential for such applications due to the high strength and high melting point of molybdenum; however, the oxidation resistance of molybdenum metal is extremely poor. Although several molybdenum silicides exhibit excellent oxidation resistance, these intermetallics are too brittle for practical use as structural materials. D. Brokken, et al (7) exhibited the large, localized deformations were handled by a combination of an Operator Split Arbitrary Lagrang Euler (OS-ALE) method and full remeshing was done. Transport of the state variables between subsequent meshes for the OS-ALE and remeshing methods was achieved by the Discontinuous Galerkin (DG) method and an interpolation procedure, respectively.Ductile fracture is incorporated using a discrete cracking approach. The calculated product shapes are compared to experimental observations, showing an overall good agreement. The element elimination procedure is capable of modelling separation, it is inherently mesh dependent. Furthermore, the adoption of rigid plastic material behaviour obstructs the modelling of spring back effects. It is concluded that a mesh independent finite element procedure to predict the shape of blanked products is not yet available. Yanshan Lou, et al.(8) implements a ductile fracture criterion was proposed to model fracture behavior of sheet metals for nucleation, growth and shear coalescence of voids during plastic deformation. In the new ductile fracture criterion, void nucleation is described as a function of the equivalent plastic strain, void growth is a function of the stress triaxiality and void coalescence is controlled by the normalized maximal shear stress. The paper presents nucleation; growth and coalescence of voids are analyzed comprehensively to develop reasonable models to describe these processes. These models are combined to construct a new ductile fracture criterion. Parametric study is carried out to investigate the effect of the normalized maximum shear stress and the stress triaxialility on the shape of FFLDs. The new criterion is applied to construct the FFLD of DP780 as well as the fracture locus of Al 2024-T351 to validate their performance on prediction of the equivalent plastic strain to fracture in a wide range of stress states from the uniaxial compression to the balanced biaxial tension of sheet metals.A ductile fracture criterion is newly proposed for prediction of FFLDs with efficient procedure to obtain the material constants in the criterion. The criterion is constructed with consideration of damage accumulation induced by nucleation, growth and shear coalescence of voids. These three processes are described as functions of the equivalent plastic strain, the stress triaxiality, and the normalized maximal shear stress to be multiplied to represent a fracture model. The model endows a cut-off value of _1/3 for the stress triaxiality for appropriate application to ductile materials. Jun Zhou a, Xiaosheng et al (9) focused the plasticity and ductile fracture behaviors of an aluminum alloy 5083-H116 are studied through a series of experiments and finite element analyses. A recently developed stress state dependent plasticity model, the I1–J2–J3 plasticity model, is implemented to describe the plastic response of this material. Furthermore, a ductile failure criterion based on a damage parameter defined in terms of the accumulative plastic strain as a function of the stress triaxiality and the Lode angle is established. The calibrated I1–J2–J3 plasticity model and ductile failure model are utilized to study the residual stress effect on ductile fracture resistance. A local out-of-plane compression approach is employed to generate residual stress fields in the compact tension specimens. Fracture tests of C(T) specimens having zero, positive and negative residual stresses are conducted. The numerical results, such as load–displacement curves and crack front profiles, are compared with experimental measurements and good agreements are observed. Both experimental and finite element results show significant effect of residual stress on ductile fracture resistance.

3. Problem Definition

From the literature survey various researchers identified blanking process is the major manufacturing process of various automotive as well as aerospace components in the very large quantity. In those stated industries, application of sheet metal functions has major impact to meet the requirements. Focusing on the prediction of failure, the enlargement of the crack in the sheet metal is characterized by the researchers concentrated only on the critical parameters. The exact fracture points due to the crack propagation during the application were not presented. If the exact nucleation of the void, void-coalescence and crack propagation is available, the fracture can be predicted before the failure happens. This project work aims to predict the fracture condition of the material in which crack propagates at its critical value by means of introducing ductile fracture criteria for a specific application. This criterion is also applicable to various other working conditions. The behaviour of materials is mainly based on the ductile properties of the sheet metal and so the ductile fracture has been chosen for prediction.

4. Proposed Work

The objective function of the work deals with the predictions of ductile fracture in the sheet metal blanking process by modelling ductile fracture initiation criteria which is suitable for various working conditions of the materials. Proposed work plan will be as follows,

- Study of blanking process
- Identifying the controllable and non-controllable factors of blanking process.
- Proposing the process conditions through literature review
- Choosing the assumptions and measuring devices
- Material selection
- Punch and die specifications
- Design of Experiments
- Results from DOE

- Applications of ANN
- Finding Optimized parameters for numerical simulation- Result-I
- Finite Element Analysis
- FEA Simulation planned to identify the crack propagation and fracture conditions – Result-II
- Comparison of Results I and II
- Validation of the model

5. Conclusion:

From the literature survey identified that blanking process is the major manufacturing process of various automotive as well as aerospace components in the very large quantity. Focusing on the prediction of failure, the enlargement of the crack in the sheet metal is characterized by the researchers concentrated only on the critical parameters. The exact fracture points due to the crack propagation during the application were not presented. If the exact nucleation of the void, void-coalescence and crack propagation is available, the fracture can be predicted before the failure happens. On the completion of implementation of the proposed work, the fracture behavior can be predicted by ANN as with the help of DOE techniques.

REFERENCES

1. A.M.Goijaerts, L.E.Govaert and F.P.T.Baaijens August 2000 "Prediction of ductile fracture in metal blanking" Journal of Manufacturing Science and Engineering,Vol 122, 476-483.
2. Emad Al-Momani, Ibrahim Rawabdeh 2008 "An Application of Finite Element Method and Design of Experiments in the Optimization of Sheet Metal Blanking Process" Jordan Journal of Mechanical and Industrial Engineering,Vol 2,2008,53-63.
3. David Hunt, September 2008 "Finite Element Assisted Prediction of Ductile Fracture in Sheet Bulging of Magnesium Alloys", DCU, M.Eng Thesis.
4. I.M. Gunter, J.H. Schneibel, J.J. Kruzic, (2007)"Ductility and fracture toughness of molybdenum with MgAl2O4 additions", Materials Science and Engineering A 458 275–280.
5. A.M. Goijaerts, L.E. Govaert, F.P.T. Baaijens (2001), "Evaluation of ductile fracture models for different metals in blanking", Journal of Materials Processing Technology 110,313-323.
6. Ahmad Rafsanjani, et.al(2009), "Investigation of the viscous and thermal effects on ductile fracture in sheet metal blanking process", Int J Adv Manuf Technol (2009) 45:459–469.
7. Jun Zhou, et.al(2012) "Modeling the ductile fracture behavior of an aluminium alloy 5083-H116 including the residual stress effect", Elsevier Ltd., Engineering Fracture Mechanics 85 103–116.
8. Ozturk, F. and Lee, D. (2004)"Analysis of Forming Limits Using Ductile Fracture Criteria" J. Mater. Process. Tech., Vol. 147 (, pp. 397- 404.
9. Yanshan Lou,et.al(2012)"New ductile fracture criterion for prediction of fracture forming limit diagrams of sheet metals", International Journal of Solids and Structures 49 3605–3615.

10. D. Brokken, W.A.M. Brekelmans and F.P.T. Baaijens,1999 "Numerical modelling of the metal blanking process", European Conference on Computational Mechanics August 31 – September 3, Munchen, Germany.
11. Takuda, H., Mori, K., Fujimoto, H., Hatta, N.1999 "Prediction of Forming Limit in Bore-Expanding of Sheet Metals Using Ductile Fracture Criterion", J. Mater. Process. Tech., Vol. 92-93 433-438.
12. Y.W. Stegeman, et.al 1999"An experimental and numerical study of a planar blanking process" Journal of Materials Processing Technology 87 266–276.

CFD Analysis of Shell and Tube Heat Exchanger to Study the Effect of Baffle Cut on the Pressure Drop

Avinash D Jadhav[1], Tushar A Koli[2]

[1]*Department of Mechanical Engineering, GF's Godavari College of Engineering Jalgaon, avinashjadhav444@gmail.com*
[2]*Department of Mechanical Engineering, GF's Godavari College of Engineering Jalgaon, kolitushar09@gmail.com*
Author Correspondence: Khote Nagar Jalgaon 425001

Abstract

The shell side design of a shell and tube heat exchanger; in particular the baffle spacing, baffle cut and shell diameter dependencies of the heat transfer coefficient and the pressure drop are investigated by numerically modelling a small heat exchanger. The flow and temperature fields inside the shell are resolved using a commercial CFD package. A set of CFD simulations is performed for a single shell and single tube pass heat exchanger with a variable number of baffles and turbulent flow. The results are observed to be sensitive to the turbulence model selection. The best turbulence model among the ones considered is determined by comparing the CFD results of heat transfer coefficient, outlet temperature and pressure drop with the Bell–Delaware method results. For two baffle cut values, the effect of the baffle spacing to shell diameter ratio on the heat exchanger performance is investigated by varying flow rate.

Keywords: CFD, Heat Exchanger, Shell and Tube, Baffle Cut, Turbulence Models.

1. Introduction

Shell and tube heat exchangers are known as the work-horse of the chemical process industry when it comes to transferring heat. These devices are available in a wide range of configurations as defined by the Tubular Exchanger Manufacturers Association. The applications of single-phase shell-and-tube heat exchangers are quite large because these are widely in chemical, petroleum, power generation and process industries. In essence, a shell and tube exchanger is a pressure vessel with many tubes inside of it. One process fluids flows through the tubes of the exchanger while the other flows outside of the tubes within the shell. The tube side and shell side fluids are separated by a tube sheet. In these heat exchangers, one fluid flows through tubes while the other fluid flows in the shell across the tube bundle. The design of a heat exchanger requires a balanced approach between the thermal design and pressure drop. The performance parameters include heat transfer, pressure drop, effectiveness etc.

The determination of pressure drop along with heat transfer in a heat exchanger is essential for many applications for at least two reasons:

(1) The fluid needs to be pumped through the exchanger, which means that fluid pumping power is required. This pumping power is proportional to the exchanger pressure drop.

(2) The heat transfer rate can be influenced significantly by the saturation temperature change for a condensing/evaporating fluid in case of multiphase flow if there is a large pressure drop associated with the flow. Ideally most of the pressure drop available should be utilized in the core and a small fraction in the manifolds, headers, or other flow distribution devices.

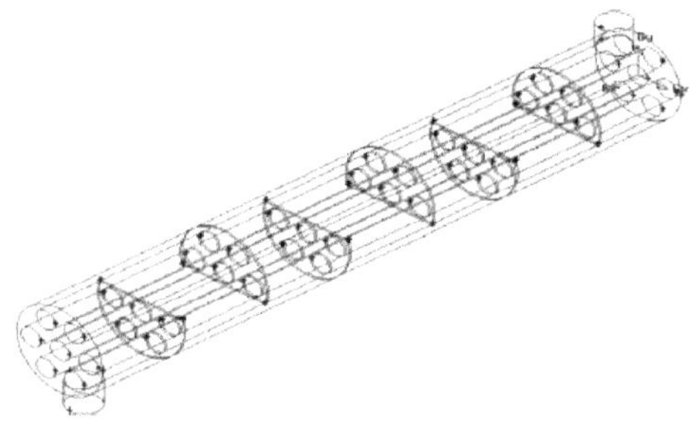

Figure 1: The Model with 6 Baffles

To be able to understand the causes of the shell side design weaknesses, the flow phenomenon inside the shell must be well understood. For that purpose, numerous analytical, experimental and numerical studies have been performed. Most of these studies were concentrated on the certain aspects of the shell-and-tube heat exchanger design. Among others, Gay et al. [4] worked on heat transfer, while Halle et al. [5], Pekdemir et al. [6], Gaddis and Gnielinski [7] investigated pressure drop. Some of the researchers concentrated only on certain parts of the shell-and-tube heat exchanger. It can be particularly useful in the initial design steps, reducing the number of testing of prototypes and providing a good insight in the transport phenomena occurring in the heat exchangers [14]. In all of these simplified approaches, the shell side pressure drop and heat transfer rate results showed good agreement with experimental data.

2. Modelling Details

In this study, a small heat exchanger is selected in order to increase the model detail and to make solid observations about the flow inside the shell. Some of the design parameters and the predetermined geometric parameters are presented here. The geometric model with 6 baffles is shown in Fig. 1. Two different baffle cut values are selected: 25% baffle cut value is very common in shell-and-tube heat exchanger designs; whereas, 36% baffle cut value is selected to place the cut just below or above the central row of tubes. The working fluid of the shell side is water. Since the properties of water are defined as constants in the Fluent database, to improve the accuracy, they are redefined using piecewise-linear functions of temperature by using the "Thermo-Physical Properties of Saturated Water" tables available in the literature [15].

2.1 Boundary conditions

The desired mass flow rate and temperature values are assigned to the inlet nozzle of the heat exchanger. The shell inlet temperature is set to 300 K. Zero gauge pressure is assigned to the outlet nozzle, in order to obtain the relative pressure drop between inlet and outlet. The inlet velocity profile is assumed to be uniform. No slip condition is assigned to all surfaces. The zero heat flux boundary condition is assigned to the shell outer wall, assuming the shell is perfectly insulated outside. Since the tube side flow is easy to resolve, the present study is concentrated on the shell side flow. After modelling the tubes as solid cylinders, the constant wall temperature of 450 K is assigned to the tube walls.

The pressure model is selected from the paper published by Kapale U C, Chand S. [16]

Design parameters and fixed geometric parameters.

Shell size (D_s) = 90 mm, Tube outer diameter (d_o) = 20 mm,

Tube bundle geometry and pitch = Triangular, 30 mm

Number of tubes (N_t) = 7, Heat exchanger length (L) = 600 mm,

Shell side inlet temperature (T) = 300 K, Baffle cut (B_c) = 25% & 30% alternate,

Central baffle spacing (B) = 86 mm, Number of baffles (N_b) = 6

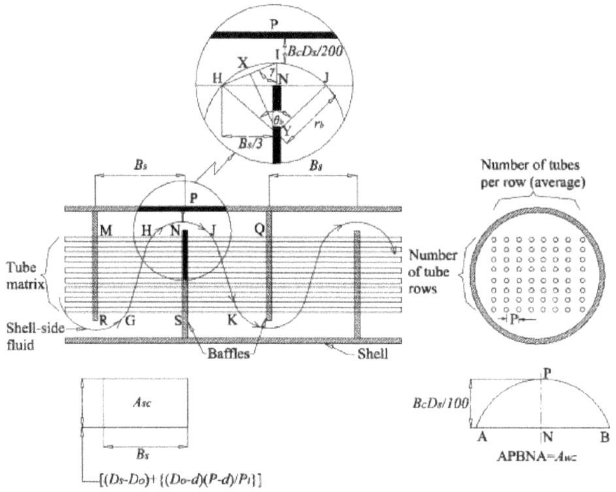

Figure 3: Showing the actual flow pattern in shell-side

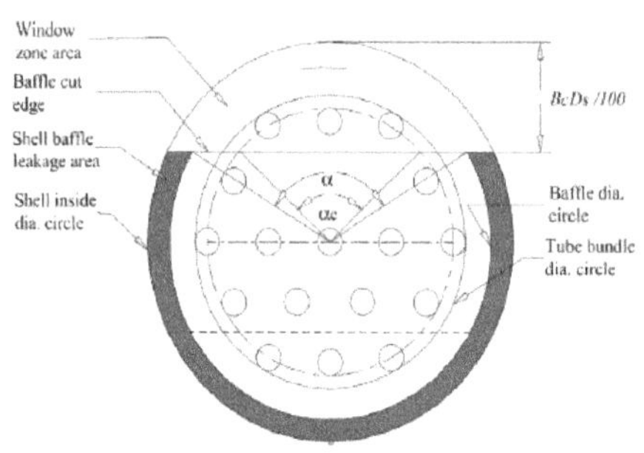

Figure 4: Illustration of baffle cut angles, leakage area.

3. Result and Discussion

3.1. Sensitivity of results to turbulence model and discretization order

In all of the preliminary simulations, the flow inside the shell is observed to be turbulent. Therefore, the viscous model is selected as turbulent. The sensitivity of the results to the turbulence model and the discretization order is investigated using the heat exchanger model with 6 baffles which is shown in Fig. 1. The first and second order discretization schemes and three different turbulence models are tried for two different mesh densities.

Sr.No.	M	G_s	Shell inlet temp	μ_s	ρ_s	Re_s	$\dot{V}$	u_n	ΔP_n	f	u_{sc}	ΔP_c	ΔP_{cdn}	u_{wz}	ΔP_b	ΔP_{wz}	ΔP_{ic}	ΔP_{ec}	ΔP_s
	kg/sec	kg/sec-m^2	^0C	kg/m.s	kg/m^3		m^3/sec	m/sec	Pa		m/sec	Pa	Pa	m/sec	Pa	Pa	Pa	Pa	Pa
1	0.5	193.7984	27	0.00083	996.05	1728.776	0.000501983	1.02264	1041.66	0.52	0.17638	28.656	636.442	0.81425	108.153	711.535	4412.49	26.2545	5480.4
2	1	387.5969	27	0.00083	996.05	3457.551	0.001003966	2.04528	4166.64	0.47	0.35275	103.603	2545.77	1.62849	422.547	2836.52	17537.1	94.9199	21798.7
3	2	775.1938	27	0.00083	996.05	6915.102	0.002007931	4.09056	16666.6	0.35	0.7055	308.603	10183.1	3.25699	1593.57	11253.8	69065.6	282.74	86014.9

Figure 5: Analytical Result for baffle cut at 25%

Sr.No.	M	G_s	Shell inlet temp	μ_s	ρ_s	Re_s	$\dot{V}$	u_n	ΔP_n	f	u_{sc}	ΔP_c	ΔP_{cdn}	u_{wz}	ΔP_b	ΔP_{wz}	ΔP_{ic}	ΔP_{ec}	ΔP_s
	kg/sec	kg/sec-m^2	^0C	kg/m.s	kg/m^3		m^3/sec	m/sec	Pa		m/sec	Pa	Pa	m/sec	Pa	Pa	Pa	Pa	Pa
1	0.5	167.8416	27	0.00083	996.05	1497.228	0.000501983	1.02264	1041.66	0.52	0.17638	20.5015	27.887	0.35153	44.4994	69.1724	517.542	26.2545	1585.46
2	1	335.6831	27	0.00083	996.05	2994.455	0.001003966	2.04528	4166.64	0.47	0.35275	74.1208	111.548	0.70306	176.595	275.349	2022.7	94.9199	6284.26
3	2	671.3662	27	0.00083	996.05	5988.91	0.002007931	4.09056	16666.6	0.35	0.7055	220.785	446.191	1.40611	692.91	1088.53	7635.08	282.74	24584.4

Figure 6: Analytical Result for baffle cut at 30%

Sr.No.	Mass flow rate	25% ΔP_s	30% ΔP_s	% Decrease
1	kg/sec	Pa	Pa	
2	0.01666	0.559493	0.539655	3.545710134
3	0.03333	1.98434	1.903972	4.05011238
4	0.05	4.303975	4.11122	4.478534378
5	0.06666	7.500554	7.165568	4.4661501
6	0.08333	11.508868	11.008043	4.351644315
7	0.1	16.518	15.611549	5.487655891
8	0.11666	22.0399	21.101822	4.25627158
9	0.13333	28.646683	27.444256	4.197438845
10	0.15	36.135445	34.566093	4.342971285
11	0.16666	44.467651	42.515915	4.389114235

Figure 7: Comparison of CFD results

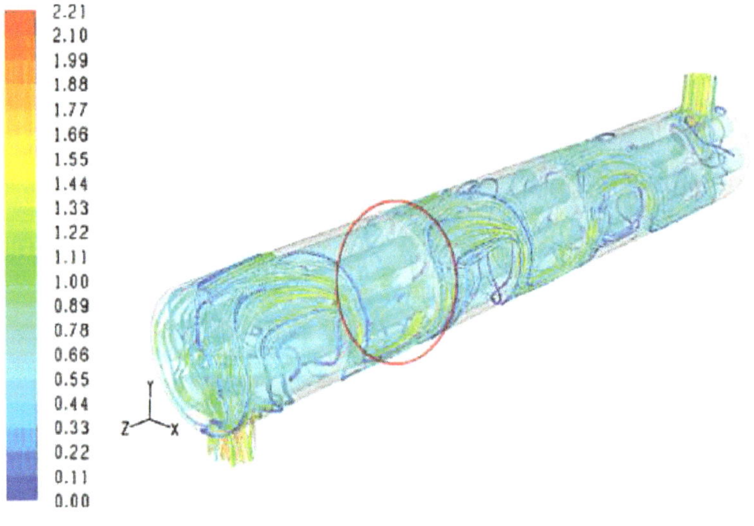

Figure 8: Velocity (m/s) path lines for 6 baffles (1 kg/s mass flow rate)

3.2. Effects of baffle cut on pressure drop

Here, the simulations are repeated for the baffle cut value of 25% and effects of the baffle cut on the heat transfer and the pressure drop are investigated. The percent differences between the analytical calculations and the CFD analysis results are presented in Figure 5. The analytical calculations are taken as the base values for the percent difference calculations.

4. Conclusion

The shell side of a small shell-and-tube heat exchanger is modelled with sufficient detail to resolve the flow and temperature fields. From the CFD simulation results, for fixed tube wall and shell inlet temperatures, shell side heat transfer coefficient, pressure drop and heat transfer rate values are obtained. The sensitivity of the shell side flow and temperature distributions to the mesh density, the order of discretization and the turbulence modelling is observed. By varying baffle cut values of 25% and 30%, for 0.5, 1 and 2 kg/s shell side flow rates, the simulation results are compared with the results from the Kern and Bell–Delaware methods. Using CFD, together with supporting experiments, may speed up the shell-and-tube heat exchanger design process and may improve the quality of the final design. In the near future, improvements in the computer technology will make full CFD simulations of much larger shell-and-tube heat exchangers possible

References

1. Gaddis D, editor. Standards of the Tubular Exchanger Manufacturers Association. Tarrytown (NY): TEMA Inc.; 2007.

2. Kern DQ. Process heat transfer. New York (NY): McGraw-Hill; 1950.

3. Bell KJ. Delaware method for shell side design. In: Kakaç S, Bergles AE, Mayinger F, editors. Heat exchangers: thermal–hydraulic fundamentals and design. New York: Hemisphere; 1981. p. 581–618.

4. Gay B, Mackley NV, Jenkins JD. Shell-side heat transfer in baffled cylindrical shell-and-tube exchangers – an electrochemical mass transfer modeling technique. Int J Heat Mass Transfer 1976;19:995–1002.

5. Halle H, Chenoweth JM, Wabsganss MW. Shell side water flow pressure drop distribution measurements in an industrial-sized test heat exchanger. J Heat Transfer 1988;110:60–7.

6. Pekdemir T, Davies TW, Haseler LE, Diaper AD. Pressure drop measurements on the shell side of a cylindrical shell-and-tube heat exchanger. Heat Transfer Eng 1994;15:42–56.

7. Gaddis ES, Gnielinski V. Pressure drop on the shell side of shell-and-tube heat exchangers with segmental baffles. Chem Eng Process 1997;36:149–59.

8. Li HD, Kottke V. Visualization and determination of local heat transfer coefficients in shell-and-tube heat exchangers for staggered tube arrangement by mass transfer measurements. Exp Therm Fluid Sci 1998;17:210–6.

9. Li HD, Kottke V. Visualization and determination of local heat transfer coefficients in shell-and-tube heat exchangers for in-line tube arrangement by mass transfer measurements. Heat Mass Transfer 1998;33:371–6.

10. Karno A, Ajib S. Effect of tube pitch on heat transfer in shell-and-tube heat exchangers – new simulation software. Heat Mass Transfer 2006;42:263–70.

11. Sparrow EM, Reifschneider LG. Effect of interbaffle spacing on heat transfer and pressure drop in a shell-and-tube heat exchanger. Int J Heat Mass Transfer 1986;29:1617–28.

12. Eryener D. Thermoeconomic optimization of baffle spacing for shell and tube heat exchangers. Energy Convers Manage 2006;47:1478–89.

13. Karno A, Ajib S. Effects of baffle cut and number of baffles on pressure drop and heat transfer in shell-and-tube heat exchangers – numerical simulation. Int J Heat Exchangers 2006;7:299–322.

14. Sunden B. Computational heat transfer in heat exchangers. Heat Transfer Eng 2007;28:895–7.

15. Incropera FP, Dewitt DP. Fundamentals of heat and mass transfer. 4th ed. New York: John Wiley; 1996.

16. Kapale U C, Chand S. Modeling for shell-side pressure drop for liquid flow in shell-and-tube heat exchanger. Int J Heat Mass Transfer 2006;49:601–10.

17. Taborek J. Thermal and hydraulic design of heat exchangers. In: Hewitt GF, editor. Heat exchangers design handbook, vol. 3. New York: Begell House Inc.; 2002.

18. Mukherjee R. Effectively design shell-and-tube heat exchangers. Chem Eng Prog 1998;94:21–37.

A Brief Author Biography

Avinash D Jadhav is Lecturer in GF's Godavari College of Engineering, Jalgaon, India. He has Graduation degree in Mechanical engineering from Dr. Babasaheb Ambedkar Marathwada University Aurangabad 2009 and Perusing Post-Graduation degree ME specialization in Thermal Engineering. He has Two years of Industrial Experience.

Tushar A Koli is Assistant Professor in GF's Godavari College of Engineering, Jalgaon, India. He received degree in BE Mechanical in 2001, Post-Graduation ME with specialization in Design in 2010 from North Maharashtra University Jalgaon.

EVALUATING THE INFLUENCE OF FIBER ORIENTATION AND FILLER CONTENT ON TENSILE, HARDNESS, AND IMPACT STRENGTH OF HYBRID LAMINATED COMPOSITES.

Vithal Rao Chavan[1], Dr. K R Dinesh[2], Dr. K Veeresh[3], Veerabhadrappa Algur[4], S.P. Jagadish [5], C. Madhu Mohan[6]

[1]*Asst. Professor, Department of Mechanical engineering, Rao Bahadur Y Mahabaleshwarappa Engineering College, Bellary, India. nrvithalchauhan@gmail.com*

[2] *Principal and Professor, Department of Mechanical Engineering, Government Engineering College, Raichur, India. dineshkrpet@gmail.com*

[3] *Principal and Professor, Department of Mechanical engineering, Rao Bahadur Y Mahabaleshwarappa Engineering College, Bellary, India. kuppagalveeresh@yahoo.co.in*

[4]*Asst. Professor, Department of Industrial Production Engineering, Rao Bahadur Y Mahabaleshwarappa Engineering College, Bellary, India. veereshalgur@gmail.com*

[5]*Asst. Professor, Department of Mechanical engineering, Rao Bahadur Y Mahabaleshwarappa Engineering College, Bellary, India. sp_jagadish1@yahoo.co.in*

[6]*P.G. Scholar, Department of Industrial & Production engineering, Rao Bahadur Y. Mahabaleshwarappa Engineering College, Bellary, India*

Abstract

The paper describes some aspects concerning the effect of the Hematite ore (Fe_2O_3) filler and fiber orientation on some mechanical characteristics of hybrid composites. Hybrid composite were fabricated with ± 45° and ±90° oriented reinforcement as E glass fiber with Hematite ore (Fe_2O_3) filler material and Polyester resin matrix. The filler content was varied as 6wt%,8 wt% and 10 wt%. The laminates were prepared by hand layup technique, cured at ambient temperature, laminates are peeled according to ASTM standards and the experiments were conducted to determine tensile strength, impact strength and hardness. The result showed that Hematite filled composite better than unfilled composite. The ±90° oriented composite exhibits better result compared to ±45° oriented composite. The 6wt% hematite filled glass/polyester has shown excellent to the other studied composite both in ± 45° and ±90° oriented composite.

Keywords: Composites; Impact Strength; FRPC; Hematite ore.

1. Introduction

Fiber reinforced polymer composites (FRPCs) have generated wide interest in various engineering fields because of the with different physical/chemical properties at the macroscopic or microscopic scale. The basic idea of the composite is to optimize material properties of the composite. Fibers are the principal load-carrying constituents while the surrounding matrix helps to keep them in desired location and orientation and also act as a load transfer medium between them [1]. The effective properties of the fiber reinforced composites strongly depend upon the geometrical arrangement of the fibers within the matrix [2]. This arrangement is characterized by the volume fraction, the fiber aspect ratio, fiber spacing parameters and orientation angles of fibers. Thermoplastic composites reinforced with long fibers, short fibers and mat (fabric) of natural and synthetic fibers like hemp, jute banana, glass, carbon, Kevlar etc are used in a variety of applications such as aerospace elements, automotive parts, marine structures, structural members and anti vibration applications due to their combined properties of resilience, creep resistance, high strength to weight and stiffness to weight ratios, corrosion resistance and good damping properties.

The fiber reinforcement provides non-isotropic in–plane strength but produces weak interlaminar resin-rich regions, where under loading extensive damage is generated, especially between plies of different orientation.

Fiber –reinforced composites show strong anisotropic mechanical behavior due to their fiber orientations. These orientations cause a variety of failure mechanisms, which are more complex under multi-axial loading conditions. Therefore, continuous effort has been made to make quasi-isotropic composite materials with controlling parameters, such as the orientation adjacent plies, stacking sequence and the properties of the constituents [3]. K.M.Kaleemulla et.al [4] focused the study on influence of fiber orientation and fiber content of epoxy resin components on mechanical properties. The main aim of the present investigation was to study the influence of fiber orientation on mechanical properties and also the influence on varying weight percentage of filler material in the composites.

2. Material details

E-glass fabric (300 GSM) of plain weave construction, procured from High tech suppliers of polymers, Bangalore, was used for the study Orthophthalic polyester resin matrix with methyl ethyl ketone peroxide catalyst and cobalt octet accelerator were used. The filler were Hematite is one of the most common minerals. The sandstone is most red and brown in color because of hematite presence. Non-crystalline forms of Hematite may be transformations of the mineral Limonite that lost water, possibly due to heat chemical formula Fe_2O_3. The filler used was hematite passed through 75-150 μm. In this work there are two types of composite is being prepared.

3. Specimen preparation and test details

A hand lay-up method is used to prepare the glass-GP composites with filler. To prepare glass GP with hematite filler composites, filler is mixed with a known amount of GP resin. The laminate was cured at ambient conditions for a period of about 24 hours. The laminate so prepared has a 300mm X 300mm X 3mm.

The composites are fabricated and cured as reported by Suresha et al [11] and basavarajappa et al. [12]. The cured materials are cut to yield test specimens in accordance of ASTM standards. Tensile test has been carried out according to ASTM D 3039, Impact test has been conducted ASTM E23 and Hardness has been measured in terms of B.H.N. value accordance of ASTM E10.

In this present work the three variables are varied.

1) Orientation of fiber material (90°,45°)

2) Volume fraction of filler material.

3) Matrix volume.

Totally the eight laminates are prepared by varying above variables and designated as (A,B,C,D) for 90° and (A1,B1,C1,D1) for 45°.

Table1:- shows the proportionate constitutes of the laminates.

composites ±90°	Composites± 45°	% of filler	Matrix volume %		Reinforcement volume %	
A	A1	0	polyester	50	Glass fiber	50
B	B1	6	polyester	44	Glass fiber	50
C	C1	8	polyester	42	Glass fiber	50
D	D1	10	polyester	40	Glass fiber	50

3.1 Ultimate tensile strength

Tensile test was performed in accordance with ASTM D3039 as shown in figure 1, under displacement with resolution of the piston movement of 0.01mm .Test specimen were well filed to attain overall length and gauge length of 250 and 140mm respectively and an appropriate cross sectional area of 25×3 mm^2 and aluminum tabs with dimensions of 55×25×2 mm with 45deg filing is done at the one end is glued as shown in Fig1

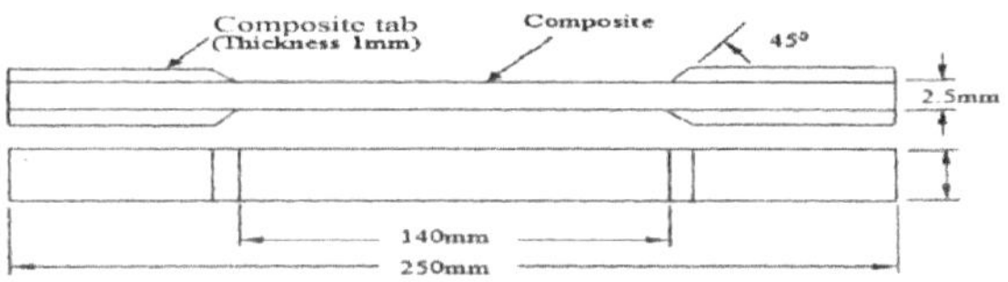

Figure 1 Tensile Test ASTM standard

3.2 Hardness test

Hardness is closely related to strength. It is the capability of a material to withstand scratching, abrasion, indentation, or penetration. It is directly proportional to tensile strength and is measured on special hardness testing machines by measuring the resistance of the material against penetration of an indenter of special shape and material under a given load. The different scales of hardness are Brinell hardness, Rockwell hardness, Vicker's hardness, etc. The test conducted on Brinell hardness machine, as per the ASTM standards the specimen prepared as per ASTM E10. A load of 100 kg was applied on the specimen for 30 sec using 1.6 mm diameter ball indenter and the indentation diameter was measured by using a microscope. The hardness was measured at three different locations of the specimen and the average value was calculated. The indentation was measured and hardness was calculated using equation (1).

$$BHN = \frac{2P}{\pi D[D - \sqrt{(D^2 - d^2)}]} \quad \text{-------- (1)}$$

Where

P= Applied load in (Kgf); D= Diameter of the indenter (mm); d= Diameter of indentation (mm).

3.3 Impact strength

Ability of the cracked (notched) material to resist the impact loading is the measure of its impact strength. It is the ratio of the energy absorbed by the specimen to cross-sectional area under the crack tip. Impact tests are designed to measure the resistance to failure of a material to a suddenly applied load. The test measures the impact energy, or the energy absorbed prior to fracture. The common methods used for measuring impact energy are

- Charpy Test
- Izod Test

Impact energy is a measure of the work done to fracture a test specimen.

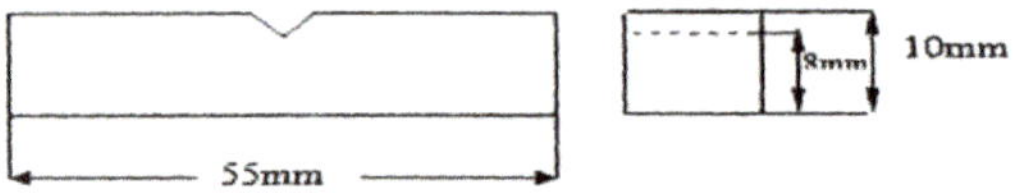

Figure 2 Impact Test ASTM standard

When the striker hits the specimen, the specimen will absorb energy until the specimen yields. At this point, the specimen will begin to undergo plastic deformation at the notch. The test specimen continues to absorb energy and work hardens at the plastic zone at the notch. When the specimen can absorb no more energy, fracture occurs. Figure 2 shows Charpy test specimens normally measure according to ASTM E23 i.e., 55x10x10mm and have a notch machined across one of the larger faces. The notches may be V-notch – A V-shaped notch, 2mm deep, with 45° angle and 0.25mm radius along the base. Impact strength was calculated by using equation (2).

$$Impact\ strength = \frac{EI}{A} \quad (\text{joules/mm}^2)\text{---------}(2)$$

Where EI= Impact energy in joules recorded on the scale.

A= Area of the specimen in mm^2.

4. Result and discussion

In all, eight different types of Hybrid laminates were fabricated, four of the Hybrid was ± 45° oriented composites and other four was ±90° oriented composites. Among four one laminate was fabricated without filler and rest other with varying filler. Various characterization tests were conducted and their results are depicted in table 2 and are discussed in the following sections.

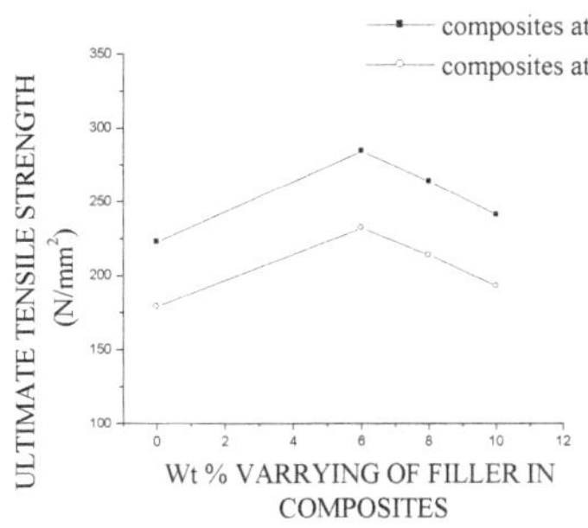

Figure 3
Comparative graph of Tensile Strength

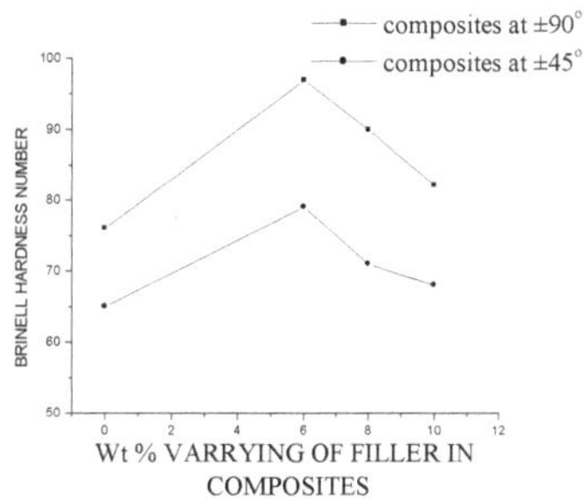

Figure 4
Comparative graph of Hardness No

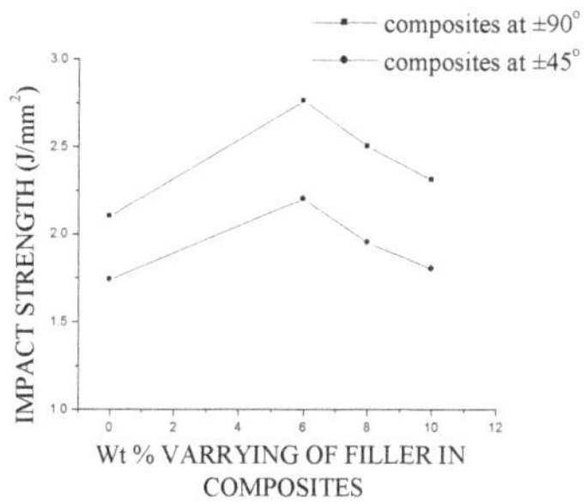

Figure 5
Comparative graph of Impact Strength

Table:- 2 Experimental Values

Sl.No	% of filler	±90° composites Mechanical properties			±45° composites Mechanical properties		
		Uts (N/mm²)	BHN	IS (J/mm2)	Uts (N/mm2)	BHN	IS (J/mm2)
01	0	222.66	76	2.10	178.96	65	1.74
02	6	283.84	96	2.76	231.71	79	2.20
03	8	263.45	89	2.50	213.4	71	1.95
04	10	240.561	82	2.31	192.45	68	1.80

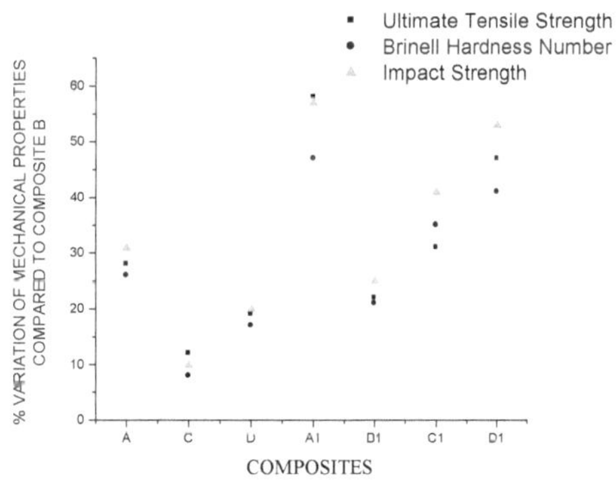

Figure 6 Comparative graph of Mechanical Strength with composite B

4.1 Tensile Strength

The values of eight different composites under consideration are shown in figure 3. The composite developed with ±90° oriented E-glass fiber reinforced and 6wt% of Fe_2O_3 filled and polyester resin matrix shows more Tensile strength compared to other composites. Hence when the composite B is compared with other composite A,C,D,A1,B1,C1,D1, the strength of composite B is increased by 28%,12%,19%,58%,22%,31%,47% respectively as shown in figure6.

4.2 Brinell hardness number

The values of eight different composites under consideration are shown in figure 4. The composite developed with ±90° oriented E-glass fiber reinforced and 6wt% of Fe_2O_3 filled and polyester resin matrix shows more Hardness Number compared to other composites. Hence when the composite B is compared with other composite A,C,D,A1,B1,C1,D1, the hardness of composite B is increased by 26%,8%,17%,47%,21%,35%,41% respectively figure 6.

4.3 Impact strength

The values of eight different composites under consideration are shown in figure 5. The composite developed with ±90° oriented E-glass fiber reinforced and 6wt% of Fe_2O_3 filled and polyester resin matrix shows more Impact Strength compared to other composites. Hence when the composite B is compared with other composite A,C,D,A1,B1,C1,D1, the strength of composite B is increased by 31%,10%,20%,57%,25%,41%,53%,respectively figure 6.

5. Conclusion

In the present research endeavor, various characterization tests were conducted over GFR-polyester-Hematite ore filled Hybrid composites. In general, the addition of fillers leads to cost and weight reduction of the regular glass fiber reinforced composites. The effect of the addition of natural filler has been studied and it is concluded that, the Tensile Strength, Hardness, Impact Strength of the composite B is better compared to other composites and at the 6wt% of filler content in both the orientations exhibits the good mechanical characterization.

Refrences

[1] Ever J. Barberoi and Liliana de Vivo, "A constitutive model for elastic damage in fiber-reinforced PMC laminate, international journal of damage mechanics, 10 (2001) 73 -93.

[2] Jane Maria Faulstich de paiva, sergio mayer, mirabel cerqueira rezende, "evaluation of mechanical properties of four different carbon/epoxy composites used in aeronautical field", materials research, 8 (2005) 91 – 97.

[3] Prashanth banakar, H.K Shivananda and H.B Niranjan "Influence of fiber orientation and thickness on tensile properties of laminated polymer composites"int.J.pure Appl. Sci.Technol., 9(1) (2012),pp.61-68.

[4] K.M.Kaleemulla and B.siddeswarappa, "Influence oif fiber orientation on the in-plane mechanical properties of laminated hybrid polymer composites", Journal of reinforced plastics and composites,29(12) (2009).

[5] Fatai Olufemi Aramide, Isiaka Oluwole Oladele, and Davies Oladayo Folorunso, "evaluation of the effect of fiber volume fraction on the mechanical properties of a polymer matrix composite", Leonardo electronic journal of practices and technologies, 14 (2009) 134 -141.

[6] S. Chauhan, Anoop Kumar, Amar Patnaik, Alok Satapathy, I. Singh, "Mechanical and wear characterization of GF reinforced vinyl ester resin composites with different co-monomers", Journal of Reinforced Plastics, 28 (2009) 2675 – 2684.

[7] Smrutisikha Bal, "Experimental study of mechanical and electrical properties of carbon nanofiber/ epoxy composites", 5 (2010) 2406-2413.

[8] R. Chatys, "Mechanical properties of polymer composites produced by resin injection molding for applications under increased demands for quality and repeatability", 64 (2009) 35 – 39.

[9] Manoj Singla and Vikas Chawla, "Mechanical Properties of Epoxy Resin – Fly Ash Composite", Journal of Minerals & Materials Characterization & Engineering, 9 (2010) 199-210.

[10] Djoković, V. and Nedeljković, J. M., "Stress relaxation in hematite nanoparticles-polystyrene composites", Macromolecular Rapid Communications, 21 (2000) 994 – 997.

[11] B. Suresha, G. Chandramohan, Siddaramaiah, P. Samapthkumaran, S. Seetharamu, Three-body abrasive wear behaviour of carbon and glass fiber reinforced epoxy composites", Materials Science and Engineering A, 443 (2007) 285–291.

[12] S. Basavarajappa, Ajith G. Joshi, K. V. Arun, A. Praveen Kumar, M. Prasanna Kumar, "Three-Body Abrasive Wear Behaviour of Polymer Matrix Composites Filled with SiC Particles", Polymer Plastics Technology and Engineering, 49(2010)8 - 12.

[13] M. N. Channapagoudra, Ajith G Joshi, Sunil Thaned, Mahantesh Patil, "Effect of Hematite Filler Material on Mechanical Properties of Glass/Epoxy Composites" International Journal of Innovative Research in Science,Engineering and Technology Vol. 2, Issue 11, November 2013

[14] K.Devendra, T.Rangaswamy, "Strength Characterization Of E-Glass Fiber Reinforced Epoxy Composites With Filler Material" Journal of Minerals Characterization and Engineering,2013,1,353-357.

Experimental study of effect of spiral tube heat exchanger in domestic refrigerator: A Literature Review

Sagar W.Matte[1], Umesh G. Telote[2]

[1] *P. G. Scholar, Mechanical Engineering, BIT, Ballarpur- India, sagarwmatte@gmail.com*
[2] *P.G guide & Assistant professor, Mechanical Engineering, Ballarpur Institute of Technology, Ballarpur*
Author Correspondence: C/o. W. S. Matte, Wadgaon road, near dhanoje kunbi samaj mandir, Chandrapur-442401, sagarwmatte@gmail.com

Abstract

This paper presents a critical review of spiral tube heat exchanger to compute COP of domestic refrigerator. Studies aimed for spiral tube in the heat exchanger device to carry an experimental study for COP increment in any heat transfer applications; a review. The purpose of this article is to compare the COP of refrigerator by using air cooled condenser & hollow spiral tube heat exchanger of same length & same diameter. It was observed from the past research that the effectiveness of the heat exchanger for other varying shaped coil is more than that for the U- shape coil. In this context, it is essential to carry out the research for COP increment; this spiral tube air cooled heat exchanger will be attached in the domestic refrigeration test rig to find out the COP of the system.

***Keywords**:* Spiral tube air cooled heat exchanger; heat transfer; domestic refrigerator, COP.

1. Introduction

Energy recovery is the prime requirement today to optimize energy consumption. The maximum utilization of thermal energy is achieved by properly designed Heat Exchangers, and selection of temperature program. The selection of heat exchanger geometry is also important for particular heating /cooling application. There are various heat exchanger geometries exist and to be selected based on applications. However, the spiral tube heat exchanger is a unique geometry of compact heat exchange for gas/liquid; vapor/liquid heat transfer applications. The secondary flow of fluid through curved tubes and in particular horizontal spiral tube heat exchanger ensure the higher rate of heat flow through the fluid. The mechanical agitating is induced due to fluid motion hence higher heat transfer rate achieved without any external mechanical induced devise in tube.

The heat exchanger is the process heat transfer equipment. The type of heat exchanger governs the effectiveness of heating and cooling. The different geometries of heat exchanger are possible however the proper selection required based on application. More performance and/or reduced the size of heat transfer devices can be achieved by heat transfer enhancement techniques. In general, these techniques can be divided into two groups: active and passive techniques. Tubes with spiral have been used as one of the passive heat transfer enhancement techniques and are the most widely used tubes in several heat transfer applications, for examples, heat recovery processes, air conditioning and refrigeration systems and chemical reactors.

2. Literature Review

Use of spiral tube heat exchanger in the refrigeration system is a new technique altogether. The focus is placed on a need to develop effective, less costly and maintenance-free auxiliary integrated with main system to achieve enhancement in COP of the domestic refrigerator. If this idea is implemented at system design level, then there would be considerable saving of energy. Following researchers contributed to the area of heat transfer enhancement significantly.

Cavallini et al. [**1**] presents a critical review of correlations to compute heat transfer coefficients and pressure drop, for refrigerants condensing inside commercially available tubes with enhanced surfaces of various types, and a theoretical analysis of the condensation phenomenon. Predictions from some of the above equations are compared with experimental data. In addition, information is presented about the influence of small amounts of compressor oil on the condensation of refrigerants in enhanced tubes.

Shinde D. & Dange M. [**2**] experimental investigation of heat transfer in cone shaped helical coil heat exchanger is reported for various Reynolds number. The purpose of this article is to compare the heat transfer in cone shaped helical coil and simple helical coil. The pitch, height and length of both the coils are kept same for comparative analysis. The calculations have been performed for the steady state condition and experiments were conducted for different flow rates in laminar and turbulent flow regime. The coil side flow rate is kept varying while the coil side flow rate is kept constant. It was observed that the effectiveness of the heat exchanger for the cone shaped helical coil is more than that for the simple helical coil. Results show that the heat transfer rates for the cone shaped helical coil are comparatively higher than that of the simple helical coil. It was found that the heat transfer rates are 1.18 to 1.38 times more for the cone shaped helical coil than that of simple helical coil.

Wisam A. Kattea [**3**] experimental study is carried out on the effect of vortex generators (Circular and square) on the flow and heat transfer at variable locations at (X = 0.5, 1.5, 2.5 cm) ahead of a heat exchanger with Reynolds number ranging from $62000<Re<125000$ and heat flux from $3000 \leq q \leq 8000$ W/m2. In the experimental investigation, an apparatus is set up to measure the velocity and temperatures around the heat exchanger. The results show that there is an effect for using vortex generators on heat transfer. Also, heat transfer depends on the shape and location. The circular is found to be the best shape for enhancing heat transfer at location [X_m=0.5 cm] distance before heat exchanger is the best location for enhancing heat transfer. The square is the best shape for enhancing heat transfer at location [X_m=2.5 cm] distance before heat exchanger is the best location for enhancing heat transfer. The results of flow over heat exchanger with vortex generators are compared with the flow over heat exchanger without vortex generators. Heat transfer around heat exchanger is enhanced (56%, 50%, 36%) at location (X=0.5, 1.5, 2.5cm) respectively by using circular vortex generators without turbulator and heat transfer around heat exchanger is enhanced (39%, 42%, 51%) at location (X=0.5, 1.5, 2.5cm) respectively by using square shape vortex generators without turbulator.

P. Naphon & S. Wongwises [**4**] the heat transfer characteristics and the performance of a spiral coil heat exchanger under cooling and dehumidifying conditions are investigated. The heat exchanger consists of a steel shell and a spirally coiled tube unit. The spiral-coil unit consists of six layers of concentric spirally coiled tubes. Each tube is fabricated by bending a 9.27mm diameter straight copper tube into a spiral-coil of five turns. Air and water are used as working fluids. The chilled water entering the outermost turn flows along the spirally coiled tube, and flows out at the innermost turn. The hot air enters the heat exchanger at the center of the shell and flows radially across spiral tubes to the periphery. A mathematical model based on mass and energy conservation is developed and solved by using the Newton–Raphson iterative method to determine the heat transfer characteristics. The results obtained from the model are in reasonable agreement with the present experimental data. The effects of various inlet conditions of working fluids flowing through the spiral coil heat exchanger are discussed.

Yanuar et al. **[5]** presents an experimental investigation of convective heat transfer coefficient for nanofluids using spiral pipe heat exchanger. The aim of this study is to investigate experimentally flow and convective heat transfer characteristics of water-based nanofluids flowing through a spiral pipe. The test section consist of spiral pipe with ratio pitch per diameter is 7.0 and mean hydraulic diameter of 30 mm. The straight spiral tube with 1600 mm length is used as the test section. At the inner of spiral pipe installed a circular copper pipe with 10 mm diameter. Measurements of pressure drop and convective heat transfer are carried out for $Al2O3$, $TiO2$ and CuO at 1% and 3%, particle volume with pure water. The convective heat transfer coefficient of the nanofluids increases by up to 28% at a concentration of 3 vol. % compared with that of pure water.

Bibin Prasad et al. **[6]** helical pipes are universally used for heat transfer enhancement in heat exchangers. In the present work, CFD simulations are carried out for a counter flow double pipe helical heat exchanger by varying the flow rates of a single fluid (water). The heat transfer characteristics of the same are compared with that of a counter flow double pipe straight tube heat exchanger for the same flow rates. The results were interpreted by developing correlations between Nusselt number and Dean Number for both the inner and outer helical pipes which shows a strong linear relationship.

R. J. Yadav & A. S. Padalkar [7] CFD investigation was carried out to study the heat transfer enhancement characteristics of air flow inside a circular tube with a partially decaying and partly swirl flow. Four combinations of tube with twisted-tape inserts, the half-length upstream twisted-tape condition (HLUTT), the half-length downstream twisted-tape condition (HLDTT), the full-length twisted tape (FLTT), and the plain tube (PT) with three different twist parameters ($\lambda = 0.14$, 0.27, and 0.38) have been investigated. 3D numerical simulation was performed for an analysis of heat transfer enhancement and fluid flow for turbulent regime. The results of CFD investigations of heat transfer and friction characteristics are presented for the FLTT, HLUTT, and the HLDTT in comparison with the PT case.

Xie Pengchenget et al. **[8]** single-phase pressure drop and heat transfer in a rotor-assembled strand inserted tube were measured using water as the working fluid. Experiment using a smooth tube was carried out to calibrate the experimental system and the data reduction method. In the experiment, fixed mounts were used to eliminate the entrance effect. The experimental results of smooth tube show that employment of fixed mounts leads to a visible bias of friction factor at relative low Reynolds numbers, although it does not significantly affect the Nusselt numbers. The measured data of inserted tube reveal that rotor-assembled strand can significantly improve heat transfer with the Nusselt number increased by 101.6%-106.6% and the overall heat transfer coefficient increased by 58.1%-67.4% within the Reynolds number range of 20000 to 36000. Meanwhile, friction factor increases by 52.2%-84.2% within the same Reynolds number range. The correlations of Nusselt number and friction factor as function of the Reynolds number and Prandtl number were determined through multi variant linear normal regression.

Kumbhar G. & Sane K. **[9]** in last decade on heat transfer enhancement in a circular tube. In the present paper emphasis is given to works dealing with twisted tape inserts because according to the recent studies, these are known to be economic heat transfer enhancement tool. The thermo hydraulic performance of twisted tape insert depends on the flow conditions i.e., laminar or turbulent flow apart from the insert configuration. The present review is organized separately for laminar & turbulent flow.

D. R. Waghole et al. **[10]** augmentation techniques refer to different methods to enhance rate of heat transfer without affecting much the overall performance of the system. These techniques are used in various applications such as solar power plants, thermal power plants, refrigeration and air conditioning, process industries, radiators for automobile systems etc. These techniques are of three types which are active, passive and compound techniques. The present paper is review of passive augmentation technique used in recent past.

J.P. Meyer & S. Vyver **[11]** potential of a very simple and inexpensive method of heat augmentation that could be used by small manufacturing companies is investigated. It consists of an aluminium heat exchanger comprising a round tube inside a twisted square tube. The aim is to increase the heat transfer between the flow in the inner tube and the flow in the annulus by increasing the heat transfer coefficient in the annulus. This investigation was conducted with measurements on five different heat exchanger configurations: the first, without any twist and the rest, with the square tube twisted at angles of 45°, 60°, 90° and 105° per metre respectively. The nusselt number and friction factor in the annulus of each, case study where measured as functions of the Reynolds number. While the Prandtl numbers were kept constant; and for each heat exchanger configuration an equation was deduced that describes this functional relationship. The relationships are shown in graphical format. It 'was found in general that the heat transfer und pressure drop increase with twist, angle and Reynolds number. Furthermore that for a twist angle of 60° and a Reynolds number lower than4000, the heat transfer rate per unit pumping power is the highest ,while for higher Reynolds numbers a twist angle of90° gives the highest heat transfer per unit pumping power.

Kevin Stone [**12**] discussion on the application of enhanced heat transfer surfaces to compact heat exchangers. The motivation for heat transfer enhancement is discussed, and the principles behind compact heat exchangers are summarized. Next, various methods for evaluating and comparing different types of heat transfer enhancement devices using first and/or second law analysis are presented. Finally, the following plate-fin enhancement geometries are discussed: rectangular and triangular plain fins, offset strip fins, louvered fins and vortex generators.

P. Naphon and S. Wongwises[**13**] heat transfer characteristics and the performance of a spiral coil heat exchanger under cooling and dehumidifying conditions are investigated. The heat exchanger consists of a steel shell and a spirally coiled tube unit. The spiral-coil unit consists of six layers of concentric spirally coiled tubes. Each tube is fabricated by bending a 9.27mm diameter straight copper tube into a spiral-coil of five turns. Air and water are used as working fluids. The chilled water entering the outermost turn flows along the spirally coiled tube, and flows out at the innermost turn. The hot air enters the heat exchanger at the centre of the shell and flows radially across spiral tubes to the periphery. A mathematical model based on mass and energy conservation is developed and solved by using the Newton–Raphson iterative method to determine the heat transfer characteristics. The results obtained from the model are in reasonable agreement with the present experimental data.

Paisarn Naphon et al. [**14**] results of the heat transfer and pressure drop characteristics in horizontal double pipes with helical ribs are presented .Nine test sections with different characteristic parameters of: helical rib height to diameter, v/di = 0.12, 0.15, 0.19, and helical rib pitch to diameter, p/di = 1.05, 0.78, 0.63, are tested. Cold and hot water are used as working fluids in the shell side and tube side, respectively. Experiments are performed under conditions of mass flow rates varying from0.01 to 0.10 kg/s for cold water and from 0.01 to 0.10 kg/s for hot water, respectively. The inlet cold and hot water temperatures are between 15 and 20 °C and between 40 and 50°C, respectively. The results obtained from the tubes with helical ribs are compared with those without helical ribs. It is found that the helical ribs have a significant effect on the heat transfer and pressure drop augmentations. Based on fitting the experimental data, non-isothermal correlations of the heat transfer coefficient and friction factor are proposed. The majority of the data falls within ±15%, ±15% of the proposed correlations for heat transfer coefficient and friction factor, respectively.

An extensive literature review of all types of augmentation technique with external inserts till the year 2005 has been discussed by Waghole [**10**]. In the following subsections literature involving recent work on passive methods by employing air cooled spiral tube heat exchanger is replace in place of air cooled condenser in domestic refrigerator it has been reviewed.

3. Experimental method

3.1 Experimental setup

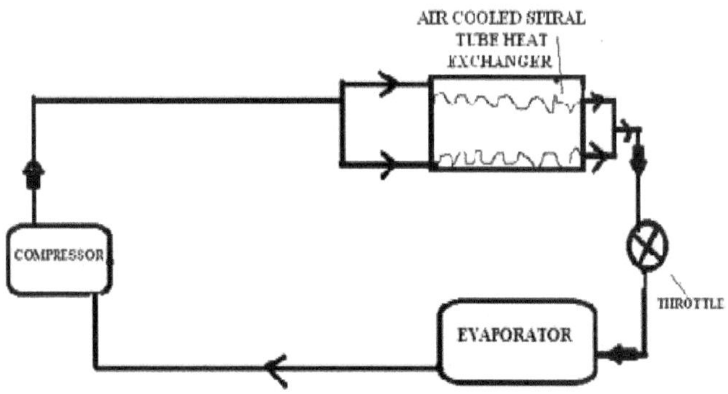

Figure 1: Flow diagram of experimental set up

The apparatus for experimental studies on COP increment in domestic refrigerator with using air cooled spiral tube heat exchanger and control panel shown in Figure 1 & 2. It consists of evaporator, air cooled spiral tube heat exchanger and thermostatic expansion valve with the knob for on/off operation. In this heat exchanger R134a refrigerant used; this unit is connected to refrigerator test rig by replacing air cooled condenser. Each unit of refrigerator system are placed as per the refrigeration cycle only the addition of the spiral tube air cooled heat exchanger with replacement of air cooled condenser in the path of refrigerant. The thermostatic expansion valve are attached in the path to find the optimum increment of COP of Vapour compression refrigeration system. Control panel consist of On/Off switch for evaporator, Dimmerstat, energy meter, digital temperature indicator connected with 9 k- type thermocouple.

To study the calculation of refrigeration system control panel is most important for calculation part for experimentation.

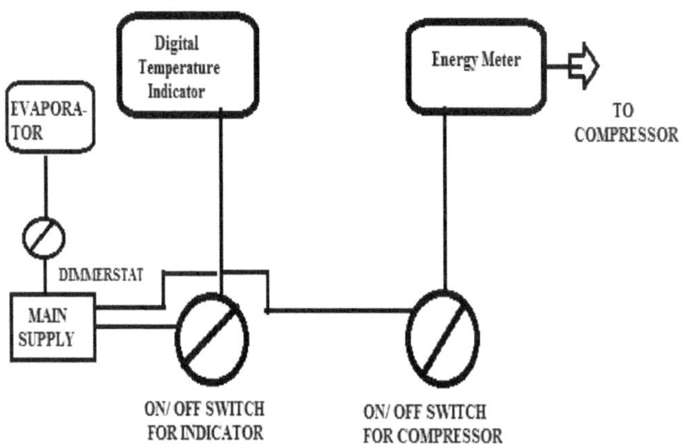

Figure 2: Photo image of Control Panel

3.2 Test procedure

Fill the tank in vapour compression test rig with the water. Switch on the evaporator, temperature indicator and compressor switch. Measure the Suction and Delivery pressure of Refrigerant with the help of Low pressure and High pressure gauge. Take initial energy meter reading of both compressor and stirrer. Before starting the experiment measure the atmospheric pressure. Wait for about 40 minutes so that the system gets stabilized. After the 40 minute completion, take the reading of nearly 5 minute of time span measure the various temperature of the setup from digital temperature indicator i.e. T_1, T_2, T_3, T_4; Measure evaporator (L.P) and condenser (H.P) pressures of the refrigerant. Experiment is conducted at no load condition at evaporator for this work. Then note down the measure temperature in tabular format along with the pulse rate per sec time. Repeat test procedure for after 5min spam.

4. Result & discussion

The literature comparison with current research on the basis of different geometry has been taken by replacing air cooled condenser in domestic refrigerator with reported heat exchanger whereas the all the author who work on spiral tube geometry included in evaporator in refrigeration system[1-4].The hydrodynamic study of spiral tube air cooled heat exchanger suggest that the pressure drop varies with the flow; however the pressure drop is maximum at a velocity indicating the maximum turbulence exists at certain velocity. The detail investigation with varied number of coil turns is under investigation. Use of such type of heat exchanger is beneficial to include in domestic refrigerator; for increasing COP.

Experiments are carried out with air cooled spiral tube heat exchanger. Mass flow rate of water circulation is changed and its effect on system performance is studied. It is found out that by using air cooled spiral tube heat exchanger the cycle, discharge pressure is reduced and work required to drive the compressor is reduced. For the analysis of cycle, software "Cool Pack" is used. Table-1 show outcomes of system. Table-1 shows the refrigerating effect QE, condenser heat rejection QC, power consumed in kW, actual COP, and theoretical COP* with air cooled spiral tube heat exchanger. Figure-3 shows p-h chart obtained by using Cool-Pack software.

Table 1: Results with air cooled spiral tube heat exchanger

Time	Q_E	Q_C	Q_S	KW	COP	COP*
12.10	13.9	16	1.5	3.2	4.34	4.8
12.50	6.93	14.1	1.56	3.4	2.03	3.8
1.30	5.12	13.9	1.96	3.4	1.50	3.2
2.10	4.43	12.7	2.12	3.4	1.30	3
2.50	8.98	12.3	2.12	3.4	2.64	3
3.30	9.48	11	2.12	3.4	2.78	2.8
4.10	10.02	13.3	2.40	3.4	2.94	3.6

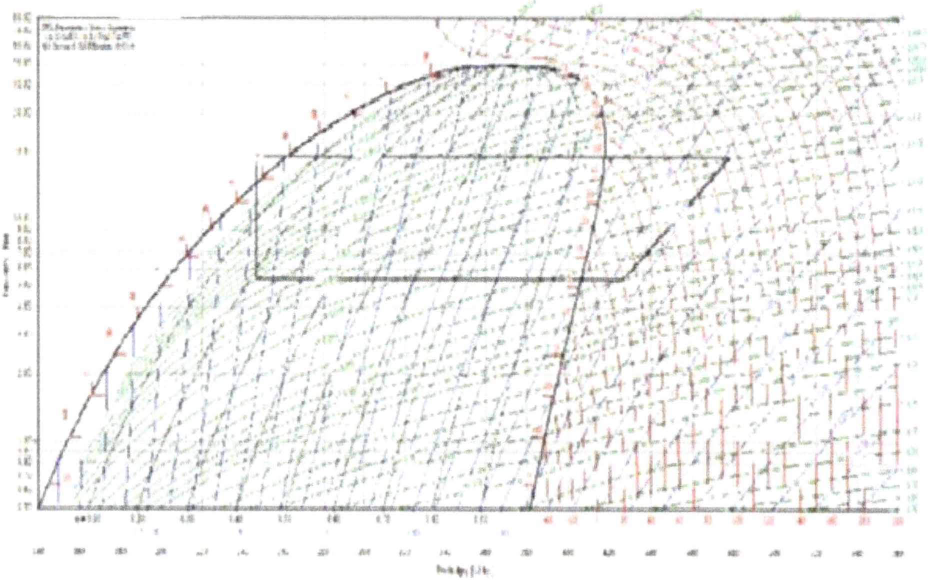

Figure 3: p-h chart obtained by using Cool-Pack software.

4. Conclusion

New experimental data for the measurement of the COP and the performance of a spiral tube heat exchanger are calculated. The results obtained from the developed experimental set up need to validate comparing with the measured data of the literature. The effects of the inlet conditions of the working fluids flowing through the spiral tube heat exchanger are discussed.

The following conclusions can be given on the basis of extensive review:

• There is reasonable agreement between the results obtained from the experiment and those from the literature data.

• Air mass flow rate and inlet-air temperature have significant effect on the increase of the outlet-air temperatures.

• The enthalpy effectiveness and humidity effectiveness decrease as the air flow rates increase.

• The enthalpy effectiveness and humidity effectiveness increase as the inlet-air temperature increases.

The horizontal spiral tube heat exchanger has higher pressure drop as compared to double pipe heat exchanger; however there is an enhancement in heat transfer rate, which off sets the disadvantage of pressure. The compact construction is suitable for heat transfer enhancement in vapour compression system.

By installing air cooled spiral tube heat exchanger in the refrigerant cycle; COP of the cycle is increase in initial level and more effects has to be under process on refrigeration cycle in deeply. This result is only from the simple calculation and experiment work carried on the basis of no load condition. Extensive experimentation result will be validate with the literature this work has to carry out.

References

[1] A. Cavallini, D. Del Col, L. Doretti, G.A. Longo, L. Rossetto, 2000, "Heat transfer and pressure drop during condensation of refrigerants inside horizontal enhanced tubes", International Journal of Refrigeration,Vol. 23, p. no. 4-25.

[2] Shinde D. & Dange M, 2013, "Heat Transfer Analysis of a Cone Shaped Helical Coil Heat Exchanger", International Journal of Innovations in Engineering and Technology,Vol. 3 Issue 1, p. no. 56-62.

[3] Wisam Abed Kattea, 2012, "An Experimental Study on the Effect of Shape and Location of VortexGenerators Ahead of a Heat Exchanger", J Al-Khwarizmi Engineering Journal, Vol. 8, No. 2, PP12- 29.

[4] P. Naphon & S. Wongwises, 2005, "A study of the heat transfer characteristics of a compact spiral coil heat exchanger under wet-surface conditions", Experimental Thermal and Fluid Science 29, 511–521.

[5] Yanuar, N. Putra, Gunawan & M. Baqi, 2011, "Flow and convective heat transfer characteristics of spiral pipe for nanofluids", International Journal of RRAS, vol. 7, p. no. 3.

[6] Bibin Prasad, Sujith V, Mohammed Shaban K, Saju Haneef, Sandeep N, Vishnu Raj, 2013, "Comparison of Heat Transfer between a Helical and Straight Tube Heat Exchanger", International Journal of Engineering Research and Technology, Vol. 6, Number 1, pp. 33-40.

[7] R. J. Yadav and A. S. Padalkar, 2012, "CFD Analysis for Heat Transfer Enhancement inside a Circular Tube with Half-Length Upstream and Half-Length Downstream Twisted Tape", Journal of Thermodynamics, Hindawi Publishing Corporation,12 pages.

[8] XIE Pengcheng, LI Fengxiang, DING Yumei, YAN Hua, GUAN Changfeng and YANG Weimin, 2008, "Experimental Study of Heat Transfer Enhancement and Friction Loss Induced by Inserted Rotor-assembled Strand (I) Water,"Chinese Journal of Chemical Engineering, 16(6), 849—855.

[9] Kumbhar G. & Sane K, 2010, "Heat Transfer Enhancement in a Circular Tube Twisted with Swirl Generator: A Review", Proc. of the 3rd International Conference on Advances In Mechanical Engineering, SVNIT, Surat, January 4-6.

[10] D.R.Waghole, R.M.Warkhedkar, V.S. kulkarni, N. K.Sane, 2012, "A Review on Heat Transfer Augmentation using Twisted Tape inserts inAbsorber/Receiver of PTC", IOSR Journal of Mechanical and Civil Engineering, PP: 33-36.

[11] J.P. Meyer &S. Vyverl, 1997, "Heat transfer augmentation in the annulus of atube inside a twistedheat exchanger consisting of a round square tube", R & D Journal, I3(3), pp.77-82.

[12] Kevin Stone, 1996, "Review of Literature on Heat Transfer Enhancement in Compact Heat Exchangers", Air Conditioning and Refrigeration Center University of Illinois, Mechanical & Industrial Engineering Dept., 1206 West Green Street, Urbana.

[13] P. Naphon and S. Wongwises, 2005, "A study of the heat transfer characteristics of a compact spiral coil heat exchanger under wet- surface conditions square tube", Experimental Thermal and Fluid Science., Vol. 29,pp. 511–521.

[14] Paisarn Naphon, Manachai Nuchjapo, Jutarat Kurujareon, 2005, "Tube side heat transfer coefficient and friction factorcharacteristics of horizontal tubes with helical rib", Experimental Thermal and Fluid Science, Vol. 21, pp. 215–220.

A Brief Author Biography

Sagar W. Matte – *is* a P.G. Student, perusing his master degree in Heat power Engineering, Department of Mechanical Engineering, Ballarpur Institute of Technology, Ballarpur. His research interest in the field of refrigeration and heat transfer enhancement techniques and its application. He completed his undergraduate program from Rajiv Gandhi college of Engineering and Research technology, chandrapur in 2009.

Umesh G. Telote – is an assistant professor and P.G. guide in Department of Mechanical Engineering, Ballarpur Institute of Technology, Ballarpur.

INTERNATIONAL JOURNAL OF RESEARCH IN AERONAUTICAL AND MECHANICAL ENGINEERING

Study of forced convection heat transfer With DAQ & ANSYS

First Authors

Moopanar karthikeyan[1], Raote Sarvesh[2], Sawant Rahul[3], Bhoi Sandesh[4]

Department of Mechanical Engineering
Saraswati college of engineering, Navi Mumbai, Maharashtra, India

Second Author
Prof. Madan Jagtap
Department of Mechanical Engineering
Saraswati college of engineering, Navi Mumbai, Maharashtra, India

Abstract

The aim of this work is to validate the Dittus-Boelter equation by experimental, correlation and Simulation method. It used to find the value of heat transfer coefficient '**h**' for turbulent flow in many fluid transfer systems. This work discusses how the Dittus-Boelter equation is applied to the problem of circular pipe. In experimental verification a DATA ACQUISTION (DAQ) SYSTEM is used to measure the temperature accurately & reliably. This DAQ system also serves the purpose of condition monitoring system, which records the temperature changes of the system continuously. In CFD simulation ICEM CFD for modeling and FLUENT 14 for analysis are used. Results of CFD simulation will be obtained by CFD-POST. Here heat transfer coefficient value is compared by correlations, experiment and CFD simulations, finally the aim of this work is to validate Dittus-Boelter equation.

Keywords: ANSYS; Heat Transfer; CFD; FLUENT; DAQ

1. Introduction

In the study of thermodynamics the average heat transfer coefficient, is used in calculating the convection heat transfer between a moving fluid and a solid. This is the single most important factor for evaluating convective heat loss or gain. Knowledge of h is necessary for heat transfer design and calculation and is widely used in manufacturing processes, oil and gas flow processes and air-conditioning and refrigeration systems. The heat transfer coefficient is critical for designing and developing better flow process control resulting in reduced energy consumption and enhanced energy conservation.

It is also influenced by flow velocity and surface geometry. It may be noted that the physical or thermal properties of the surface material play no part in the process of convective heat transfer. As the fluid properties vary with the temperature and locations the value of convective heat transfer coefficient vary from point to point, this leads to the situation that analytically derived equation are applicable only to a limited extent

1.1 DITTUS-BOELTER RELATION

The conventional expression for calculating the heat transfer coefficient in fully developed turbulent flow in smooth pipes is the Dittus-Boelter equation.

$$N_u = C\, R_e m P_r n$$

Where C, m and n are constant determined experimentally.

We will use these values -

C = 0.023, m = 0.8 and

n = 0.4 for heating of the fluid

n = 0.3 for cooling of the fluid

The properties of this relation have been calculated at the average fluid bulk temperatures. Equation is valid for single phase heat transfer in fully developed turbulent flows in smooth pipes for fluids with Prandtl number ranging from 0.6 to 100 at low heat fluxes. At high fluxes the fluid properties changes resulting in higher errors. At higher heat fluxes the Sieder-Tate equation is used to reduce the error. The work presented is about designing and acquiring data from an experimental setup to verify the Dittus-Boelter empirical relation by finding the heat transfer coefficient. The value of the heat transfer coefficient has been calculated from theory, experiment and CFD simulation to compare the results.

2. Methodology

The below flow chart represents the step by step working procedure of this work. Thestandard procedure for ICEM-CFD work follows the above components. For the present work, in case of pipe creating the geometry then computational domain is constructed around the pipe geometry, later set the mesh parameters then generates the mesh. Later this mesh is exported in the fluent readable format

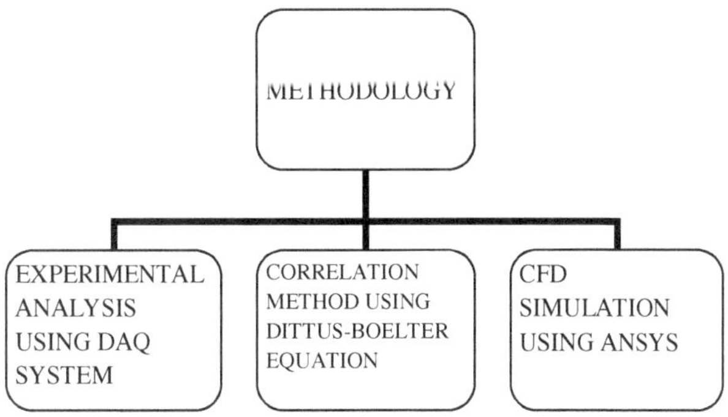

Fig 2.1: Methodology

3. Experimental Set Up Of Forced Convection

Fig. 3.1: Experimental Set Up Of Forced Convention

The experimental set up consists of a blower unit fitted with the rest pipe. The test section pipe is grey cast iron surrounded by nichrome heater. Four temperature sensors are embedded on the test section and two temperature sensors are placed in the air stream at the entrance and exit of the test section to measure air inlet and outlet temperature. Test pipe is connected to the delivery side of blower along with orifice to measure flow of air through the pipe. Input to the heater through dimmer stat and measured by voltmeter and ammeter. Air flow is controlled by gate valve and is measured with the help of orifice meter and the manometer fitted on board.

Specifications:

Length of hallow pipe L = 0.6 m

Pipe inner diameter Di = 0.032 m

Orifice diameter D_o= 0.016 m

3.2 Lay Out Of Experimental Setup

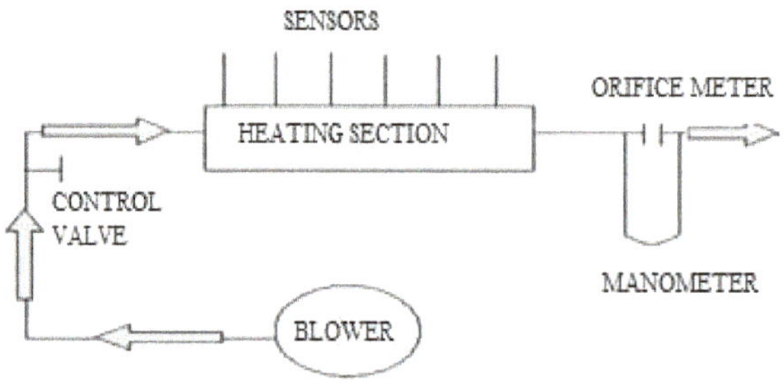

Fig 3.2 Lay Out Of Experimental Setup

3.3 Data Acquisiton (DAQ) System

Our DAQ system consists of an Arduino Duemilanove microcontroller and six DS18B20 sensors. The sensors are connected in Daisy chain arrangement and connected to the microcontroller using one wire bus. Another serial port application "COOL TERM" is used to display & save the data sent by these sensors.

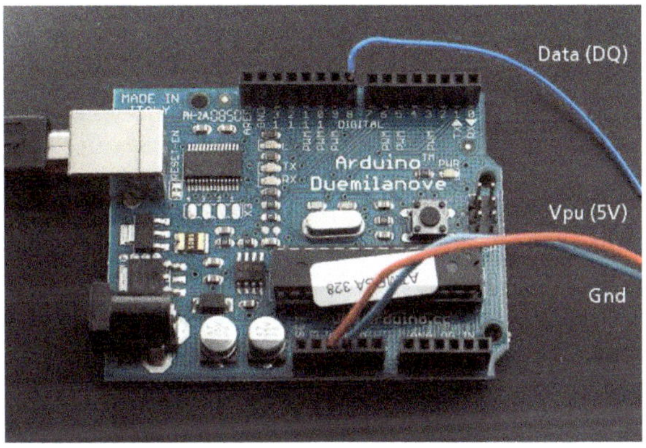

Fig 3.3 Sensors Connected to Arduino

The main advantage of using the DAQ system to measure temp is the accuracy & flexibility. They have a very quick response time compared to thermocouple based system & can also function as a condition monitoring system. The data can be saved to an excel file as shown in fig 3.4.

	A	B	C	D	E	F
1	37.56	52.88	57.81	60.63	58.44	42.56
2	37.56	52.88	57.88	60.63	58.5	42.63
3	37.56	52.88	57.88	60.69	58.5	42.63
4	37.56	52.94	57.88	60.69	58.56	42.56
5	37.56	52.94	57.94	60.69	58.56	42.63
6	37.56	52.94	57.94	60.75	58.56	42.63
7	37.5	52.94	57.94	60.75	58.56	42.63
8	37.56	52.94	57.94	60.75	58.56	42.69
9	37.56	53	57.94	60.75	58.63	42.69
10	37.56	53	58	60.81	58.69	42.69
11	37.56	53	58	60.81	58.63	42.69
12	37.63	53	58	60.81	58.69	42.69
13	37.63	53	58	60.81	58.69	42.69

Fig 3.4: Recorded Temperature

3.4 Calculation of Heat Transfer Coefficient

1. Results obtained by experimental set up:

$T_1 = 37.56$ c, $T_2 = 52.88$ c, $T_3 = 57.81$ c, $T_4 = 60.63$ c,

T_5 = 58.44 c, T6 = 42.56 c, H = .12 m

$T_s = \frac{T2 + T_3 + T_4 + T_5}{4} = 57.44$ c

$T_a = \frac{T_1 + T6}{2} = 40.1$ c

Properties of air: Density = 1.2 kg/m^3, $\gamma = 16.9 \times 10^{-6}$,

K = 0.027, P_r = 0.711, C_p = 1.005 kJ/kg

A_i=area of inlet is 8.04 * 10^{-4} m^2

Discharge through orifice:

$$q = C_d A_o \sqrt{2gh_a}$$

Where C_d= *0.65,* A_o *is the area of orifice which is* *2.01×10*$^{-4}$*m*^{2}

$$h_a = \frac{\text{(Density of water * head of water)}}{\text{Density of air}} = \frac{1000 * 0.12}{1.2}$$

= 100 m

q = 0.65 * 2.01 * 10^{-4} *√ (2*9.81*100) = 5.79*10^{-3} m^3/s

Velocity V = (q/A) = 7.2 m/s

Mass flow rate m = q * density of air = 6.948 * 10^{-3} kg/s

Heat Carried by air:

$Q_a = m * C_p * dT = 34.91$ J/s

$A = \pi * D_i * L = 0.06\ m^2$

Now, $Q_a = h A (T_s - T_a)$

Therefore,

$$h = \frac{Q_a}{A (T_s - T_a)} = \frac{34.91}{0.06 * (57.44 - 40.1)} = 33.55\ W/m^2 K$$

$h = 33.55\ W/m^2 K$

2. Result by correlation method

In the correlation method, we use Dittus-Boelter eqn to obtain the value of Nusselt (Nu) number for the given system & with the eqn for Nu we obtain the heat transfer coefficient.

$$R_e = \frac{VD}{\gamma} = \frac{7.2 * 0.032}{16.9 \times 10^{-6}} = 13585$$

γ is kinematic viscosity obtained from data hand book properties of air which is16.96×10^{-6}.

Using Dittus-Boelter eqn,

$$N_u = 0.023\ R_e^{0.8} P_r^{0.4} = 40.64$$

We know,

$N_u - \frac{hD}{K}$, where h is heat transfer coefficient, D is pipe inner diameter and k is thermal Conductivity of air at mean temperature obtained from properties

Therefore,

$$h = \frac{Nu * K}{D} = \frac{40.64 * 0.027}{0.032} = 34.29\ W/m^2 K$$

$h = 34.29\ W/m^2 K$

Experimental h= $33.55\ W/m^2 K$. Now validate the correlation using CFD simulation.

4. CFD Simulation

4.1 CFD Analysis of Pipe

There are four basic steps involved in the CFD analysis of pipe. They are given as below:

1. Creating Geometry & meshing
2. Pre-processing
3. Solver
4. Post-processing

4.2 Geometry Model

ANSYS ICEM is used to build a 3D geometry for CFD analysis. The geometry is created and Meshed using ANSYS ICEM CFD. The length of pipe is 600mm and inner diameter is 0.032m.

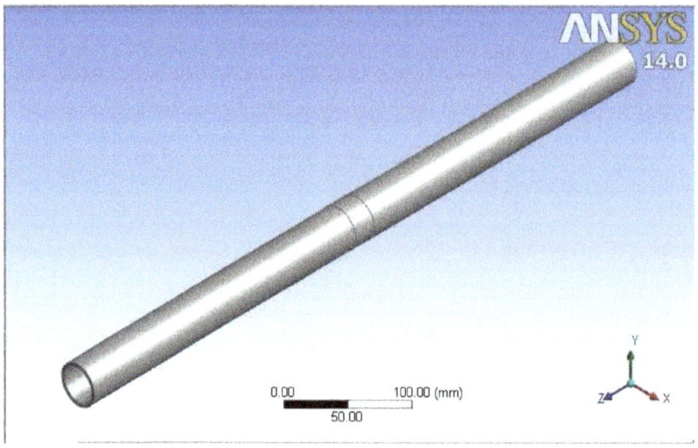

Fig. 4.1: Modelling of Pipe Using ICEM CFD

4.3 Meshing

The computational model is created in ICEM-CFD as per the dimensions shown in Figure 4.2. Later the Tetrahedral Mesh with Smooth Transition grid is generated.

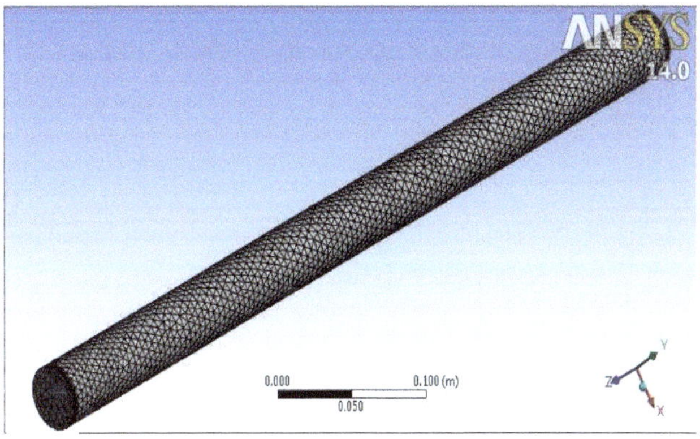

Fig. 4.2: Mesh Generated Using ICEM CFD

4.4 CFD Simulation in FLUENT

The computational model is created in ICEM-CFD as per the dimensions shown in Figure 4.1. Later the Tetrahedral Mesh with Smooth Transition grid is generated.The solver and the boundary conditions are specified in FLLUENT i.e. used segregated solver with asymmetric space and k-epsilon for turbulent modelling.

The boundary conditions and reference values are shown below.

Boundary Conditions:

1. Location = inlet

   Boundary details, subsonic

   Normal speed = 7.2 m/sec

   Medium intensity heat transfer

   Static temp = 310 k

2. Location outlet

Boundary details

Relative pressure = 1 Pa

Wall Option: no slip wall ,Smooth wall

4.5 Results

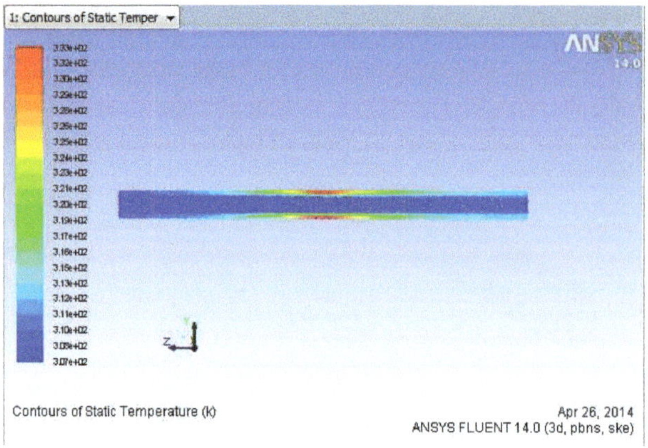

Fig 4.3 : Temperature Plane

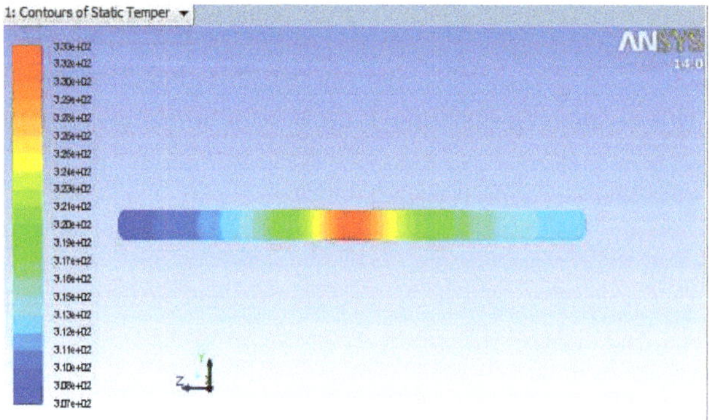

Fig 4.4: Temperature Plane

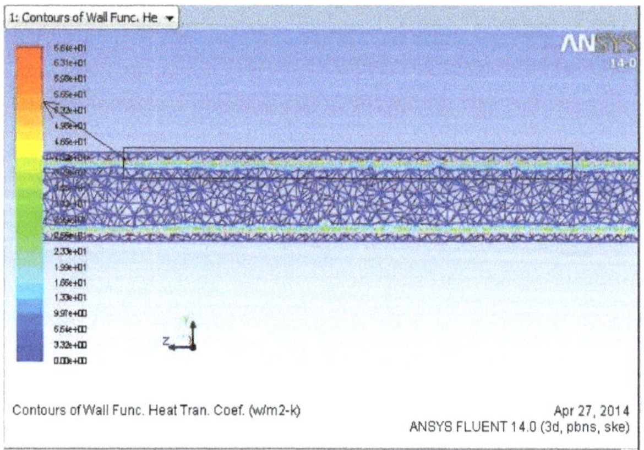

Fig 4.5: Heat Transfer Coefficient

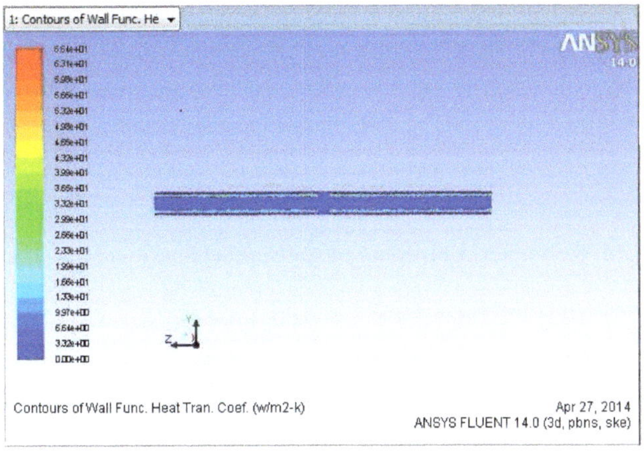

Fig 4.6: Heat Transfer Coefficient

5. Validation of Experimental, Cfd And Correlation

Experimental h= 33.55 W/m^2 K

Correlation h= 34.29W/m^2 K

CFD simulation h= 36.5 W/m^2 K (from fig 4.6)

From this, we observe that the value obtained by CFD is very nearer to correlation and experimental hence Dittus-Boelter equation is validated.

The primary objective of this work is to determine heat transfer coefficients (local and average) for different flow geometries and this heat transfer coefficient (h) may be obtained by experimental or theoretical methods.

Theoretical methods involve solution of the boundary layer equations to get the Nusselt number. On the other hand the experimental methods involve performing heat transfer measurement under controlled laboratory conditions and correlating data in dimensionless parameters. Many correlations for finding the convective heat transfer coefficient are based on experimental data which needs uncertainty analysis, although the experiments are performed under carefully controlled conditions. The causes of the uncertainty are many. Actual situations rarely conform completely to the experimental situations for which the correlations are applicable. Whatever the value obtained by experimental, correlation and CFD simulation agree with each other with an error less than 10 %. Hence Dittus-Boelter has been validated

6. Conclusion

By carrying out experiment, simulation and calculation, it is conclude that these correlations give an error well within the range 10 % and usually less in controlled environment. The results clearly validate the use of Dittus-Boelter correlation in many industrial applications. There are many reasons which contribute to the errors. The heat loss does account for errors, but the value of heat loss is negligible as compared to total heat flux. The heaters are local made and do not provide constant surface heat flux although it has been tried to control wattage of each heater.

Secondly all sensors (temperature) are attached in a close proximity of 220 V AC lines coming from heaters. Therefore signals from sensors are distorted and give an error. If somehow these problems are overcome results may improve. Another thing is correlation results must be more but here it is less because this has been taking the property of air at mean temperature of inlet and out let obtained by experiment.

References

[1] Dittus, F. W., and L. M. K. Boelter. University of California (Berkley) Pub. Eng., vol. 2. p. 443, 1930.

[2] Holman, J. P., Heat Transfer, 9th edition, McGraw Hill, p. 267-271.

[3] Diane L. Linne, Michael L. Meyer, Tim Edwards, and David A. Eitman. Evaluation of Heat Transfer and Thermal Stability of Supercritical JP–7 Fuel. NASA Technical Memorandum 107485.

[4] Incropera, Fundamentals of Heat and Mass Transfer, 5th edition, John Wiley and Sons

[5] J. R. Lamarsh& Anthony B., Intro to Nuclear Engineering, 3rd edition, Wiley

[6] Munson, Young, Okishi, Fundamentals of Fluid Dynamics, 4th edition, Wiley

[7] Incropera F. and D. De Witt. Fundamentals of Heat and Mass Transfer, 5th Edition, USA. 2002

[8] Fand, R. M. and K. K. Keswani. "A Continuous Correlation Equation for Heat Transfer from cylinders to Air in Crossflow for

[9] Reynold's Numbers from 10-2 to 2(10)5," International Journal of Heat and Mass Transfer, 15:559-562. 1972. CFD Simulation & Experimental Verification of Air Flow through heated pipe. Jamuna A.B., Somashekar V.

Solar Water Heating System

Abhishek Verma[1], Vishal Kumar[2]

[1]Student, Mechanical Engg. Deptt. M.I.E.T, KUK, (vermaabhishek1009@gmail.com)

[2] Assistant Professor in Mechanical Engg. Deptt. MIET, KUK, (vishal2106481@gmail.com)

Abstract

In solar heater water is heated by the use of solar energy. Solar heating systems are generally composed of solar thermal collectors, fluid system to move the heat from the collector to its point of usage. The systems may be used to heat water for a wide variety of uses, including home, business and industrial uses. In many climates, a solar heating system can provide up to 85% of domestic hot water.

Solar energy provides power security, enabling you to continue getting power even when utility power is disrupted.

Business can save 40% to 80% on electric or fuel bills by replacing their conventional water heater with a solar water heating system.

Home solar heating can lead to saving of 85% on utility bills over the costs of electric water heating.

Using solar energy is free and it is an environmentally friendly source of power. Solar energy can be integrated with existing energy methods, such as wood stove and demand boiler, to produce 100% of the required thermal loads.

Keywords: Solar Energy, Solar Thermal collectors, Passive Solar water heater system (PSWHS), Active Solar water heater system (ASWHS).

1. Introduction

A low cost pipe type solar water heater having a capacity of 100 liter for the storage use in urban as well as rural areas has been developed. To minimize the cost, the absorber and the storage tank has been integrated together i.e. it is built storage tank. The unit works on push through principle i.e. hot water will come out from the outlet when the cold water in it through the outlet. The unit consist of ten number of pipe fabricated from 34SWG copper pipe. The entire ten pipes are placed parallel to each other with a net spacing of 1201 μm between them. The free edge at the sites not to exceed 60 mm from centre of the extreme and riser tube.

The unit as a whole is housed in aluminum sheet cover box with ordinary window glass in front. The whole unit has been placed in an inclined position, depending upon the latitude of the place and oriented towards south direction. A number of tests has been conducted for it . The maximum temperature of water at 1500hrs in winters is 60-70 degree Celsius. The hot can be withdrawn at the rate of 15-30 liters hourly or two hourly.

This water heater can be easily coupled with electric geyser. This combined system will conserve electricity and result in overall economy. Feasibility studies carried in this respect show that about 50% electrical energy consumed in the geyser can easily be saved by integration of solar water heater with electric geyser. Considerable amount of fuel can be saved by using this water heater for daytime use for taking bath and washing clothes, utensils etc.

2. PRINCIPLE:

This is passive type solar water heater. This requires no pumping for circulation of water. Water is circulated with the function of natural convection. Thus this solar heater is based on the principle of natural convection. In this solar water heater. Collector absorbs the radiation from the sun. Cold water comes from the storage tank flows through collector pipe and gets heated there. Now the density of hot water is low as compared to cold water .So the hot water start flowing in upward direction and goes to storage tank .The hot water stored in the tank can be used for various applications.

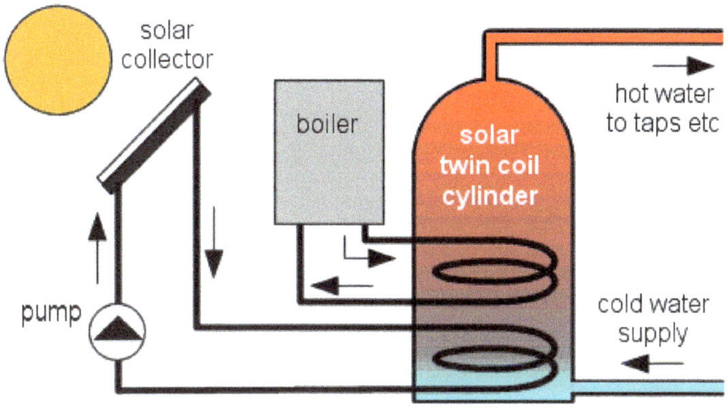

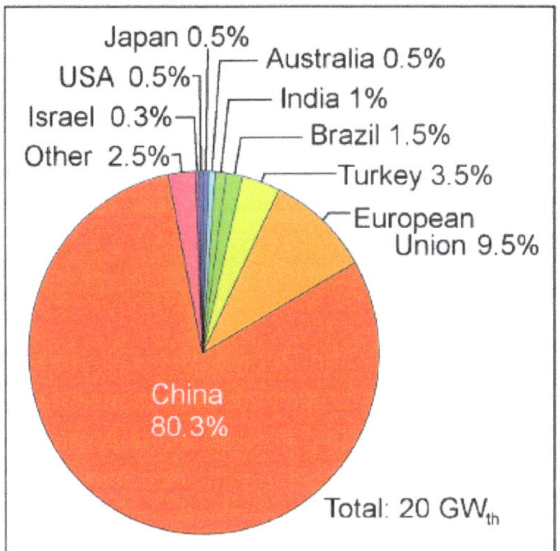

Source: REN21 2009. GTZ Gmbh

Solar Water Heater:

Solar water heater heats the water by the use of energy. These are generally composed of solar thermal collectors ,a fluid system to move the heat from the collector to its point of usage . The system may use for pumping the fluid, and have a reservoir tank for heat storage and subsequent use. The system may be used to heat water for a wide variety of uses including homes, business and industrial uses. Heating swimming pool, under floor heating or energy input for space heating or cooling are more specific examples.

In many climates, a solar heating system can provide up to 85% of domestic hot water energy. This can include domestic hot water energy.

System design requirements

The type, complexity, and size of a solar water heating system is mostly determined by:

• Changes in ambient temperature and solar radiation between summer and winter.
• The changes in ambient temperature during the day-night cycle.
• The possibility of the potable water or collector fluid overheating.
• The possibility of the potable water or collector fluid freezing.

The minimum requirements of the system are typically determined by the amount or temperature of hot water required during winter, when a system's output and incoming water temperature are typically at their lowest. The maximum output of the system is determined by the need to prevent the water in the system from becoming too hot.

Freeze protection

Freeze protection measures prevent damage to the system due to the expansion of freezing transfer fluid. Drainback systems drain the transfer fluid from the system when the pump stops. Many indirect systems use antifreeze (e.g.Propylene glycol) in the heat transfer fluid. In some direct systems, the collectors can be manually drained when freezing is expected. This approach is common in climates where freezing temperatures do not occur often, but is somewhat unreliable since the operator can forget to drain the system. Other direct systems use freeze-tolerant collectors made with flexible polymers such as silicone rubber. A third type of freeze protection is freeze-tolerance, where low pressure polymer water channels made of silicone rubber simply expands on freezing. One such collector now has European Solar Keymark accreditation, following extra durability testing.

Overheat protection

When no hot water has been used for a day or two, the fluid in the collectors and storage can reach very high temperatures in all systems except for those of the drainback variety. When the storage tank in a drainback system reaches its desired temperature, the pumps are shut off, putting an end to the heating process and thus preventing the storage tank from overheating. One method of providing over heat protection is to dump the heat into a hot tub. Some active systems deliberately cool the water in the storage tank by circulating hot water through the collector at times when there is little sunlight or at night, causing increased heat loss. This is most effective in direct or thermal store plumbing and is virtually ineffective in systems that use evacuated tube collectors, due to their superior insulation. No matter the collector type, however, they may still overheat. High pressured sealed solar thermal systems versions ultimately rely on the operation of temperature and pressure relief valves. Low pressure, open vented ones have simpler, more reliable safety controls, typically an open vent.

Passive and active systems

Passive systems rely on heat-driven convection or heat pipes to circulate water or heating fluid in the system. Passive solar water heating systems cost less and have extremely low or no maintenance, but the efficiency of a passive system is significantly lower than that of an active system. Overheating and freezing are major concerns.
Active systems use one or more pumps to circulate water and/or heating fluid in the system.
Though slightly more expensive, active systems offer several advantages:

• The storage tank can be situated lower than the collectors, allowing increased freedom in system design and allowing pre-existing storage tanks to be used.
• The storage tank can always be hidden from view.
• The storage tank can be placed in conditioned or semi-conditioned space, reducing heat loss.

• Drainback tanks can be used.
• Superior efficiency.
• Increased control over the system.

Modern active solar water systems have electronic controllers that offer a wide range of functionality, such as the modification of settings that control the system, interaction with a backup electric or gas-driven water heater,calculation and logging of the energy saved by a SWH system, safety functions, remote access, and informative displays, such as temperature readings.

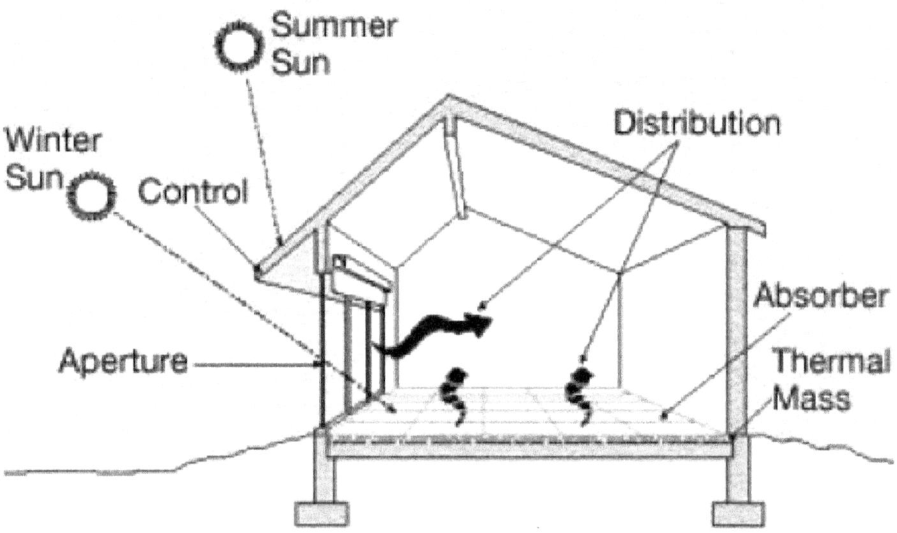

Elements of passive solar design, shown in a direct gain application

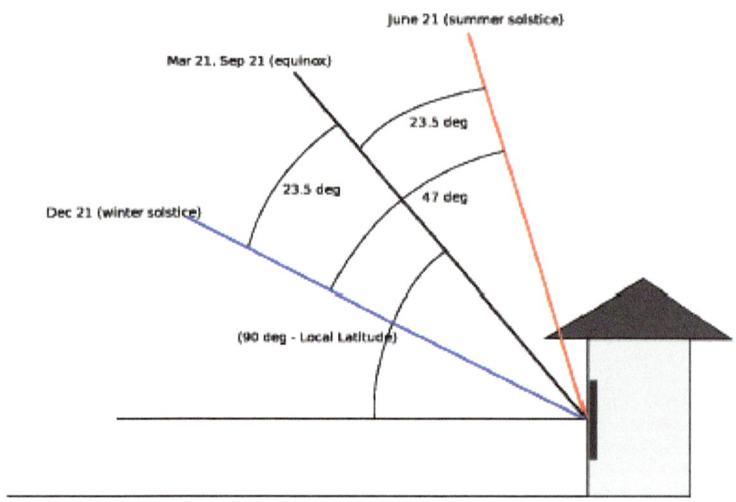

Solar altitude over a year; latitude based on New York, New York

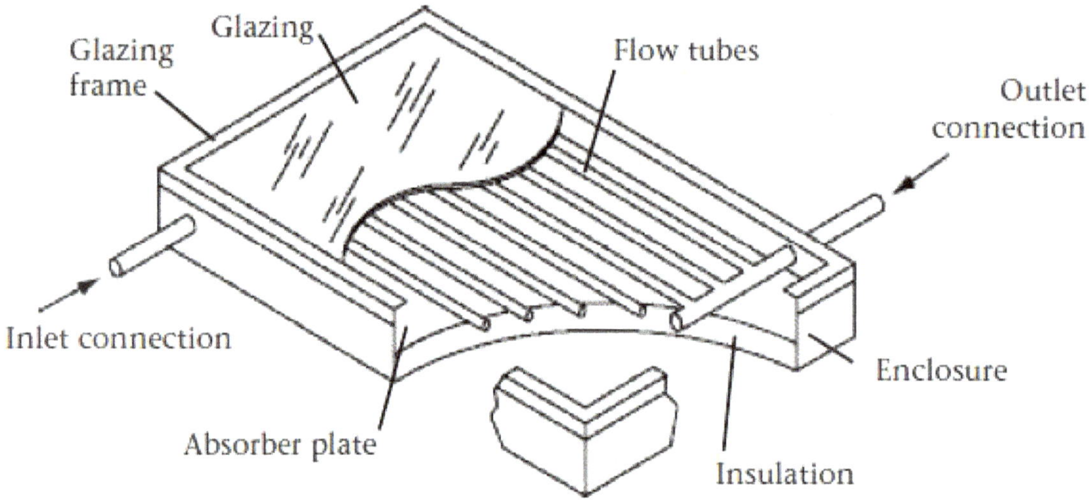

Glazed flat-plate collector

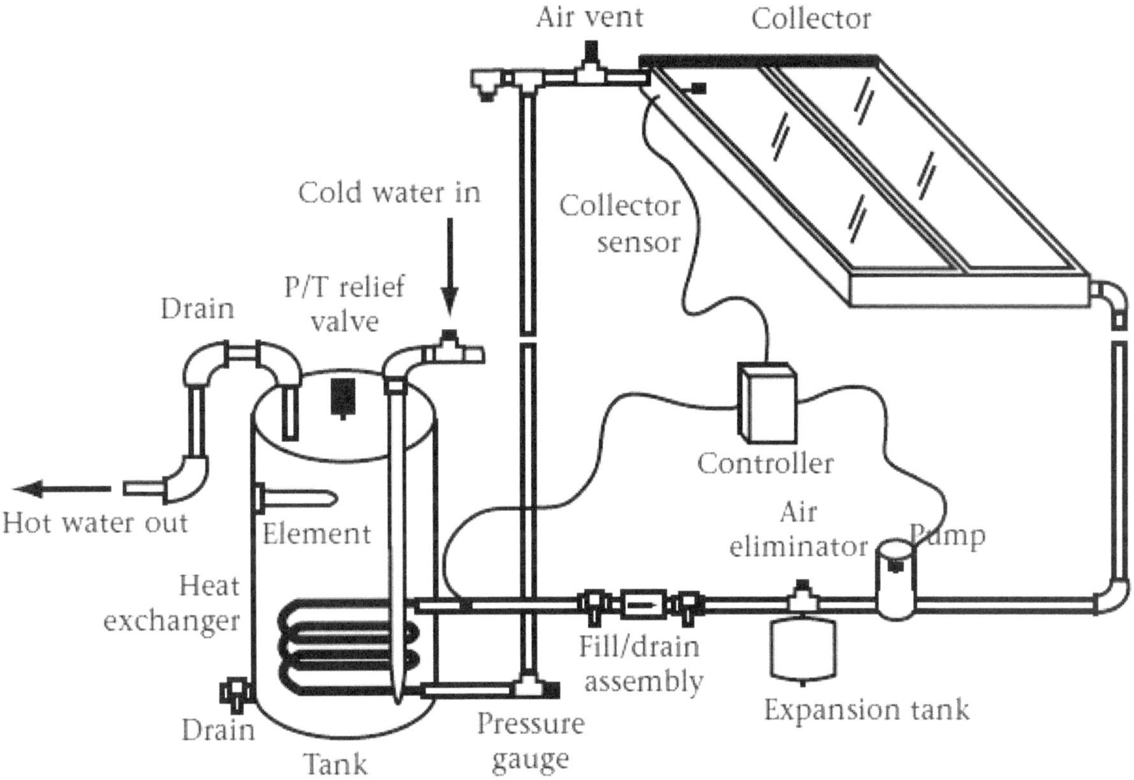

Technique and design:

In order to heat water using solar energy, a collector is fastened to the roof of a building, or on a wall facing the sun. In some cases, the collector may be freestanding. The collector could be made of a simple glass topped insulated box with a flat solar absorber made of sheet attached to copper pipes and painted black, or a set of metal tubes surrounded by an evacuated glass cylinder. A simple water heating system would pump cold water out to a collector to be heated water flows back to a collection tank.

The typical 50 gallion electric water heater uses 11.1 barrels of oil a year, which translates into the same amount of oil used by a typical 4-door sedan driven by the average consumer. Electrical utility companies often provide electricity by burning and releasing energy from fuels such as oil, coal and nuclear energy. An electrical home hot water heater sits on an electrical grid and may be driving the use of unclean fuels on the other end of the grid.

So using the solar water heaters to heat water is much reliable and appropriate thing than using our preciously non-renewable fuels.

SPECIFICATION OF COMPONENTS

Solar panel: Specification of different parts of solar panel is as below. This specifications is taken according to INDIAN STANDARD (IS 12933 PART : 5) for solar panel.

1 Absorber material: Absorber shall be of copper sheet and copper tube
Thickness of copper sheet : 34SWG
Riser specification:
Diameter of riser tube: 12.7+0.5mm
Thickness: 0.56mm
Number of riser tubes: 9
Header specification:
Diameter: 25.4+0.5mm
Thickness: 0.71mm
Projection outside: 40mm+0.5mm including flanges of the collector box
Number of header tubes:2
Space between riser tubes: 120mm maximum for centre to center of the riser. The free edges at the sides not to be exceed 60 mm from the centre of the extreme end riser tubes. For independent tins or joints in the sheet an overlap of minimum 2mm shall be provided.
coating of copper tubes and sheets: contain of this part is done with absorbability>0.92 and emmissivity<0.2. This is achieved by black powder coating.
Collector box materials:
Minimum aluminum extruded sections of size approximately 100mm*25 mm channel section and thickness 1.6 mm for the sides.
sheet for bottem :
Thickness:0.71nun
Material: aluminium

Angles for cover box:
Thickness: 1.5mm
Dimensions:25mm*25mm*1.2mm
Material:aluminium

Front glazing:

Material: single piece glass plain.
Thickness: 4.8+0.3mm
Transmittance: 80'1"0(min)
General appearance of glass: free from bubbles/rough surface.
spacing between : 20 to 40 mm.

Collector box insulation:

Black insulation:
Material: rock wool
Thickness: 50 mm

Side insulation:
Material : rock wool
Thickness 25mm

Aluminium foil of thickness 0.01 mm to 0.015 mm is used of covering the side insulation

Header flange:
Material: brass
Diameter: 75 mm
Thickness: 4 mm

Grommet:
It is used for sealing between the inlet and outlet of header.
Materials : silicon rubber or butyl rubber
Outer diameter: 40mm
Inner diameter: 25mm
Grommet shall be suitable for a temperature up to 200°c.

Gasket:
It is in between flange and pipe.

Storage tank
The specification of different part of storage tank is as below.
Material : stainless steel with epoxy coating.
Shape : cubic
Length: 32cm
width: 32cm
Height: 102cm
Volume of tank: 32*32*102cm3= 104448 cm3
Now 1liter= 10UO cm3
So that the tank having capacity about 104.4 liters
Thickness of stainless steel sheet:20SWG

Insulation:

Material: glass wool
Thickness:1no. mm approx.
Density:48kg/cu.m.
Temperature limit: upto 250 °c
Chicken mesh is used as retaining material

Cladding of insulation:
Material: aluminium sheet
Thickness: 24SWG

Wooden Frame
It gives support to cladding. Aluminium sheet cover wooden frame along with glass wool
Length:18 inch
Width:19 inch
Weight: 47inch

Nipples:
Material: grey iron
Diameter:0.75mm
Length: 6 inch

Valves:

Gun metal valves are used.

Elbows:

Diameter: 0.75 inch
Material: mild steel.

Angles for fitting aluminium sheet:

Material: aluminium
Dimensions: 0.75* 0.75 inch

Instrumentation:
Temperature gauge: It is for hot water storage tank/outlet
Gun metal stainer at cold water inlet.piping:
Diameter:1inch
Material: grey iron.
Cock valve and tap:

Cock valve at cold water inlet and tap at hot water outlet.

The storage tank has five opening.
1 For cold inlet(from cold water tank)
2 For hot water outlet (for use)
3 ventilation.
4 Cold water to solar panel.
5 Hot water from solar panel
For cold water circulation black flexible pipes can be used

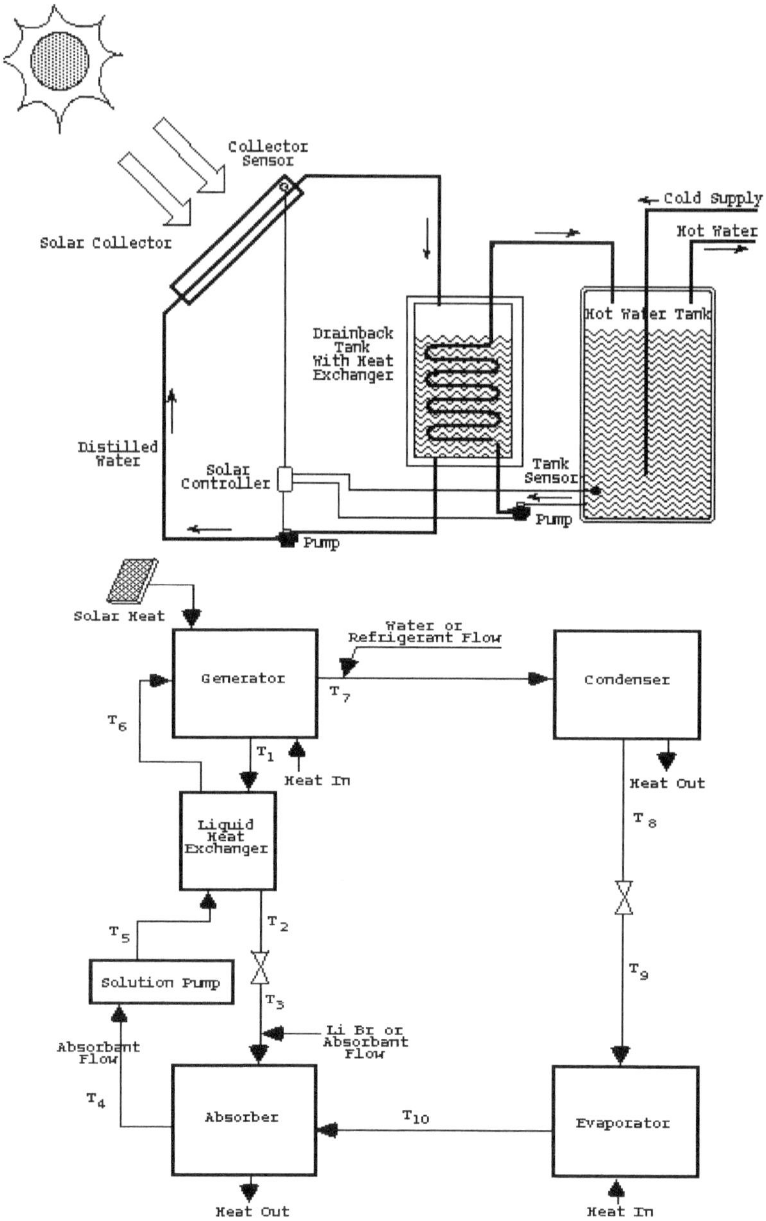

WORKING:

As we discussed in the principle, our solar water heater is based on the principle of natural convection. We supply old water to storage from external tank or water supply. Then cold water from tank flows through solar collector. As radiation fall on the collector, so the water flows through collector tube get heated and shirts flowing in upward direction through outlet valve of panel and goes to storage tank for storage. We can use this water for various applications as discussed earlier. The flow of cold water from tank to solar panel is due to gravitational forces as height of tank provides head. Hot water flows from panel to storage tank due to the conventional process. As hot water has lower density than cold water, so hot water lift up and cold water occupy its position of hot water.

This is the working of solar water heater system. After assembling of different parts i.e. after completion of project, we made it work. We mount it at suitable place. We found that it heats the water. We measure the temperature of water. We found that it heats the water up to 70 °c.

References

1. http:/ / www. sunnovations. com/ content/ how-sunnovations-system-works
2. Israeli Section of the International Solar Energy Society (http:/ / www. ises. org. il/ assets/ files/ ISES Info/ IsraelSectionISESfinal. pdf)
3. Your Home - Orientation (http:/ / www. yourhome. gov. au/ technical/ fs43. html)
4. Your Home - Insulation (http:/ / www. yourhome. gov. au/ technical/ fs47. html)
5. Your Home - Passive Cooling (http:/ / www. yourhome. gov. au/ technical/ fs46. html)
6. Your Home - Shading (http:/ / www. yourhome. gov. au/ technical/ fs44. html)
7. Your Home - Thermal Mass (http:/ / www. yourhome. gov. au/ technical/ fs49. html)
8. Chiras, D. The Solar House: Passive Heating and Cooling. Chelsea Green Publishing Company; 2002.
9. Earthships (http:/ / earthship. com/)
10. Annualized Geo-Solar Heating, Don Stephens (http:/ / greenershelter. org/ TokyoPaper. pdf)- Accessed 2009-02-05
11. http:/ / www. energy. ca. gov/ 2006publications/ CEC-500-2006-067/ CEC-500-2006-067.pdf
12. http:/ / www. ornl. gov/ sci/ solar/
13. Brian Norton (2011) Solar Water Heaters: A Review of Systems Research and Design Innovation, Green. 1, 189–207, ISSN (Online)1869-8778
14. http:/ / en. wikipedia. org/ w/ index. php?title=Template:Sustainable_energy& action=edit
15. Brian Norton (2011) Solar Water Heaters: A Review of Systems Research and Design Innovation, Green. 1, 189–207, ISSN (Online)1869-8778
16. 2011 global status report by Renewable Energy Policy Network for the 21st Century (REN21)

21

EXPERIMENTAL STUDY OF ALTERNATIVE REFRIGERANTS TO REPLACE R134a IN A DOMESTIC REFRIGERATOR

Balakrishnan.P[1], Dr.K.Karuppasamy[2], Ramkumar.J[3], Anu Nair.P[4]

[1,4]PG Scholar, Department Of Mechanical Engineering, Regional Centre Anna University: Tirunelveli Region, Tirunelveli -627 007,krishna234431@gmail.com,anunair67@gmail.com

[2] Assistant Professor, Department Of Mechanical Engineering, Regional Centre Anna University: Tirunelveli Region, Tirunelveli -627 007,kkssurabu2001@yahoo.co.in

[3]Teaching Faculty, Department Of Mechanical Engineering, Regional Centre Anna University: Tirunelveli Region, Tirunelveli-627007,rammech07@gmail.com

Author Correspondence:144/3Pillaiyar Kovil Street,Singilipatti,Tenkasi,Tirunelveli,Tamilnadu-627765

Email address- krishna234431@gmail.com

Abstract

This paper explores an experimental investigation of an alternative eco-friendly refrigerant for R134a with a better Coefficient Of Performance (COP), reduced Global Warming Potential (GWP) and Ozone Depletion Potential(ODP). This investigation has been accessed using a hydrocarbon refrigerant mixture composing of R32/R600a/R290 in the ratio of 70:5:25 by weight. The performance characteristics of the domestic refrigerator were predicted using continuous running tests under different ambient temperatures and cyclic running (On/Off) tests at the fixed temperatures i.e., evaporation temperature (-5°C) and condensation temperature (30°C). The obtained results showed that the hydrocarbon mixture has lower values of energy consumption; pull down time and ON time ratio also higher Coefficient of Performance (COP). Thus, the performance of the alternate refrigerant derives the better choice than R134a.

Keywords: Domestic refrigerator, Alternative refrigerant, R32, R600a, R290, R134a

1. Introduction

Recently, The Ozone Depleting Potential (ODP) and Global Warming potential(GWP) have become the most important criteria in the development of new refrigerants apart from the refrigerant CFCs and HCFCs, both of which have high ODP and GWP due to their contribution to ozone layer depletion and global warming(Dalkilic et al,2010). Results from researchers show that the ozone layer is being depleted due to the presence of chlorine in the stratosphere. The general consensus for the cause of this that CFCs and HCFCs are large class of chlorine containing chemicals, which migrate to react with ozone. This leads to the strict prohibition of

CFCs. The R134a was the first chlorine-free refrigerant discovered (Bolaji,2010). R134a is used nowadays as the working fluid in domestic refrigerators. But it was found that the R134a increasing significantly to the world's greenhouse warming problem. This caused scientists to investigate more environmentally friendly refrigerants than HFC refrigerants for the protection of the environment such as hydrocarbon mixtures as working fluid in refrigeration and air-conditioning systems. The possibility of using hydrocarbon mixtures as working fluid to replace R134a in domestic refrigerators has been evaluated through the simulation analysis. This simulation concludes that the hydrocarbon refrigerants offer desirable environmental requirements i.e., zero ODP and approximately zero GWP (Fatouh et al ,2006). A hydrocarbon blend of difluoromethane (R32), propane (R290) and iso-butane (R600a) is recommended to avoid the stratospheric ozone depletion. The presence of R600a in the mixture is miscible with both mineral oil and synthetic lubricants (Mohanraj et al,2009).The hydrocarbon mixture (R290 and R600a) in the ratio 45.2:54.8 by weight is investigated in the domestic refrigerator in which the COP is improved upto 3.6% and also reduce the indirect global warming due to its higher energy efficiency (Mohanraj et al,2009). The propane/iso-butane mixture at 0.6 mass fraction used as the refrigerant has a 3-4% higher energy efficiency and faster cooling rate. (He et al,2014).The energy analysis using Taguchi method indicates that the charge amount required was 50g which brings economic advantages and also reduces the risk of flammability of the hydrocarbon refrigerant. The amount of total energy destruction in optimum condition is 45.05% of the base refrigerator one (Joybari et al,2013). Researches on R290 and R290/R600a for substituting R134a in a large capacity chest freezer were done. This work concluded that the data of power consumption of R290 is lower by 26.7% and its charging amount is lesser than that of R134a.The COP of R290 is lower by about 4.7% than that of R600a, so that the working fluid R290/R600a in an appropriate proportion can improve the energy conversion effect. Further, this hydrocarbon mixture has an elevated effect on energy conservation when a specialized compressor is designed (Wongwises et al,2005).

In the present work, the problem of R134a is identified from environmental site and the alternative refrigerants are selected. Then, their mass flow ratio are mixed and a better refrigerant is chosen with better mixing ratio. Thereby, the COP of the alternate refrigerant and R134a are compared. Hence, the the main objective is to select an alternate refrigerant with an increased Coefficient Of Performance(COP) and to reduce Global Warming Potential (GWP) and Ozone Depletion Potential (ODP) hydrocarbon mixture R32/R600a/R290 in the ratio of 70:5:25 by weight.

2. Experimental Methodology

The first step is to study the alternative refrigerant to replace R134a in a domestic refrigerator. The details of the experiment setup is shown in Table-1

Table 1 Details of Experimental Set-up

SI NO	Description	Dimension/Range
1	Refrigerator Capacity	200 litres
2	Capillary Tube	0.031mm φ
3	Compound Gauge	-30-220psi
4	Pressure guage	0-300 psi
5	Vaccum Pump	-30PSIG
6	R32/R600a/R290	100gm

The R134a domestic refrigerator setup consist of a hermatically sealed compressor ,natural convection air cooled condenser having a cooling capacity level of 5.67KW/hr, an evaporator and copper capillary tube whose schematic diagram and photographic view of the experimental set up is given in the fig 1 and fig 2.Sensor is attached at the inlet and outlet of compressor, condenser and evaporator. Pressure gauge is attached at the compressor inlet and outlet. Compound gauge is fitted at the condenser outlet. Evacuation of moisture takes place with the help of service port service port .Vaccum pump is used for evacuation and through the charging system refrigerant was filled in the refrigeration system. Properties of R32/R600a/R290 refrigerant from ASHRAE handbook are given in the Table 2.

Table 2 Properties of R32/R600a/R290

Refrigerant	R32	R600a	R290
Safety level	A1	A1	A1
Boiling point	-52^{0}C	-11^{0}C	-41^{0}C
T_{con}	78.1^{0}C	137.7^{0}C	66^{0}C
P_{cond}	5.78 bar	3.78 bar	3.62bar
ODP	0	0	0
GWP	715	-20	-20
COP	2.11	1.99	2.01

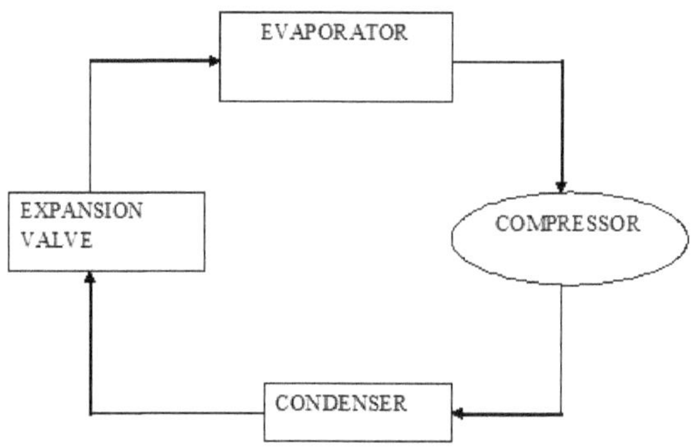

Figure 1 Schematic diagram of the Experimental setup

Figure 2 Photographic View of the Experimental setup

2.1 Experimental Procedure

• As a first step, R-134a is used as a working fluid at room temperature of 32^0C. Pressure and temperature in the inlet and outlet of the compressor, condenser pressure and temperature, evaporator outlet temperature are monitored in every one hour. The Compound pressure gauge and temperature sensors are mounted on the both ends of the compressor and condenser to measure the pressure and temperature of the refrigerant respectively. Finally COP and Mass flow rate is evaluated for the working fluid R134a.

• Then working fluid is released to the ambient and it can be fully evacuated by using vaccum pump running for three hours.

• R32/R600a/R290's are mixed in the proportions of 70%/5%/25% used as refrigerant in the refrigerator.

• The Same procedures are repeated to calculate the COP and Mass flow rate for those mixtures of gases.

3. RESULTS AND DISCUSSION

The assumptions used in the R134a domestic refrigerator are

- Steady state condition.
- No pressure drop through the pipe line in the refrigerator.
- No sub cooling and super heating takes place.
- No heat losess and no heat gain from or to the system.

.

Experimental results for COP and Mass flow rate are determined using REFPROP software are shown in Table 3. The various equations used for calculating the parameters under the study are given below

Coefficient of Performance (COP), Mass flow rate, cooling capacity is calculated using some given values.

COP is ratio between Refrigeration effect and work done.

Refrigeration effect = h_1-h_4,

h_1 is compressor side enthalpy.

h_4 –expansion enthalpy.

$$\text{Work done} = (h_2\text{-}h_1)$$

h_2 – compressor enthalpy.

$$COP = (h_1\text{-}h_4) / (h_2\text{-}h_1)$$

Mass flow rate is ratio between cooling capacity and refrigeration effect.

$$\text{Capacity} = \text{mass flow rate} * (h_{fg})$$

Capacity = 5.67 kW.

h_{fg} – is vapour enthalpy.

Super-heated vapour refrigerant at compressor enthalpy exit (h_2)

$$h_2 = h_1 + (h_{2,is} - h_1) / (\eta_{is,com})$$

Where,

$\eta_{is,\ com}$ is the isentropic compressor efficiency.

$h_{2\ is}$ the superheated vapour enthalpy at compressor exit.

Table 3. COP and Mass flow rate of (R32/R600a/R290) using Refprop Software

Refrigerants	Mixing ratios	COP	Mass flow rate (kg/s)
R32/R600a/R290	45/5/50	1.90	0.0303
R32/R600a/R290	35/9/56	2.01	0.0286
R32/R600a/R290	30/10/60	1.91	0.0299
R32/R600a/R290	60/5/35	2.01	0.0217
R32/R600a/R290	50/6/44	1.98	0.0203
R32/R600a/R290	65/4/21	2.03	0.0213
R32/R600a/R290	70/5/25	2.21	0.0311
R32/R600a/R290	45/10/45	2.07	0.0295
R32/R600a/R290	50/5/45	1.33	0.0219
R32/R600a/R290	44/6/64	1.93	0.0301
R32/R600a/R290	30/6/64	2.08	0.0233
R32/R600a/R290	72/8/20	1.73	0.0212
R32/R600a/R290	63/7/30	2.04	0.0242
R32/R600a/R290	20/7/73	2.02	0.0227

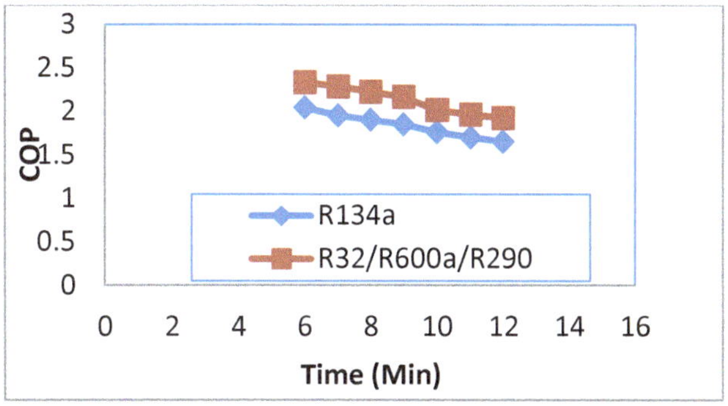

Figure 3 Time vs COP in Half loading condition

Figure 3 shows that the variation of Cop with time in half loading ON condition The pressure and temperature at the suction and discharge of the compressor and that of condenser are measured under half loading condition by switching the refrigerator on. COP decreases with increases in time due to suction pressure and temperature decreases in the compressor i.e, work done in the compressor decrease. R32/R600a/R290 refrigerant mixture has high COP compared with R134a.

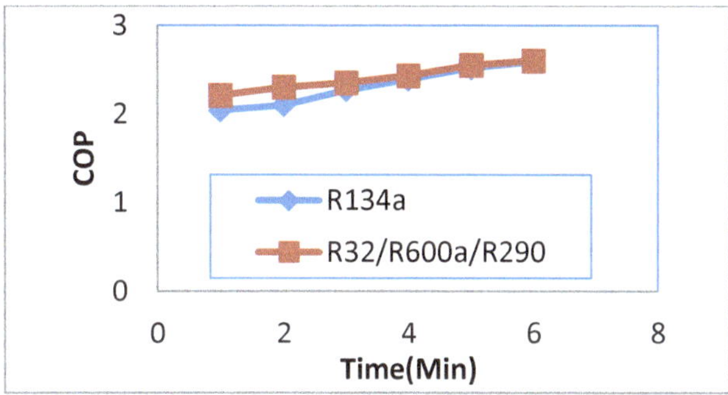

Figure 4 Time vs COP in full loading condition. Figure 4 shows that the Variation of COP with time in Full loading condition. The pressure and temperature at the suction and discharge of the compressor and that of condenser are measured under loading condition by switching the refrigerator on. COP increases with increase in time due suction pressure and temperature increases in the compressor i.e. work done in the compressor increases.

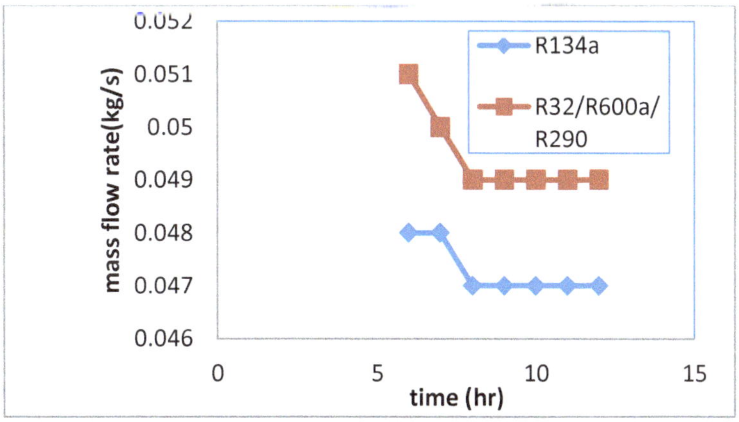

Figure 5 Time vs COP loading ON condition

The mass flow rate of the R134a refrigerant measured at different time intervals by switching the refrigerator on. Figure 5 shows that the Variation of mass flow rate with time. Mass flow rate decreases with increases in time, after that it remains constant. Initially Refrigerant takes more time to flow through the compressor.

4. CONCLUSION

The problem of R134a refrigerant is identified from the environmental site. An alternative refrigerant is chosen with better COP and is compared with R134a. The mass flow rate and COP of the chosen refrigerant is calculated and is compared with R134a.

- Mass concentration ratio is (70/5/25) mixer time that COP and Mass flow rate calculated.
- Condensation temperature is fixed (33 C) and evaporator temperature is also fixed (-5 C) that time calculated COP and Mass flow rate.
- R134a at unloading ON condition time, the COP and mass flow rate were lower than (R32/R600a/R290).
- Using (R32/R600a/R290) at loading ON condition time, the COP and mass flow rate values are higher than R134a.
- (R32/R600a/R290) is starting cooling process lower than R134a.
- Mixing refrigerants GWP (global warming potential) is lower than R134a.

Nomenclature

c specific heat (kJ/kg k)

C- Clearance ratio

COP coefficient of performance

h specific enthalpy (kJ/kg)

m mass flow rate (kg/s)

n specific heat ratio

N compressor speed (rpm)

P power (kW)

Q heat transfer rate (kW)

T - Temperature (^{o}C)

T_{cond} condenser temperature(^{o}C)

P_{cond} condenser pressure(bar)

R134a 1,1,1,1 tetraflouroethane

R32 difluoro methane

R600a isobutene

R290 propane

ODP ozone depletion potential

GWP global warming potential

References

A.S. Dalkilic, S. Wongwises,2010,A performance comparison of vapour-compression refrigeration system using various alternative refrigerants, *International Communications in Heat and Mass Transfer*, vol. 37 ,pp. 1340–1349.

B.O. Bolaji, 2010, Experimental study of R152a and R32 to replace R134a in a domestic refrigerator *Energy* ,vol .35 .pp.3793-3798.

M. Fatouh, M. El Kafafy, 2006,Assessment of propane/commercial butane mixtures as possible alternatives to R134a in domestic refrigerators,*Energy Conversion and Management* ,Vol. 47,pp.2644–2658.

M. Mohanraj, 2013,Energy performance assessment of R430A as a possible alternative refrigerant to R134a in domestic refrigerators, *Energy for Sustainable Development* ,vol 17,pp . 471–476.

M. Mohanraj, S. Jayaraj, C. Muraleedharan, P. Chandrasekar,2009,Experimental investigation of R290/R600a mixture as an alternative to R134a in a domestic refrigerator, *International Journal of Thermal Sciences* , Vol. 48,pp. 1036–1042.

M.G. He, X.-Z. Song, H. Liu, Y. Zhang,2014,Application of Natural Refrigerant Propane and Propane/ Iso butane in Large Capacity Chest Freezer, *Applied Thermal Engineering,*Vol.55,pp.25-35.

Mahmood Mastani Joybari, Mohammad Sadegh Hatamipour, Amir Rahimi Fatemeh Ghadiri Modarres,2013,Exergy analysis and optimization of R600a as a replacement of R134a in a domestic refrigerator system, *International Journal of Refrigeration,* Vol.20,pp.1-10.

Somchai Wongwises, Nares Chimres, 2005,Experimental study of hydrocarbon mixtures to replace HFC-134a in a domestic refrigerator, *Energy Conversion and Management*,Vol.46,pp.85–100.

22

COMPARITIVE STUDY OF BAYESIAN APPROACH AND LEAST SQUARE RESIDUAL OPTIMIZATION METHOD IN HORIZONTAL HEATED PLATE FACING UPWARD-AN EXPERIMENTAL APPROACH

Anu Nair.P[1], Dr.K.Karuppasamy[2]

[1] *PG Scholar, Department Of Mechanical Engineering, Regional Centre Anna University: Tirunelveli Region,Tirunelveli -627 007,anunair67@gmail.com*

[2] *Assistant Professor, Department Of Mechanical Engineering, Regional Centre Anna University: Tirunelveli Region, Tirunelveli -627 007,kkssurabu2001@yahoo.co.in*

Author Correspondence: Madhusadhanam,Thrikkodithanam P.O,Changanacherry,Kottayam,Kerala

Email address- anunair67@gmail.com

Abstract

Free convection heat transfer is encountered in many engineering applications such as Cooling of electronic equipment, pollution, materials processing, energy systems, and safety in thermal processes and geophysical flows. The present work discusses about the result of an experimental study of Steady, Laminar, Free convection heat transfer in a horizontal plate facing upwards in which two walls are adiabatic and other two ends are open to the ambient .The aim is to estimate the value of constant C in the Nusselt number of available correlation using least square residual method and Bayesian approach where air is used as a working medium. The steady temperature time history is mainly used to estimate constant of the Nusselt number in both optimization method. The experimental setup has been designed and fabricated. Finally, Estimated Parameter is compared with Experimental benchmark.

Keywords**:** Free Convection, Least Square Residual Method, Bayesian Approach

1. Introduction

Natural Convection is mainly due to the density differences between the solid and fluid surface ($T_s > T_a$). The density difference gives rise to buoyancy forces due to which the flow is generated. A heated body cooling in ambient air generates such a flow in the region surrounding it. ($T_s > T_a$). The buoyant flow arising from heat or material rejection to the atmosphere, heating and cooling of rooms and buildings, recirculating flow driven by temperature and salinity differences in oceans, and flows generated by fires are other examples of natural convection.There has been growing interest in buoyancy-induced flows and the associated heat and mass transfer over the past several decades, because of the importance of these flows in many different areas, such as cooling of electronic equipment, pollution, materials processing, energy systems, and safety in thermal processes.

(Elenbass,1942) conducted experimental work in laminar natural convection heat transfer in smooth parallel plate vertical channel was investigated and reported a detailed study of the thermal characteristics of cooling by natural convection.(R.A. Wirtz et al.,1982) have considered a geometry with constant heat sources placed over the entire length of the wall. Since the geometry cannot simulate discrete placement of chips, a number of discrete heat sources placed over a wall was considered by, (Chen Linhui et al., 2006).(Osterle,1962) conducted numerical analysis on free convection heat transfer for development of boundary layer between parallel isothermal vertical plates and get result for velocity, temperature and pressure variation throughout the flow field .The numerical method used is hybrid finite difference method.(Yousef et al.,1982) conducted an experimentally study of free convection heat transfer in air from isothermal horizontal surfaces heated and facing upward by using a Mach-Zehnder interferometer. The local and the average heat-transfer coefficients and the temperature distributions were determined and compared with available experimental and theoretical results.. The nature of the free convection flow over the heated surface and the separation of the boundary layer were inferred from these random changes in the local and average Nusselt numbers.(Oztop et al., 2004) conducted numerical investigation of natural convection heat transfer in a square cavity with a heated plate placed in vertical and horizontal manner. The governing equations were solved with TDMA using finite difference equation based on the finite control volume approach with non- staggered grid arrangement and SIMPLEM algorithm. Computation was done in Rayleigh number ranging from10^4 to 10^6 at different aspect ratios and position of heated plate. Air was used as a working fluid (Pr=0.71).The effect of the position and aspect ratio of heated (Vertical and horizontal) plate on heat transfer and flow in square cavity were investigated. The result showed that as Rayleigh number increases with increase in mean Nusselt number at both vertical and horizontal locations position. At higher Ra numbers, when the plate is located horizontally heat transfer is decreased about 80% less than for vertical located position.(Krishnan and Balaji,2004) conducted a synergistic approach to parameter estimation in multimode heat transfer. This paper reports the efficacy of the least square residual method in parameter estimation when more than one mode of heat transfer is encountered. (Parthasarathy and Balaji, 2008) determined multiple parameter estimation in 2d conduction using Bayesian inference. (Venugopal and Balaji, 2008) conducted experimental analysis of transient heat transfer to recall the constant c in the Nusselt number using hybrid optimization techniques.

From literature survey it can be inferred that free convection in vertical channel geometry with discrete heat source and optimization techniques used in parameter estimation has considerable attention. This geometrical configuration has physical relevance with respect to electronic chip placement. The goal of the study is to compare the value of Nusselt number constant C with standard correlation of vertical plate using least square residual method and Bayesian approach.

2. Experimental Methodology

As a first step, a validation experiments has been conducted by studying the heat transfer in an isolated heated horizontal plate. The details of the experiment setup is shown in Table 1

Table 1 Details of Experimental Set-up

SI NO	Description	Dimension/Range
1	Rectangular Horizontal SS Plate (1 no.)	250×50× 3mm
2	Rectangular Box with Thermocol (2 no.)	500 x 500x 150 mm

3	Heat Input	0-300 W
4	Single Phase Closed Type Dimmerstat	0-1 Amps
5	Ammeter	0-1 Amps
6	Multimeter	350 V_{AC}
7	K-type Thermocouple with indicator	0-1000°C

Experimental apparatus has been specially planned and formulated to carry out investigations on electronics devices. The experimental setup consist of an apparatus, temperature indicator, K –type thermocouple, Multimeter, Ammeter, and AC power supply whose schematic diagram is as shown in Figure1.

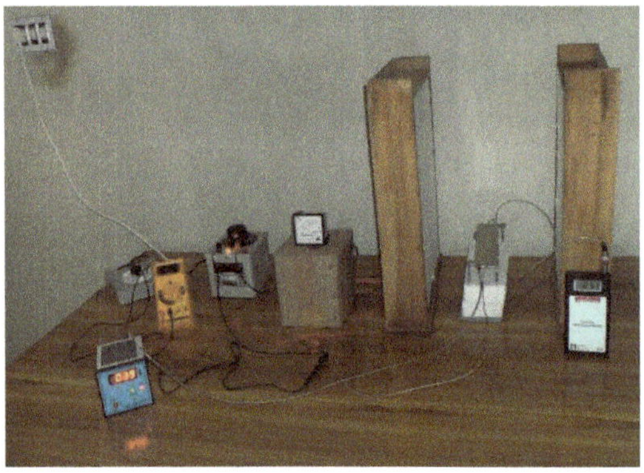

Figure 1 Photographic View of the Experimental setup

The apparatus consists of a stainless steel plate and flat heater fitted on a support structure in a horizontal fashion. The heat input to the heater is measured by an ammeter and a multimeter and is varied by a dimmerstat.Ammeter are connected in series manner and multimeter are connected in Parallel manner. The two rectangular box are kept as adiabatic walls which is filled with thermocol to avoid the external disturbance on both sides and only heat is supplied to the horizontal plate Temperatures is measured using K-type thermocouple. Thermocouples is fixed in between the flat heater and stainless steel plate with the help of blind holes by using the nuts and bolt on the support structure and taken out of the plate. The K type thermocouples are connected to the temperature indicator to measure the temperature of the plate. The stainless steel plate has been polished to minimize the radiation losses. As it could be seen there are 2 holes on the plate which house the screws that are used in fastening the plate and the heater together. The parameter varied during the experimentation is heat input to the heater. The whole assembly was highly polished on its outer surface to obtain an emissivity of 0.17

2.1 Procedure for the least Square Residual Optimization Method and optimization method

The experimentation consists of plate heater assembly which was given definite amount of power supply that was switched off after steady state was reached. Temperature measurements were done at regular interval of five minute. The experiment was carried out under controlled condition; simulating natural convection .

In this method, typically all the properties of the system will be known beforehand or a priori and the temperature response of the system will be sought. However, in this case the experimentally measured temperature response is available and to estimate or retrieve the value of C.

3. Results and Discussion

3.1 Least Square Residual Method

The Yousef et al correlation for Horizontal plate facing upward losing heat by free convection is

$$Nu = C\,Ra^{n} = 0.62 Ra^{0.25}$$

The constant C in the above relation was determined by least square residual method and Bayesian Approach.

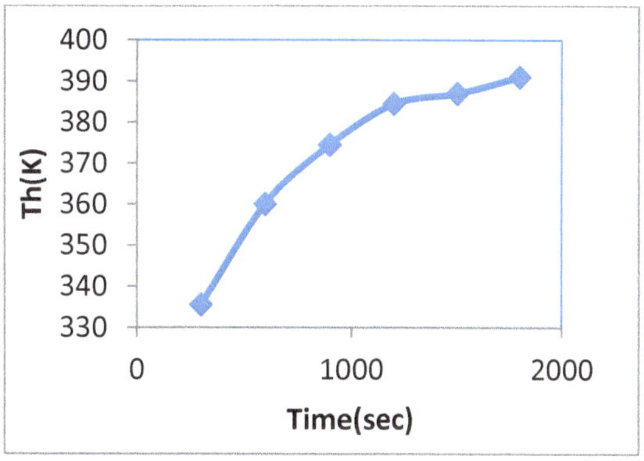

Figure 2 Plate temperature Vs Time

Figure 2 shows the Variation of Plate temperature with Time. Time increases with increase in plate temperature due to the lattice vibration of molecules between plate and heater.

Assumptions

- There is no heat loss from/to the stainless steel plate .
- The temperature of the enclosure remains constant throughout,
- The properties of the SS plate do not change with temperature
- The foil is spatially isothermal(lumped capacitance formulation)

For the above assumptions, the heating of the SS Plate can be mathematically represented as

$$E_{stored} = E_{lost} + E_{gen}$$

$$mc_p \frac{dT}{dt} = -hA(T - T_a) \qquad (1)$$

Equation (1), the left hand side represent the rate of change of enthalpy and the right hand side represents the heat transfer by convection. For substituting Nusselt number for h, Equation(1) becomes

$$\rho v C_p \frac{dT}{dt} = \frac{-Nu.K}{L}.A(T - T_a)$$

$$\rho v C_p \frac{dT}{dt} = \frac{-CRa^{0.25}KA}{L}(T - T_{a)}$$

$$\frac{dT}{dt} = \frac{-CRa^{0.25}KA}{L\rho v C_p}(T - T_a)$$

Rayleigh Number, $Ra = \frac{g\beta\Delta TL^3}{\gamma\alpha}$

$$\frac{dT}{T - T_a} = \frac{-CRa^{0.25}KA}{L\rho v C_p}dt$$

Initial Condition T=T_i at t=0

$$\int_{T_I}^{T}\frac{dT}{T - T_a} = \frac{-CRa^{0.25}KA}{L\rho v C_p}\int_{0}^{t}dt$$

$$\ln\left(\frac{T - T_a}{T - T_a}\right)_{T_i}^{T} = \frac{-CRa^{0.25}KAt}{LC_p\rho v}$$

$$\ln\left(\frac{T - T_a}{T - T_a}\right) - \ln\left(\frac{T_i - T_a}{T_i + T_a}\right) = \frac{-CRa^{.0.25}KAt}{LC_p\rho v} \quad (2)$$

One possibility of solving the inverse problem is to substitute various values of C in Equation(2) and determine the temperature T_i at various time instants given in the problem. With these following can be calculated

$$S(C)= \sum_{i=1}^{N}\left(T_{\exp,i} - T_{calc,i}\right)^2$$

Upon doing such an exercise for C ranging from 0.53-0.67 in the steps of 0.1. From the plot we can at the best, say that 0.53<C<0.67.The residual are plotted against the constant C and are shown in Figure 3.

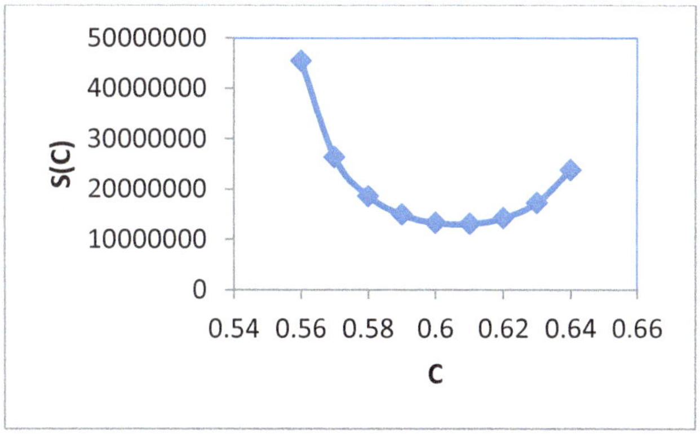

Figure 3 Residual Vs Constant C

By locally fitting a Lagragian interpolation polynomial for S(C), by employing three values of C where the residuals appear to be heading towards a minimum.These happens to be 0.57,0.6and 0.62.

$$S=\frac{(C-0.6)(C-0.62)}{(0.57-0.6)(0.57-0.62)}\times 26422412.04+\frac{(C-0..57)(C-0.62)}{(0.6-0.57)(0.6-0.62)}\times 13278078.79+\frac{(C-0..57)(C-0.6)}{(0.62-0.57)(0.62-0.6)}\times 14194282.9$$

$$S=9679092938\mathbf{C}^2-11762683179\mathbf{C}+3586414529$$

Take $\frac{ds}{dC}$ and equate it to zero to make S Stationary

$$\frac{ds}{dC}=19358185876C-11762683179=0$$

C=0.608

Therefore the best estimate of C with the level of computational intensity is 0.608.

3.2 Bayesian Approach

Bayesian inference is based on Bayes conditionality probability theorem and employs probability to characterize all forms of uncertainty in the problem. The Bayes theorem to relate the experimental data Y and the parameter is

$$P\left(\frac{x}{Y}\right)=\frac{P\left(\frac{Y}{x}\right)P(x)}{P(Y)}=\frac{P\left(\frac{Y}{x}\right)P(x)}{\int P\left(\frac{Y}{x}\right)P(x)dx} \qquad (3)$$

Where P(x/Y) is called the posterior probability density function (PPDF) for which the effect is Y and x is the cause, P(Y/x) is the likelihood density function(x) is the prior density function and P(Y) is the normalizing constant. In equation (3), the first term in the RHS represent the probability of getting Y for an assumed value of x. This can be obtained from a solution to the least square residual method problem for an assumed x and convert the $S = \sum_{i=1}^{N} \left(Y_{\exp,i} - Y_{sim,i}\right)^2$ into a PDF. The P(x) is prior knowledge belief about even before the measurements are made .Bayesian inference also helps us to inject a prior if it is known already from previous knowledge and this will hasten the retrieval and also helps us in obtaining sharper PPDFs signifying lower standard deviation of the estimates.

Steps involved in Bayesian approach

The Bayesian method to solve an inverse problem involves three steps

- Experimental data collection(The data is collected in the form of Temperature)
- Modeling Likelihood and priori
- Estimation of x

The first step is done by conducting experiments.in so far as the likelihood is concerned,we exploit the idea of measurement error in temperature as follow

$$Y_{measured} = Y_{simulated} + \omega$$

ω is a random variable from a normal distribution with mean"0" and standard deviation σ ,where σ is the standard deviation of the measuring instrument. Assuming that the uncertainty ω follows a normal or Gaussian distribution, the likelihood can be modeled as

$$P\left(\frac{Y}{x}\right) = \frac{1}{\left(\sqrt{2\Pi\sigma^2}\right)^n} \exp\left(\frac{\left(T - F(x)^T\right)\left(Y - F(x)\right)}{2\sigma^2}\right)$$

Where Y is a vector of dimension n i.e., n measurement are available and the F(x) is the solution to the least square residual method with the parameter vector x.

$$P\left(\frac{Y}{x}\right) = \frac{1}{\left(\sqrt{2\Pi\sigma^2}\right)^n} \exp\left(\frac{\left(-\chi^2\right)}{2}\right)$$

Where N is the total number of samples in one single experiment λ^2 .

Where $\chi^2 = \sum_{i=1}^{n} \frac{\left(Y_{meas,i} - Y_{sim,i}\right)^2}{\sigma^2}$

Where $Y_{sim,i}$ are the simulated values of Y for an assumed x.

The posterior PDF(PPDF) then becomes

$$P\left(\frac{x}{Y}\right)=\frac{\left[\frac{1}{\left(\sqrt{2\Pi\sigma^2}\right)^n}\exp\left(\frac{-\chi^2}{2}\right)\right]P(x)}{\int\left[\frac{1}{\left(\sqrt{2\Pi\sigma^2}\right)^n}\exp\left(\frac{-\chi^2}{2}\right)\right]P(x)dx}$$

The prior probability density(x) typically follows a uniform, normal or log normal distribution. In this case of a uniform prior,P(x) is the same for all values of x,i.e,we have absolutely no selective preference. Such a prior is called a non- informative prior.

Consider P(x) follows a normal distribution with mean μ_p and standard deviation σ_p.Mathematically P(x) is given by

$$P(x)=\frac{1}{\left(\sqrt{2\Pi\sigma^2}\right)^n}\exp\left(\frac{\left(-\left(x-\mu_p\right)^2\right)}{2\sigma_p^2}\right)$$

Hence the PPDF becomes

$$P\left(\frac{x}{Y}\right)=\frac{\frac{1}{(2\Pi)^{\frac{n+1}{2}}(\sigma^n\sigma_p)}\exp(-)\left[\frac{\lambda^2}{2}+\frac{(x-\mu)^2}{2\sigma_p^2}\right]}{\left[\int\frac{1}{(2\Pi)^{\frac{n+1}{2}}(\sigma^n\sigma_p)}\exp(-)\left[\frac{\chi^2}{2}+\frac{(x-\mu)^2}{2\sigma_p^2}\right]\right]dx}$$

Therefore for every assumed value of the data vector $X(x_1,x_2,x_3,\ldots\ldots\ldots\ldots x_n)$,P(x/Y) can be worked out.

Therefore,

$$P\left(\frac{x}{Y}\right)=\frac{\exp(-)\left[\frac{\lambda^2}{2}+\frac{(x-\mu)^2}{2\sigma_p^2}\right]}{\left[\int\exp(-)\left[\frac{\chi^2}{2}+\frac{(x-\mu)^2}{2\sigma_p^2}\right]\right]dx}$$

The mean estimate of x then becomes

$$\bar{x}=\frac{\int x\exp(-)\left[\frac{\lambda^2}{2}+\frac{(x-\mu)^2}{2\sigma_p^2}\right]dx}{\left[\int\exp(-)\left[\frac{\chi^2}{2}+\frac{(x-\mu)^2}{2\sigma_p^2}\right]\right]dx}$$

Often the integral is replaced by a summation when only discrete values of x are used,

$$\bar{x} = \frac{\sum_i x_i \exp(-)\left[\frac{\lambda^2}{2} + \frac{(x-\mu)^2}{2\sigma_p^2}\right]\Delta x_i}{\sum\left[\exp(-)\left[\frac{\lambda^2}{2} + \frac{(x-\mu)^2}{2\sigma_p^2}\right]\right]\Delta x_i}$$

$$\bar{x} = \frac{\sum_i x_i \exp(-)\left[\frac{\lambda^2}{2} + \frac{(x-\mu)^2}{2\sigma_p^2}\right]}{\sum\left[\exp(-)\left[\frac{\lambda^2}{2} + \frac{(x-\mu)^2}{2\sigma_p^2}\right]\right]}$$

$$\sigma_x^2 = \frac{\sum_i (x_i - \bar{x})\exp(-)\left[\frac{\lambda^2}{2} + \frac{(x-\mu)^2}{2\sigma_p^2}\right]}{\sum\left[\exp(-)\left[\frac{\lambda^2}{2} + \frac{(x-\mu)^2}{2\sigma_p^2}\right]\right]} \quad (4)$$

Now we can use this framework to estimate the constant C in the Nusslet number.in Equation(4) is the standard deviation of the estimated parameter, which is very diagnostic of the potency of the estimation process.Table 2 shows the Mean and Standard Deviation for single parameter using Bayesian method(With and Without Priori).Figure 4 and Figure 5 shows the typical PPDF for the constant C in the Nusslet number correlation.Table 5 shows the comparision of optimization result with standard correlation.

Table 2 Mean and Standard Deviation for single parameter using Bayesian method

	Bayesian Approach	
	Without Prior	With Prior
Mean (C)	0.608167	0.60770467
Standard Deviation	0.012452	0.00479451

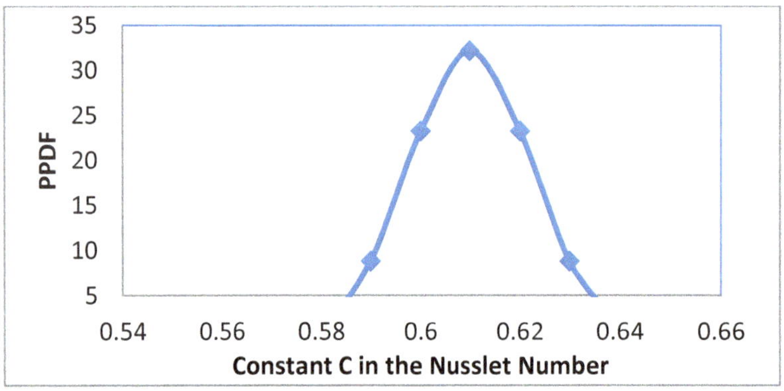

Figure 4 PPDF of Constant C in the Nusslet Number using Bayesian Method

(Without Prior)

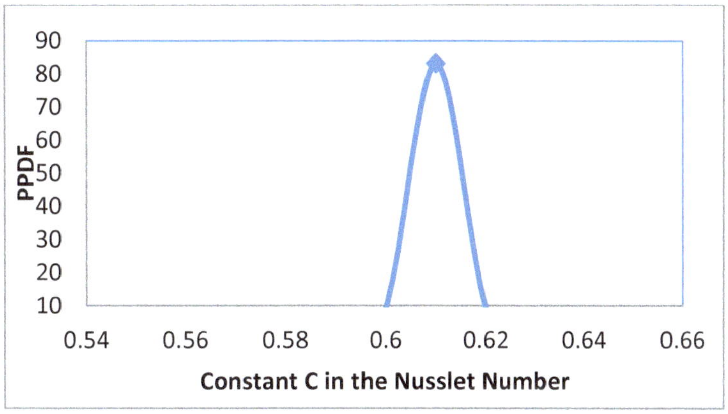

Figure 5 PPDF of Constant C in the Nusslet Number using Bayesian Method

(With Prior)

Table 3 Comparison of Optimization Results with Yousef Correlation

SI No.	Correlation	C
1	Yousef Correlation	0.62
2	Least square residual Optimization Method	0.61
3	Bayesian Approach	0.61

4. Conclusion

Steady heat transfer laminar natural convection experiment in horizontal heated plate facing upward have been done for a highly polished stainless steel plate to retrieve single parameter (constant value C in the nusslet number) by using least square residual method and Bayesian appproach and compared with the value available from yousef correlation and found to be in good agreement.

Nomenclature

A- Surface Area

C- Constant in the Nusslet Number

C_p- Specific Heat at Constant Pressure

K- Thermal Conductivity of the Fluid

L- Characteristic Dimension of the Surface

m- Mass of the Material

Nu- Nusslet Number

ρ - Density of Fluid

S(c)- Residual

T - Plate Temperature

T_a- Ambient Temperature

Ti- Instantaneous Temperature

Greek Symbols

σ - Stefan-Botzmann Constant

γ - Kinematic Viscosity of the Fluid

β -Coefficient of Volumetric Expansion of the Fluid

α - Thermal Diffusivity of the Fluid

σ_p - Standard Deviation

μ_p - Mean

References

W. Elenbaas,1942, Heat dissipation of parallel plates by free convection, *Physica* ,Vol.9, pp 1-28.

R.A. Wirtz, R.J.Stutzman,1982, Experiments on free convection between vertical plates with symmetric heating , *ASME , Journal of heat transfer*,Vol. 104, pp 501-507.

Chen Linhui,Tian Huaizhang, Li Yanzhong,Zhang Dongbin,2006, Experimental study on natural convective heat transfer from a vertical plate with discrete heat sources mounted on the back ,*Energy conservation and management*, Vol. 47,pp.3447-3455.

Osterle,1962, The development of free convection between heated vertical plates, *ASME Journal of Heat Transfer* ,Vol.84,pp. 40–44.

W.W.Yousef, J.D.Tarasuk,W.J.P McKeen,1982, Free convection heat transfer from upward facing isothermal horizontal surfaces ", *Journal of Heat Transfer*, Vol. 104,pp. 493-500.

H.F.Oztop, I.Dagtekin,A.P Bahoul,2004,Comparision of position of a heated thin plate located in a cavity for natural convection,*International Communication of Heat Mass Transfer* ,Vol. 31, pp.121-132.

A.S. Krishnan, C.Balaji,S.P. Venkateshan,2004, A synergistic approach to parameter estimation in multimode heat transfer,*International Communication 0f Heat Mass Transfer*,Vol.30,No.4,pp.515-524.

G Venugopal, C Balaji,S P Venkateshan, 2008,A Correlation For Laminar Mixed Convection From Vertical Plates Using Transient Experiment,*Journal of Heat Mass Transfer*,Vol.44., Pages 1417-1425.

S Parthasarathy,C.Balaji,2008,Estimation of parameters in Multimode heat transfer problems using Bayesian Inference-Effect of Noise and a priori, *International Journal of Heat and Mass Transfer*,Vol.51,pp.2313-2334.

OPTIMIZATION OF PROCESS PARAMETER IN TITANIUM ALLOY BY USING GENETIC ALGORITHM

R. Arularasan[1], K.Sivasakthivel[2], M.Alagesan[3], R.Anand[4], K. Arshad Ahmed[5], S.Vimalan[6]

[1,2] Faculty,[3,4,5,6] Students Dept. of Mechanical Engineering, University College of Engineering Arni, Thatchur, Tiruvannamali District- 632326, India

E-Mail: vrarasan@gamil.com[1], ksivasakthivel73@gmail.com[2].

Abstract

Turning is one of the very important machining operations in engineering industries. Optimization of turning processes parameters still remains as one of the most challenging problems because of its high complexity and non-linearity. Hence, there is a need to apply most powerful optimization techniques to get desired accuracy of optimum solution. In this paper, non-conventional optimization techniques, genetic algorithm (GA) results were compared with Taguchi Optimization technique. The process variables considered for optimization are speed, feed, and depth of cut. The objective considered in the present work is surface finish subjected to the constraints.

Keywords: Turning, Optimization, Process parameters, Taguchi Analysis, Genetic algorithm.

1. Introduction

Process parameters composed of cutting speed, feed and depth of cut (for turning operation), have essential effects on the machining productivity and cost. The selection of cutting parameters has long depended on the skills and experience of machine tool operators or handbooks, and conservative cutting parameters are usually selected. This situation would cause significant productivity loses and lead to a costly machining operation. The determination of optimum cutting parameters is a combinatorial optimization problem and is usually realized by applying optimization algorithms. These algorithms include neural network, geometric programming, simulated annealing, genetic algorithm (GA), particle swarm optimization (PSO) etc. GA was considered as a suitable algorithm for solving any type of machining process optimization problem. In this paper, Process parameters optimization by using GA and Taguchi Method were discussed comprehensively.

2. LITERATURE REVIEW

Tarng. Y.S , S.C. Juang and C.H. Chang [2] proposes the use of grey-based Taguchi methods for the optimization of the Submerged Arc Welding (SAW) process parameters in hard facing with considerations of multiple weld qualities. In this new approach, the grey relational analysis is adopted to solve the SAW process with multiple weld qualities. A grey relational grade obtained from the grey relational analysis is used as the performance characteristic in the Taguchi method.

Vijayan. P and V. P. Arunachalam[3] reported research in their work Taguchi off-line quality control method applied for determines the optimal process parameters which maximize the mechanical properties of squeeze cast LM24 aluminum alloy. For this purpose, concepts like orthogonal array, S/N ratio and ANOVA were employed.

Anshuman Sahu and Rudrajit Tapadar attempts to solve the generalized Assignment problem through genetic algorithm and simulated annealing. The generalized assignment problem is basically the N men N jobs problem where a single job can be assigned to only one person in such a way that the overall cost of assignment is minimized. While solving this problem through genetic algorithm (GA), a unique encoding scheme is used together with Partially Matched Crossover (PMX).

3. EXPERIMENT DETAILS

Work Piece & Cutting Tool:

The work piece used here is Titanium Alloy grade (II) because it's high strength and low density[1]. We are using carbide cutting tool for machining titanium CNMG PR-3215.

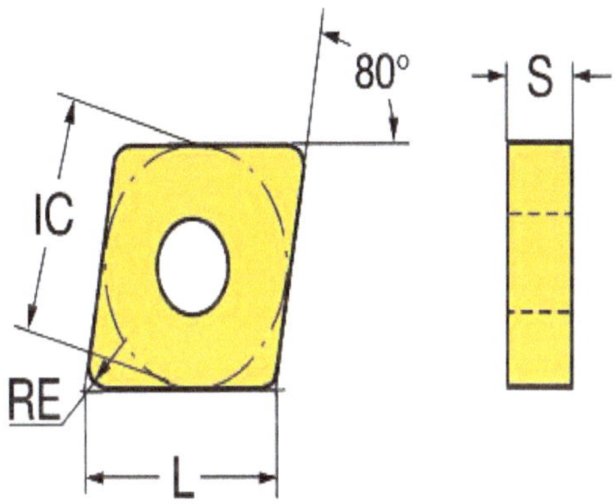

Fig.1: Insert geometry

Fig.2: CNC Machining

Tablel Cutting parameters

Parameter	Level 1	Level 2	Level 3
Cutting	60	70	80

Speed(m/min)			
Feed (mm/rev)	0.05	0.1	0.15
Depth Of Cut (mm)	0.2	0.4	0.6

4. TAGUCHI METHOD

The Taguchi method is a well-known technique that provides a systematic and efficient methodology for process optimization and this is a powerful tool for the design of high quality systems. Taguchi approach to design of experiments (DOE) is easy to adopt and apply for users with limited knowledge of statistics, hence gained wide popularity in the engineering and scientific community. This is an engineering methodology for obtaining product and process condition, which are minimally sensitive to the various causes of variation, and which produce high-quality products with low development and manufacturing costs. Signal to noise ratio and orthogonal array are two major tools used in the design.

The S/N ratio characteristics can be divided into three categories when the characteristic is continuous

a) Nominal is the best

b) Smaller the better

c) Larger is better characteristics.

For the minimum Cutting Force, the solution is "Smaller is better" and S/N ratio is determined according to the Following equation

Table 2. S/N Ratio

Cutting Speed	Depth Of Cut	Feed	Surface Roughness	S/N Ratio
60	0.2	0.05	6.415	-16.1439
60	0.4	0.10	1.804	-5.1247
60	0.6	0.15	4.594	-13.2457
70	0.2	0.10	2.714	-6.7452
70	0.4	0.15	2.946	-9.3847
70	0.6	0.05	2.145	-6.6285
80	0.2	0.15	3.632	-11.2029
80	0.4	0.05	3.084	-9.7823
80	0.6	0.10	2.298	-7.2270

ANOVA Table:

Analysis of variance (ANOVA) was introduced by Sir Ronald Fisher. This analysis was carried out for a level of significance of 5%, i.e., for 95% level of confidence. The purpose of ANOVA is to investigate which turning parameter significantly affects the performance characteristics.

Table 3. ANOVA Table

Source	Df	Ss	Ms	F-Value	P-Value	C(%)
Cutting Speed	2	5.36	2.68	1.93	0.34	30.45
Depth Of Cut	2	3.24	1.71	1.23	0.44	19.60
Feed	2	5.89	2.94	2.12	0.32	33.75
Error	2	2.78	1.39			15.73
Total	8	17.46				

According to the data obtained from the ANOVA table the following were place major and minor contribution on the surface roughness of the work piece. Here cutting speed contribution on surface is 30.40% of the total contribution, depth of cut 19.6% of total contribution and feed plays a most important role on surface roughness in the range of 33.75% of total contribution.

5. GENETIC ALGORITHM

Genetic algorithms are robust and adaptive methods, successfully used for solving optimization problems. They are powerful tools for the optimization of functions that can more easily locate the global optimum.

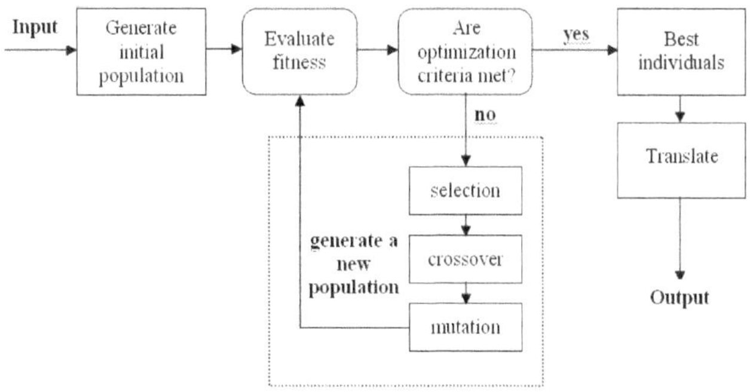

Fig.3: Genetic Algorithm procedure

The reason lies in the fact that GA seeks an optimal solution in the space of solutions, starting from groups of points, rather than a starting point. GA use only the objective function to search optimal solutions [4]. Genetic algorithm is stopped after a certain number of generations or after a certain time interval, after which the quality of the best individuals is tested. If the result is not acceptable, we can start again to search for new (better) solutions [6].

Objective function,

C = 800+8088.64/(x(1)*x(2))+6.332*10^-7*x(1)^4.55*x(2)^.67;

Constraint functions,

$V_c f^{0.3} \leq 146.65$

$V_c f^{0.75} \leq 849.15$

$F^{0.75} \leq 129.45$

$F^{0.75} \leq 55.33$

$V_C \leq 5.278$

$V_C \geq 7.345$

$F_{min} \leq 0.05$

$F_{max} \geq 0.15$

Program 1:

```
function C = turning_cost(x)
C = 800+8088.64/(x(1)*x(2))+6.332*10^-   7*x(1)^4.55*x(2)^.67;
end
```

Program 2:

```
function [c, ceq] = nonlinear_constraints(x)
c = [x(1)*x(2)^0.3-146.46;
x(1)*x(2)^0.75-849.15;
x(2)^0.75-129.63];
ceq = [];
end
```

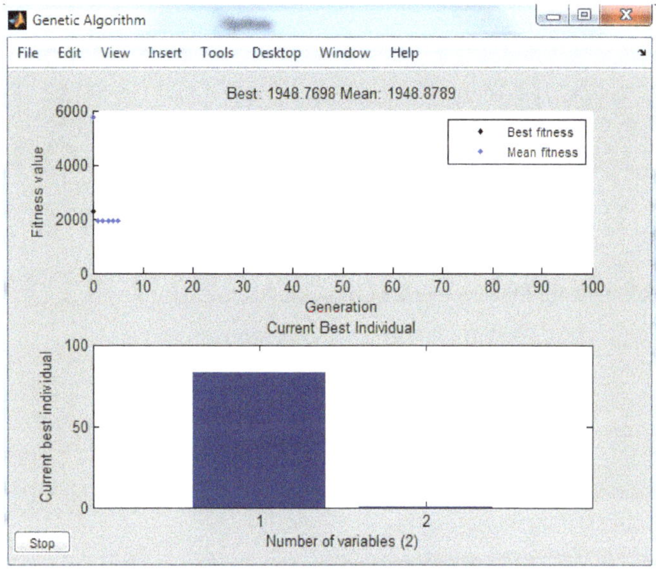

Fig.4: Genetic Algorithm plot

6. CONCLUSION

Modern methods of optimization are powerful and popular tools for solving complex engineering optimization problems. This paper shows the possibilities of using genetic algorithms for solving such problems. Cost of machining in turning process, depending on cutting speed and feed was minimized under some nonlinear constraints. The final result obtained from the GA is

Cutting speed = 83.33 rev/min Feed = 0.09 mm/rev

REFERENCE

[1] Dillibabu.R, Sivasakthivel.K , Vinoth Kumar.S Optimization of process parameters in dry and wet machining of Ti-6Al-4v ELI using Taguchi method, International journal of Design and manufacturing technology, vol-4(3), PP.15- 21,(2013)

[2] Tarng.Y.S , S.C. Juang and C.H. Chang, The use of grey-based Taguchi methods to determine submerged arc welding process parameters in hard facing. Journal of Materials Processing Technology Vol 128, Issues 1-3, 2002, pp 1-6.

[3] Vijayan. P and V. P. Arunachalam, Optimization of squeeze casting process parameters using Taguchi analysis .The International Journal of Advanced Manufacturing Technology Vol 33, Numbers 11-12, 2007, pp.1122-1127

[4] Onwubolu, G., Kumalo, T., Optimization of multi pass turning operations with genetic algorithms, International Journal of Production Research, vol. 39, no. 16, 2001, p.3727-3745

[5] Sardiñas RQ, Santana MR, Brindis EA (2006) Genetic algorithm-based multi objective optimization of cutting parameters in turning processes. Eng Appl Artif Intell 19:127–133.

[6] Dusan Petkovic, Miroslav Radovanovic 1university Of Nis, Faculty Of Mechanical Engineering, A. Medvedeva 14, 18000 Nis, Serbia Journal of Engineering Studies and Research – Volume 19 (2013) No. 1.

24

A REVIEW ON HEAT TRANSFER AND PRESSURE DROP CORRELATIONS IN SOLID CIRCULAR FINNED TUBE BUNDLES POSITIONED AT INLINE AND STAGGERED ARRANGEMENT IN CROSS FLOW

Abhishek Tripathi[1], Ajay Kumar Singh[2]

[1] *Post Graduate (PG) student, Mechanical Engineering Department, Radharaman Institute of Technology and Science, Bhopal (M.P.), India, Telephone: 09630444300, email-abhiyoho@gmail.com*

[2] *Associate professor & HOD, Mechanical Engineering Department, Radharaman Institute of Technology and Science, Bhopal (M.P.), India*

Abstract

The present review is organized in different arrangement of finned tube bundles placed on inline arrangement and staggered arrangement in cross flow. A large number of experimental and numerical works had been performed for enhancement of air-side heat transfer. A brief discussion is done on the effect of local heat transfer behaviour of circular finned tube and analysis of geometric and flow parameters included in this paper. Different parameters like fin height, fin spacing, fin thickness, tube diameter, tube spacing, effects of row and arrangement of tube bundles affect directly on the performance of solid circular finned tube. All these parameters are briefly discussed in this paper. Discussions on some important points which affect the performance of tube bundles (i.e. inline and staggered arrangement) from various authors and their problem and related issues are presented in this paper. The flow profiles and the related heat transfer characteristics in the complex geometries are still needed to be verified.

Keywords: Heat Transfer Augmentation, inline arrangement, staggered arrangement, cross flow.

1. Introduction

Finned-tube heat exchangers have been used to exchange heat between gases and liquids, which are single- or two-phase, for many years. The performance of the finned-tube heat exchanger is limited by the air side heat transfer resistance because the air side heat transfer coefficient is significantly lower than the refrigerant side heat transfer coefficient. Finned-tube heat exchangers having discrete plate fins with large fin pitches have been widely used to enhance the heat transfer performance of evaporators in refrigerators and freezers under dry and frosting conditions. Adding the fins in a heat exchanger is a very common procedure to enhance the overall heat transfer coefficient. One of the important processes in engineering is the heat exchange between flowing fluids, and many types of heat exchangers are employed in various types of installations, as power plants, petro chemical plants, building heating, ventilating, air conditioning and refrigeration (HVAC/R) systems. As far as construction design is concerned, the tubular or shell and tube type and the extended surface or finned tube type heat exchangers are widely in use. Finned-tube heat exchangers are used for the processes in which a liquid or gas is required to be either cooled or heated. Generally, a liquid flows within the tubes while gas is directed across the finned-tubes. Because of the poor thermal conductivity and thus the heat transfer coefficient of the gases, it is necessary to apply the extended surfaces on the gas side to enhance the heat transfer without losing its compactness. In addition, the problems concerned with pressure loss in flow medium must not be overlooked. A large number of experimental works has been performed for enhancement of the overall heat transfer coefficient and for enhancement of air side heat transfer; however, the flow profiles and the related heat transfer characteristics in the complex geometries are still needed to be verified. The circular finned-tube bundles are commonly used in the industries. Investigations are carried out mainly for the staggered arrangement under the cross flow conditions and numerous correlations have developed. The heat-transferring surface can be enlarged by an arrangement of annular fins or other elements. This increase of total tube surface allows transfer of a greater amount of heat from hot gas, but the demand for smaller installation sizes requires smaller fin pitch with larger fin height. On the other hand, finned tubes with segmented fins show somewhat higher turbulence than those with smooth fins since the boundary layer has to be established at each individual segment [1].An excellent overview of plate fin and circular finned tube studies is presented. Factors such as fin spacing, fin efficiency and fin configuration yield enhanced heat transfer. Yet one finned tube heat exchanger study claims to maximize the heat transfer rate while simultaneously minimizing pumping power [2].The heat transfer in a finned-tube heat exchanger is a conjugate problem. Conjugate heat transfer means computing more than one mode of the heat transfer simultaneously and it can be established efficiently by the way of numerical means. For a finned-tube heat exchanger, when the convection effect is intended to calculate for the fluid through the bundle, the conduction in the fins has to be considered as well [3].The fin efficiency of serrated fins is calculated using an analytical model, with the assumption of a heat transfer coefficient uniformly distributed over the fin surface and the segmented section, as well as a second equation for theoretical fin efficiency [4]. Investigation on seven different segmented finned tube bundle arrangements in cross flow. They ascertained that, compared to solid fins, the fin height of segmented fins can be greater than the maximum fin height because such fins are easier to laser-weld. This would increase the total outside surface area and as a result the efficiency of the heat exchanger [5]. Investigated experimentally local heat transfer behaviour in staggered and in-line arrangements with the help of optical adaptation of the naphthalene sublimation technique in order to evaluate the analytical fin efficiency and also gave the expression for the pressure drop of in-line and staggered arrangement [6].Experimentally studied the influence of rows in cross-flow heat exchangers. They compared their results with theoretical correlations. All measurements were performed in the Reynolds range between 10000 and 30000.They also present evaluation of heat transfer and pressure drop predictive method for tube bundles with stud fins [7].

2. LITERATURE SURVEY

Brauer [8] had done his investigation on the total surface area of the finned-tube bundle and the heat transfer coefficient *h* were closely linked and governed by the layout and the form of fins and tubes. Therefore, it is important to ensure that an enlargement of the fin surface area must be implemented without causing the decrease in heat transfer coefficient. It concluded that the fin surface area can be augmented by increasing the fin height and the number of fins per meter.

Zozulya et al. [9] had done his work on Effect of Flow Turbulization on Heat Transfer in a Finned Tube Bundle and concluded that the flow mode should be treated as a temperature dependent problem because the flow variables directly depend on the temperature of air. Furthermore, the heat transfer from a bundle (especially in the first row) depends on the conditions of turbulence intensity occurred at the bundle inlet.

Sparrow and Chastain [10] investigated on a single circular fin by mass transfer method. They measured the variations of the angle of attack by a thin film of a naphthalene sensor. The authors concluded that the overall heat transfer performance of a circular fin was not significantly affected by small angles of attack. However, the measurement of the heat transfer coefficient at only one side may yield erroneous results.

Neal and Hitchcock [11] carried out a comprehensive study on the local heat transfer and airflow occurring within a circular finned-tube bank in staggered tube arrangement. Three different tube spacing were examined and instrumented tubes were installed in row two and six. They found and concluded that the heat transfer upstream in the fin is considerably higher, and the heat transfer coefficient decreases at the base of the tube as the boundary layer increases.

Hu and Jacobi [12] investigated that the use of thermocouples and few sensors might be inadequate for resolving details of the flow and the heat transfer interaction and suggested that the point heating method could lead to serious errors. The author concluded that the total heating method, the thermal boundary layer will not be developed at the fin tips where both (velocity and thermal) boundary layers should have to develop. Instead, thermal boundary layer only will develop when the airflow reach to the heated area and According to this dependency on the heating models, the heat transfer distribution patterns over the fin will be dissimilar.

Watel et al. [13] had done his work on the influence of the fin spacing on the convective heat transfer from a rotating circular finned-tube in transverse airflow. The fin cooling was monitored by aid of the infrared thermography and the flow field in the mid-plane between two fins was obtained by Particle Image Velocimetry (PIV) device. Their results were compared and validated with various authors' experimental works and concluded that the heat transfer coefficients over the circumference were more uniformly distributed on a tube in the bundle rather than on a single tube.

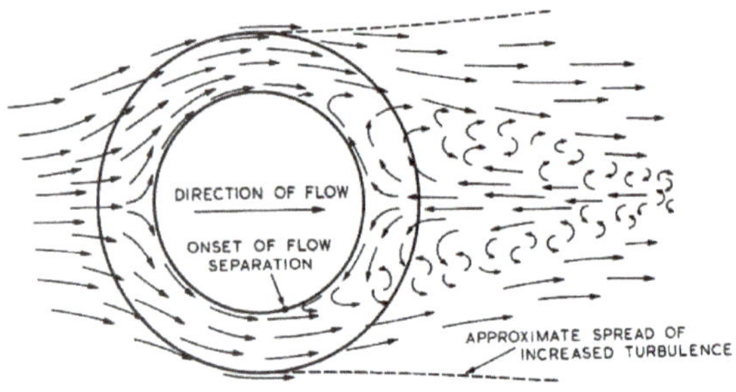

Fig -1 Schematic view of general flow pattern over a row of staggered arrangement tube bundle

Fig -2 Cross-flow circular finned-tube heat exchanger

Yudin et al. [14] had done his investigation on experimental correlation data on convective heat transfer in cross flow over finned bundles with different fins and these fins are transverse, spiral and circumferential fins and observed and concluded that an increase of the fin height of the staggered tube arrangement provides the decrease of heat transfer coefficient and the increase of pressure drop.

Zukauskas et al. [15] investigated that the highest value of the average heat transfer coefficients along the fin height, are found to be narrowest near the flow passage in the bundle exists. It can be explained that due to the construction in the flow passage, the flow velocity reaches maximum and causes a high local heat transfer. At ($\theta = 0°$) and ($\theta = 180°$) the relative heat transfer coefficient, i.e. the ratio of local heat transfer coefficient to the averaged heat transfer coefficient over a fin height, attained its maximum at the fin tip. At ($\theta = 90°$), the relative maximum heat transfer occurred in the middle of the fin height. Author concluded in his work is that the nature of the local heat transfer distribution over the fin height was changed according to the angle of fin surface.

Rabas et al. [16] had investigated that the fin density has much stronger impact on the performance for the larger diameter of tube (d = 31.75 mm). They examined for the low fins, which is less than 6.35 mm at the range of $1000 \leq Re \leq 25000$. He also gave his results for the longitudinal pitch effect to the heat transfer performance. Authors concluded that the thermal performance is almost independent of the Reynolds number for the larger fin density (0.98 fins / mm).

Mirkovic [17] investigated the heat transfer and pressure drop in an eight-row deep staggered tube bundle for the two tube diameters 38.1mm and 50.8 mm with constant transverse and longitudinal tube pitches. The tube diameter only was changed while other parameters such as the fin height and fin spacing were kept constant in his investigation. Author concluded that when the tube diameter increases, the wake region behind the tube will increase and the air-side pressure drop will rise.

Torikoshi et al. [18] investigated numerically the tube diameter effect on heat transfer and flow behaviour for a two-row staggered arrangement of plate finned-tubes. In contrast to the circular finned-tubes, the plate fin surface is decreased when the tube diameter increases. He observed no significant variation of the average heat transfer coefficient. Author concluded that the tube diameter effect may be largely ignored for the cases where the diameter is changed only slightly.

Rabas and Taborek [19] studied on the rows effect of low finned-tube bundles and presented a shallow bundle correction factor. It means that the factor increased with the number of rows for low fin density (0.393 fins/ mm) but decreased for high fin density (0.984 fins/ mm). For a high fin density tube bank in a staggered array, some performance characteristics similar to those of the in-line bank are noted. In his study he concluded that the heat transfer performances of shallow in-line tube banks always decreased with row number regardless of the fin density.

F.Halici and I. Taymaz [20] they provide experimental data for the study of air side performance of tube row spacing in finned tube. They also provided an extensive reference database for heat transfer and pressure drop of finned-tube bundles in cross flow. Their work covers 21 different seven-row finned-tube bundles. The heat transfer correlations were obtained from their own data and the experimental data for the bundles scatter over the range of ± 14 %. They concluded that the effect of bundle configuration and the fin parameters are affect directly on performance of finned tube bundle.

Sparrow and Samie [21] observed that the pressure drop increases with the longitudinal pitch. It is possible that the tubes will act as a single tube if the longitudinal pitch is too wide. They showed that the longitudinal pitch effect depends on the fin density of the low finned-tube bundle. They concluded that the heat transfer performance is identical for Sl = 37.12 mm and 50.80 mm with 0.4 fins per mm. However, for the fin density of 0.98 fins per mm, the heat transfer performance lowered for a larger Sl applied.

Jacobi and Shah [22] discussed that the air flow was likely to exhibit all of possible flow features (e.g. steady or unsteady, laminar or turbulent) in a single heat exchanger. They suggested that there were still limitations to the air-side heat transfer performance and a clear understanding of airflow in the complex passages of heat exchangers was needed so that surface design can be optimized efficiently.

Rene Hofmann et al. [23] had done his work on heat transfer and pressure drop at different transverse serrated and solid finned-tubes and investigated in cross-flow with aim of optimizing heat exchanger performance. Three different finned-tube shapes were investigated. The I-shaped and U-shaped fin geometries under consideration have varying geometrical constants, i.e. fin height, fin pitch, fin thickness, and fin width. The heat exchanger consists of eight consecutive finned-tube rows and eleven tubes on top of each other. The finned tubes are arranged in a staggered formation at equal transverse and longitudinal pitch. The experimental setup, measurement technique and measurement uncertainties are presented. The design of an optimum heat exchanger must take into account the advantages and disadvantages of geometrical factors which influence heat transfer and pressure drop. After measurement validation, the derived correlations for the Nusselt number and the pressure drop coefficient were compared with experimental results and equations from literature. Authors concluded that serrated finned tubes, three ranges could be identified: $Re < 10000$, $10000 < Re < 20000$ and $Re > 20000$. Despite varying fin height and fin pitch in the two different geometries, no substantially differing tendencies could be observed when applying these equations.

3. CONCLUSION

All studies verified that the heat transfer coefficient around the fin, and from row-to-row vary in accordance with the bundle depth. It is useful to note that only limited results with a single tube and very few rows are found, and further studies applying four and more tube row bundles should follow. On the other hand, some studies have done to resolve this situation by developing the row correction factors. A considerable amount of related data on the local and average heat transfer and the pressure drop were established and qualitative judgements on circular finned tube configurations are rendered. Despite of these earlier developments, this review indicated that further concentration on the existing problems in designing of optimum fin geometry and tube arrangements are still necessary. Some conclusion points are,

• Considerably more information on numerical simulation has been published for plate fins than circular fin tubes.

• The heat transfer coefficient and flow distribution over a tube in the bundle is different to a single tube.

• Temperature distributions over the fin surface and the flow structures between fins are of complex pattern. When the need arises to measure such effects accurately, it is an experimentally difficult task to do without disturbing the heat transfer behaviour on a fin surface. Therefore, more precise data on the local behaviour are necessary.

• Different results came out of the relevant information regarding the tube spacing adjustments and number of rows.

• Since most correlations were based on their own data, authors gave different formula for the heat transfer and pressure drop correlations. In addition, the characteristic dimension to define Reynolds number was dissimilar. Thus, it is fairly anticipated that to compare directly to experimental correlations is found to be difficult.

• Finally, all related works for the circular finned-tubes have been correlated experimental ones and respective correlations have not been verified yet under numerical simulations. Therefore, additional numerical data are needed in order to establish improved correlations.

4. FUTURE SCOPE

More work on the existing problems in designing of optimum fin geometry and fin spacing and tube arrangement are still necessary. A lot of works are still to do on these parameters and this area.

References

[1] R.G. Carranza, Spine Fin Efficiency a Three Sided Pyramidal Fin of Scalene Triangular Cross- Sectional Area: Analytical Solution, *5th WSEAS Int. Conf. on Heat and Mass transfer (HMT'08), Acapulco, Mexico, January 25-27, 2008.*

[2]. Webb, R. L., "Air-Side Heat Transfer in Finned Tube Heat Exchanger," *Heat Transfer Engineering*, Vol. 1, No. 3, pp. 33-49, 1980.

[3]. Fiebig, M., Gorsse-Georgemann, A., Chen, Y., and Mitra, N. K., "Conjugate Heat Transfer of a Finned Tube Part A: Heat Transfer Behaviour and Occurrence of Heat Transfer Reversal," *Numerical Heat transfer, Part A*, Vol. 28, pp.133-146, 1995.

[4] K. Hashizume et al, Fin efficiency of serrated fins, *Heat Transfer Engineering*, Vol.23, 2002, pp. 6-14.

[5] T.J. Rabas, P.W. Eckels, Heat transfer and pressure drop performance of segmented extended surface tube bundles, *AIChE-ASME Heat Transfer Conference, San Francisco, California, Aug.11-13, 1975, pp. 1-8.*

[6]. Kearney, S. P., and Jacobi, A. M., "Local Convective Behaviour and Fin Efficiency in Shallow Banks of Inline and Staggered, Annularly Finned Tubes," *ASME Journal of Heat Transfer*, Vol. 118, pp.317-326, 1996.

[7] H. Necula, E. Nineta, G. Darie, An experimental analysis of heat transfer for a tube bundle in a transversally flowing air, *WSEAS Transactions on Heat and Mass Transfer,* Issue 3, Vol.1, 2006, pp. 349-355.

[8]. Brauer, H., "Compact Heat Exchangers," *Chemical Progress Engineering*, London. Vol. 45, No. 8,p p.451-460,1964.

[9]. Zozulya, N. V., Vorob'yev, Y. P., and Khavin, A. A., "Effect of Flow Turbulization on Heat Transfer in a Finned Tube Bundle," *Heat Transfer Soviet Research*, Vol. 5, No. 1, pp. 154-156, 1973.

[10]. Sparrow, E. M., and Chastain, S. R., "Effect of Angle of Attack on the Heat Transfer Coefficient for an Annular Fin," *International Journal of Heat Mass Transfer*, Vol. 29, pp.1185 1191, 1986.

[11]. Neal, S. B H. C., and Hitchcock, J. A., "A Study of the Heat Transfer Process in Banks of Finned Tube in Cross Flow, Using a Large Scale Model Technique," *Proceeding of the Third International Heat Transfer Conference*, Vol. 3, Chicago, IL, pp. 290-298, 1966.

[12]. Hu, X., and Jacobi, A.M., "Local Heat Transfer Behaviour and its Impact on a Single-Row Annularly Finned Tube Heat Exchanger," *ASME Journal of Heat transfer*, Vol. 115, pp. 66-74, 1993.

[13]. Watel, B., Harmand, S., and Desmet, B., "Experimental Study of Convective Heat Transfer from a Rotating Finned Tube in Transverse Air Flow," *Experiments in Fluids,* Vol. 29, No. 1, pp. 79-90, 2000.

[14]. Yudin, V F., Tokhtarova, L. S. Lokshin, V., and Tulin, S. N., "Correlation of Experimental Data on Convective Heat Transfer in Cross Flow Over Bundles with Transverse Spiral and Circumferential Fins," *Trudy TsKTI*, No. 82, 1968.

[15]. Zhukauskas, A., Stasiulevičius, J., and Skrinska, A., "Experimental Investigation of Efficiency of Heat Transfer of a Fin Tube with Spiral Fin in Cross Flow," *Third International Heat Transfer Conference*, Chicago, Vol. 3, pp. 299-305, 1966.

[16]. Rabas, T. J., Eckels, P. W., and Sabatino, R. A., "The Effect of Fin Density on the Heat Transfer and Pressure Drop Performance of Low-Finned Tube Banks," *Chemical Engineering Communications*, Vol. 10, No.1-3, pp. 127-147, 1981.

[17].Mirkovic, Z., *Heat Transfer and Flow Resistance Correlation for Helically Finned and Staggered Tube Banks in Cross Flow,* Heat Exchangers: Design and Theory Source Book, (edited by N. H. Afgan and E. U. Schlünder), Hemisphere, Washington, D. C, pp. 559-584, 1974.

[18]. Torikoshi, K., and Xi, G. N., "A Numerical Steady of Flow and Thermal Fields in Finned Tube Heat Exchangers (Effect of the Tube Diameter)," *IMECE Proceedings of the ASME Heat Transfer Division*, HTD-Vol. 317-1, pp. 453-457, 1995.

[19]. Rabas, T. J., and Taborek, J., "Survey of Turbulent Forced-Convection Heat Transfer and Pressure Drop Characteristics of Low-Finned Tube Banks in Cross Flow," *Heat Transfer Engineering*, Vol. 8, No. 2, pp. 49-62, 1987.

[20] F. Halici, I. Taymaz, Experimental study of the airside performance of tube row spacing in finned tube heat exchangers, *Heat and Mass Transfer*, Vol.42, 2006, pp. 817-822.

[21]. Sparrow, E. M., and Samie, F., "Heat Transfer and Pressure Drop Results for One and Two row Arrays of Finned Tubes," *International Journal of Heat Mass Transfer*, Vol. 28, No. 12, pp. 2247-2259, 1985.

[22]. Jacobi, A. M., and Shah, R. K., "Air-Side Flow and Heat Transfer in Compact Heat Exchangers: A discussion of Enhancement Mechanisms," *Heat Transfer Engineering*, Vol.19, no.4, pp.29-41, 1998.

[23]. Rene Hofmann, Friedrich Frasz, Karl Ponweiser, "Heat Transfer and Pressure Drop Performance Comparison of Finned-Tube Bundles in Forced Convection," WSEAS TRANSACTIONS on HEAT and MASS TRANSFER, ISSN: 1790-5044 Issue 4, Volume 2, October 2007.

[24]. Tutar, M., and Holdo, A. E., " Computational Modelling of Flow around a Circular Cylinder in Sub-Critical Flow Regime with Various Turbulence Models," *International Journal for Numerical Methods in Fluids*, Vol.35, pp.763-784, 2001.

[25]. K. Kawaguchi, Heat transfer and pressure drop characteristics of finned tube banks in forced convection, *Journal of Enhanced Heat Transfer*, Vol.12, No.1, 2005, pp. 1-20.

OPTIMIZATION OF PROCESS PARAMETER IN WEDM FOR MONEL K-500 USING TAGUCHI AND GRA

Amit kumar[1],Vishal Gulati[2],Amitesh Goswami[3]

Amit kumar, Email add:-twinkleverma11@gmail.com

Vishal Gulati, Email add:-vishal_gulati_in@yahoo.com

Amitesh Goswami, Email add:-amitesh550@yahoo.com

Abstract

Wire Electrical Discharge machining process is extensively used in mechanical industry to manufacture geometrically complex, intricate shape and very hard material parts that are extremely difficult to machine by conventional machining process. The objective of the present work is to investigate the effects of the various WEDM process parameters like pulse on time (T_{ON}), pulse off time (T_{OFF}), peak current (IP), Servo voltage (SV) on material removal rate(MRR) and surface roughness (SR) of Monel K-500 by using Ultima-1F wire-cut EDM machine.The experimentation has been done by using Taguchi's L_{27} Orthogonal Array. Multi response optimization method using grey relation analysis is proposed for optimization of parameters.

Keywords: WEDM, Material removal rate, Surface roughness, GRA.

1. Introduction

In mechanical industry, the demands for alloy materials having high hardness, toughness and impact resistance are increasing. These materials are difficult to be machined by traditional machining methods. To satisfy the present demands of the manufacturing industries, a different class of modern machining techniques such as Wire Electrical Discharge Machining (WEDM) have emerged. Which is the best alternative for the production of small scale complex parts with the highest degree of surface finish and dimensional accuracy.

Wire EDM machining (Electrical Discharge Machining) is an electro thermal production process, which erodes and vaporize material from the surface of workpiece by using a series of discrete spark between workpiece andtool electrode separated by dielectric fluid.

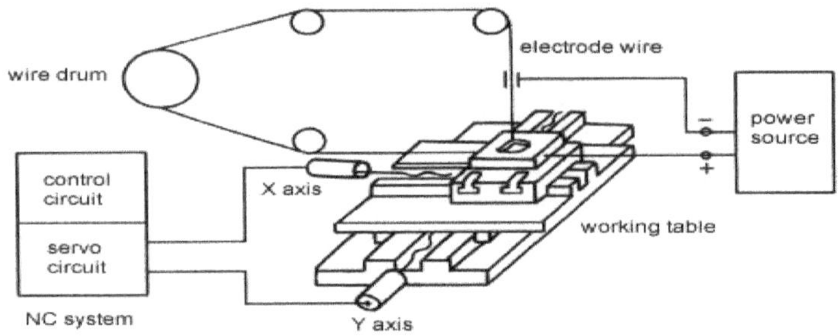

Figure1: Set up of Electric discharge machining

This fluid acts as cooling agent and it also flushes away the eroded particles from the work zone. WEDM utilizes a continuously traveling single stranded wire electrode made of thin copper, brass or tungsten of diameter 0.05-0.3 mm, against workpiece through CNC wire guide[2]. Fig.1 is shows the setup of the Electric discharge machining. Proper gap is maintained between wire and workpiece,when the voltage across the gap becomes sufficiently high spark occurs and pressure developed between work and tool. As a result of which a very high temperature is reached and at such high pressure and temperature that some metal is melted and eroded. Such localized extreme rise in temperature leads to material removal. Wire electrical discharge machining is complex, time varying stochastic process in which output of machine depends on input variables. To harness the full potential of WEDM machine toolit is priority to run the machine with optimum parameters. Many researchers have been contributed their efforts towards optimization cutting parameters.**Tosunet at. [2004]** studied the influence of open circuit voltage, pulse duration, wire speed, flushing pressure on kerf and metal removal rate on a Sodick A320D/EX21 WEDM machine tool with AISI 4140 steel(with 200mm × 40mm × 10mm) with CuZn37 Master brass wire of 0.25mm diameter. L_{18} orthogonal array was used. ANOVA shows that open circuit voltage and pulse duration were highly effective parameters on both the kerf and the MRR whereas wire speed and dielectric flushing pressure were less effective factors.**Mahapatra et al. [2006]** conducted experiment to study various significant control factors which effect the machining performance on Robofil 100 high precision five axis CNC WEDM using Zinc coated copper wire with 0.25 mm diameter and a block of D2 tool steel as work piece material. TAGUCHI method was used for experiment. The results of confirmation experiment agree well with the predicted optimal settings as an error of 4.062 % in MRR and 1.53 % SF and discharge current, pulse duration, pulse frequency play significant role in finish cutting operation was observed.**Mahapatra et al. (2007)** established a relationship between control factors and responses like MRR, SF and kerf by means of nonlinear regression analysis, resulting in a valid mathematical model. Genetic algorithm was employed to optimize the wire electrical discharge machining process with multiple objectives. Experiments were performed on ROBOFIL 100 high precision 5-axis CNC WEDM as per Taguchi method (L_{27} OA) using D2 tool steel specimen of 200mm×25mm×10mm. The confirmation experiment shows that the errors associated with MRR, SF, and kerf are 3.14, 1.95, and 3.72%, respectively**Ramakrishnan et al. [2008]** studied the influence of pulse on time, delay time, wire feed speed, and ignition current on metal removal rate and surface roughness of Inconel 718 material of thickness 20mm on five-axis Robofil 290P CNC WEDM. Experiments were planned as per Taguchi's L_9 orthogonal array. Artificial neural network (ANN) models and multi response optimization technique was used to predict and select the best cutting parameters of WEDM. ANOVA was employed to identify the level of importance of the machining parameters. Experimental result shows that the developed ANN models could predict the performance characteristics of WEDM process more accurately and the pulse on time, delay time and ignition current influenced more than wire feed speed on the performance characteristics. **H.Singh et al. (2009)** studied the effects of various process parameters of WEDM like pulse on time

(T_{ON}), pulse off time (T_{OFF}), gap voltage (SV), peak current (IP), wire feed (WF) and wire tension (WT) to reveal their impact on material removal rate of hot die steel (H-11) using one variable at a time approach on Electronica sprint cut WEDM machine. Results show that the material removal rate (MRR) directly increases with increase in pulse on time (T_{ON}) and peak current (IP) while decreases with increase in pulse off time (TO_{FF}) and servo voltage (SV).**Jangra et al. [2011]** investigated the influence of taper angle, peak current, pulse-on time, pulse-off time, wire tension and dielectric flow rate to optimize MRR and SR simultaneously on 5-axis sprint cut ELPLUS-40 WEDM machine made by Electronica with WC-Co composite material. Taguchi's L_{18} orthogonal array was used for the experimentation. GRA was used for the optimization of process parameters and ANOVA was used to show the most significant parameters affecting the multiple machining characteristics. The experimental setup showed that taper angle, pulse-on time and pulse-off time were found the most significant affecting the MRR and SR.**Jaganathanet al. [2012]** evaluatedthe influence of applied voltage, Pulse width, Pulse interval and speed on MRR. Experiment was performed on ST CNC-E3 (MCJ) WEDM machine with E 31 material. Taguchi's Mixed L_{27} orthogonal array and DOE was used. The experimentation setup resulted that applied voltage, Pulse width speed was most significant factors. **Kulkarni et al. [2014]** studied the influence of various process parameters such as pulse on time, pulse off time, wire feed, wire tension, upper flush and lower flush on MRR and SR for high carbon high chromium steel using Electronica maxicut 734 CNC Wire-cut electrical discharge machining (WEDM). Taguchi's L_{25} Orthogonal Array along with utility concept was used to optimize the response characteristics such as Material removal rate, surface finish and kerf width. Shows that the order of influencing parameters for MRR was pulse-off time, upper flush, lower flush and wire tension and for SR was pulse-on time, Lower Flush, Wire Feed, Wire tension.

2. Methodology

In the present work, Taguchi method using L_{27}orthogonal arrays (Table 4) has been employed for the design of experiments. The various control factors along with their symbols shown in Table 1. Pulse on time, pulse off time, peak current and servo voltage are set as control factors (independent variable) which vary during the experiment. With the help of machine manual, many trail experiments were done to determine the operating range of each factor finally selected.The Taguchi method tests pairs of combinations instead of testing all possible combinations. From the available literature on WEDM, total four numbers of input parameters were finally selected and range of levels of control factors is shown in Table 2.

In this present research work, orthogonal array with four control factors viz., A, B, C, D, and three interactions viz. A× B, A×C have been studied. Signal to noise ratio was obtained using Minitab 16 software. Higher is better (HB) for MRR and lower is better (LB) for SR were taken for obtaining optimum machining characteristics.

Table 1: Process parameters and their symbols

Control factor(Symbol)	Parameter kept constant
Pulse on time (A)	Wire tension = 8
Pulse off time (B)	Wire feed = 8
Peak current (C)	Servo Feed- 2050 mm/min
Servo voltage (D)	Dielectric flow rate = 10

Table 2: Levels for various control factors

Control factor	Symbol	Level		
		Level 1	Level 2	Level 3
Pulse on time	A	110	115	120
Pulse off time	B	50	55	60
Peak current	C	11	12	13
Servo voltage	D	40	50	60

Table 4: Taguchi's L_{27} Standard Orthogonal Array

Column no	1	2	3	4	5	6	7	8	RESPONSE RAW DATA			S/N
Trial no.	A	B	A×B	A×B	C	A×C	A×C	D	R1	R2	R3	RATIO
1	1	1	1	1	1	1	1	1	$Y1_1$	$Y1_2$	$Y1_3$	S/N(1)
2	1	1	1	1	2	2	2	2	.	.	.	.
3	1	1	1	1	3	3	3	3	.	.	.	.
4	1	2	2	2	1	1	1	2	.	.	.	.
5	1	2	2	2	2	2	2	3	.	.	.	.
6	1	2	2	2	3	3	3	1	.	.	.	.
7	1	3	3	3	1	1	1	3	.	.	.	.
8	1	3	3	3	2	2	2	1	.	.	.	.
9	1	3	3	3	3	3	3	2	.	.	.	.
10	2	1	2	3	1	2	3	1	.		.	.
11	2	1	2	3	2	3	1	2	.	.	.	.
12	2	1	2	3	3	1	2	3	.	.	.	.
13	2	2	3	1	1	2	3	2	.	.	.	.
14	2	2	3	1	2	3	1	3	.	.	.	.
15	2	2	3	1	3	1	2	1	.	.	.	.
16	2	3	1	2	1	2	3	3	.	.	.	.
17	2	3	1	2	2	3	1	1	.	.	.	.
18	2	3	1	2	3	1	2	2	.	.	.	.
19	3	1	3	2	1	3	2	1	.	.	.	.
20	3	1	3	2	2	1	3	2	.	.	.	.
21	3	1	3	2	3	2	1	3	.	.	.	.
22	3	2	1	3	1	3	2	2	.	.	.	.
23	3	2	1	3	2	1	3	3	.	.	.	.
24	3	2	1	3	3	2	1	1	.	.	.	.
25	3	3	2	1	1	3	2	3	.	.	.	.
26	3	3	2	1	2	1	3	1	.	.	.	.
27	3	3	2	1	3	2	1	2	$Y27_1$	$Y27_2$	$Y27_3$	S/N(27)

3. Experimental set up

The experiments were performed on Ultima-1F wire-cut EDM machine of Electronica Machine tools Ltd. The Monel k-500 (Nickel 63.0%,Copper 27.0%-23.0%,Iron 2.0%, Titanium0.35%-0.85%) plate of 100mm x 100mm x 19.85mm size is mounted on machine tool fig.2. Table 3 shows the limiting chemical composition of Monel K-500. In the present research work Mitutoyosurfest SJ- 210 instrument (fig. 3) is used for the measurement of Surface Roughness of the specimen.This (Mitutoyosurfest SJ- 210) is shop- floor type surface roughness measuring instrument.Mitutoyosurfest SJ-210 traces the surface of various machine part and calculate the surface roughness based on roughness standard and display the results.

Figure 2: Monel K- 500 Plate Mounted on WEDM Machine

Fig. 3: MitutoyoSurftest SJ-210 (Surface RoughnessTester)

Table 3: Limiting Chemical Composition, %

Nickel	63.0 min
Copper	27.0-23.0
Iron	2.0 max
Sulfur	0.01 max
Carbon	0.25max
Manganese	1.5max
Silicon	0.5 max
Sulfur	0.01 max
Aluminum	2.3-3.5 max
Titanium	0.35-0.85

Digital stop watch was used to calculate the time taken by machine to cut the particular specimen. Material removal rate is measure of material removal from workpiece which is quantitatively given as mm^2/min and is calculated as:
MRR = [Distance travelled by wire (mm) x Thickness of workpiece (mm)]/Time (min.)] (1)

4. Single response optimization

The Taguchi approach has been used for predicting the confidence intervals for the predicted mean has been applied. Three Confirmation experiments for each performance characteristics have been performed at optimal settings of the process parameters and the average value has been reported. The optimum value of material removal Rate (MRR) is predicted at the optimal levels of significant variables which have already been selected as pulse on time (A3), pulse off time (B1), peak current (C3) and spark gap set voltage (D1) shown in table fig.4 and experimental results of MRR and SR are shown in Table no.5.Predicted Optimal Values, Confidence Intervals and Results of Confirmation experiments are shown in Table 8. The estimated mean of the response characteristic (MRR) can be determined [8] as:-

$$\begin{aligned}\mu_{MRR} &= \{(A3 + B1 + C3 + D1) - 3(\mu)\} \\ &= \{(48.56+47.16+46.98+45.89)-3(39.641)\} \\ &= 188.59 - 118.923 \\ &= 69.67 mm^2/min\end{aligned}$$

The 95 % confidence intervals of confirmation experiments (CI_{CE}) is calculated as:

$$CI_{CE} = \sqrt{F_\alpha(1 \cdot f_e)\left\{\frac{1}{n_{eff}} + \frac{1}{R}\right\}V_e} = 1.830$$

Where, $F\propto (1, f_e)$ = The F ratio at the confidence level of (1-α) against DOF 1 and error degree of freedom f_e = 3.99 (from Tabulated F value)

$$n_{eff} = \frac{N}{1 + \text{Total DF involved in estimation of mean}}$$

= 81 / (1+8)

= 9

N = Total number of results = 27 x 3 = 81

R = Sample size for confirmation experiments = 3

Ve = Error variance = 1.89,fe = error DOF = 64 (Table 6)

Therefore, the predicted confidence interval for confirmation experiments is:-

$\mu_{MRR} - CI_{CE} < \mu_{MRR} < \mu_{MRR} + CI_{CE}$

$67.84 < \mu_{MRR} < 71.50$

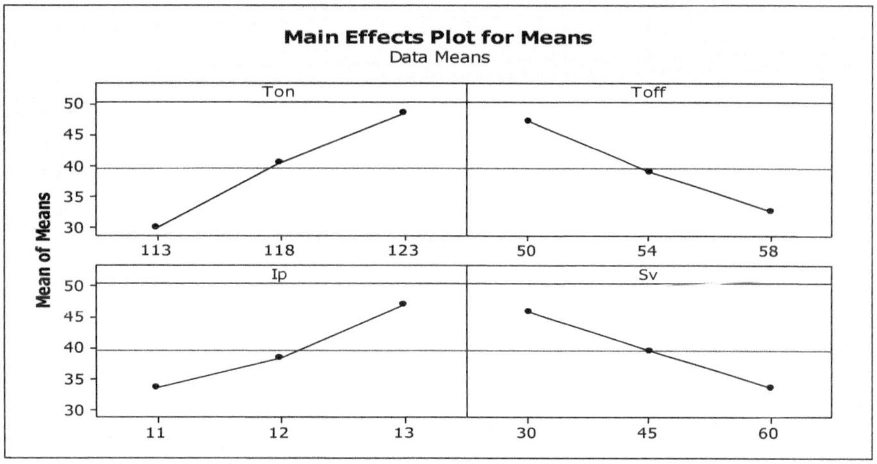

Figure 4: Main Effect Plot for Means (for Material Removal Rate)

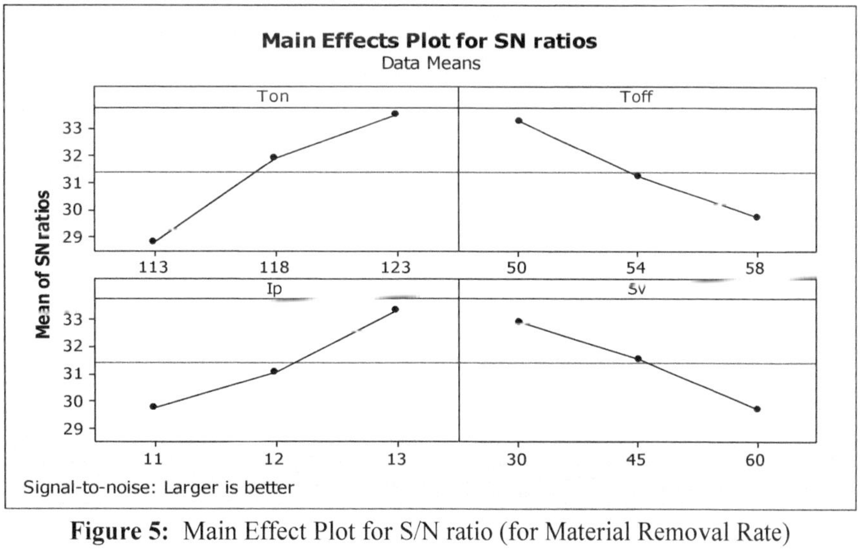

Figure 5: Main Effect Plot for S/N ratio (for Material Removal Rate)

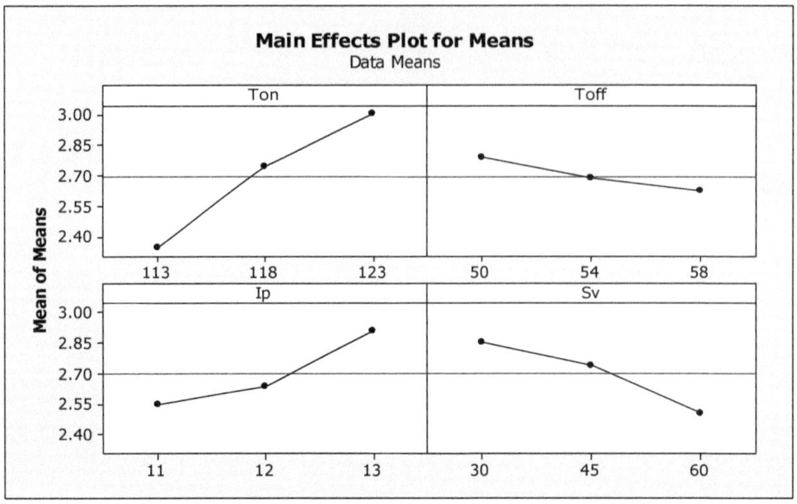

Figure 6: Main Effect Plot for Means (Surface Roughness)

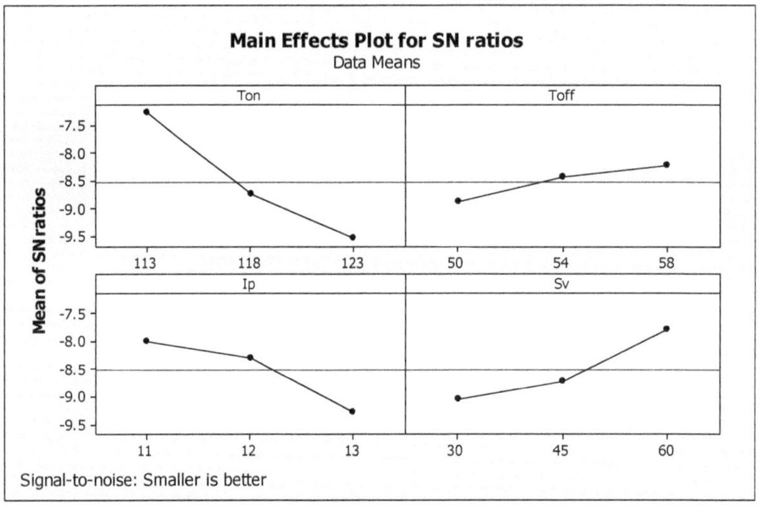

Figure 7: Main Effect Plot of Surface Roughness (S/N ratio)

Similarly for Surface Roughness (SR):

$\mu_{SR} = \{(A1 + B3 + C1 + D3) - 3(\mu)\}$

$= \{(2.345+2.621+2.547+2.504)-3(2.697)\}$

$= 10.017 - 8.091$

$= 1.926\mu m$

Where,

The 95 % confidence intervals of confirmation experiments (CI_{CE}) are calculated as:-

$$CI_{CE} = \sqrt{F_{\alpha}(1 \cdot f_e)\left\{\frac{1}{n_{eff}} + \frac{1}{R}\right\}V_e}$$

= 0.268

Where, F∝ (1, f_e) = The F ratio at the confidence level of (1-α) against DOF 1 and error degree of freedom f_e = 3.99 (from Tabulated F value)

$$n_{eff} = \frac{N}{1 + \text{Total DF involved in estimation of mean}}$$

= 81 / (1+8)

= 9

N = Total number of results = 27 x 3 = 81

R = Sample size for confirmation experiments = 3

V_e = Error variance = 0.0406 (Table 7) f_e = error DOF = 64

Therefore, the predicted confidence interval for confirmation experiments is:

$$\mu_{SR} - CI_{CE} < \mu_{SR} < \mu_{SR} + CI_{CE}$$

$$1.658 < \mu_{SR} < 2.194$$

Exp. No.	MRR			MEAN1	SNRA1	SR			MEAN2	SNRA2
	MRR	MRR 2	MRR 3			Ra 1	Ra 2	Ra 3		
1	32.19	33.168	32.67	32.6760	30.2826	2.192	2.81	2.46	2.48733	-7.9594
2	28.296	29.474	28.89	28.8867	29.2103	2.830	2.126	2.381	2.44567	-7.8290
3	42.262	45.489	43.873	43.8747	32.8325	2.227	2.841	2.638	2.56867	-8.2369
4	20.521	21.337	20.923	20.9270	26.4108	1.889	2.488	2.214	2.19700	-6.8902
5	17.308	18.008	17.652	17.6560	24.9344	1.648	1.857	1.732	1.74567	-4.8497
6	49.851	54.347	52.148	52.1453	34.3275	2.978	2.647	2.727	2.78400	-8.9045
7	12.925	13.548	13.256	13.2430	22.4349	1.605	1.680	1.629	1.63800	-4.2879
8	22.646	23.458	23.054	23.0527	27.2517	2.15	2.447	2.225	2.28467	-7.1890
9	36.942	37.061	37.004	37.0023	31.3646	2.858	3.040	2.971	2.95633	-9.4179
10	52.933	54.026	53.47	53.4763	34.5623	2.852	2.674	2.771	2.76567	-8.8390
11	48.361	48.168	48.269	48.2660	33.6728	2.734	2.801	2.789	2.77467	-8.8647
12	44.761	46.133	45.441	45.4450	33.1477	2.612	3.230	2.981	2.94100	-9.4021
13	36.291	37.646	36.962	36.9663	31.3532	2.617	2.667	2.698	2.66067	-8.5005
14	30.398	31.838	31.121	31.1190	29.8559	2.438	2.542	2.499	2.49300	-7.9357
15	49.255	50.98	50.112	50.1157	33.9969	2.971	2.968	2.987	2.97533	-9.4707
16	24.812	25.097	24.958	24.9557	27.9431	2.484	2.479	2.304	2.42233	-7.6899
17	35.895	37.219	36.551	36.5550	31.2561	2.333	3.086	2.089	2.74267	-8.8189
18	36.648	38.223	36.937	36.9360	31.3384	2.764	3.033	2.915	2.90400	-9.2662
19	55.517	54.68	53.593	54.5967	34.7406	3.066	3.215	3.161	3.14733	-9.9605
20	62.537	64.22	63.37	63.3757	36.0369	3.059	2.995	2.805	2.95300	-9.4111
21	53.846	53.913	53.872	53.8770	34.6281	2.939	3.066	3.001	3.00200	-9.5495
22	38.646	38.475	38.564	38.5617	31.7231	2.889	3.053	2.935	2.95900	-9.4253
23	45.413	44.411	44.916	44.9133	33.0464	3.030	3.053	3.151	3.07800	-9.7666

24	58.102	59.691	58.891	58.8947	35.3999	3.195	3.358	3.236	3.26300	-10.2743
25	26.361	27.068	26.711	26.7133	28.5330	2.550	2.716	2.680	2.64867	-8.4637
26	50.005	53.042	51.528	51.5250	34.2328	3.140	3.242	3.198	3.19333	-10.0856
27	43.889	45.253	44.579	44.5737	32.9795	2.805	2.716	2.865	2.79533	-8.9308

Table 5: Experimental Results of Material Removal Rate and Surface Roughne

Source	DF	Seq SS	Adj MS	F	P
Ton	2	4704.71	2352.36	1243.54	0.000
T_{off}	2	2828.09	1414.04	747.52	0.000
Ip	2	2495.32	1247.66	659.56	0.000
SV	2	2063.22	1031.61	545.35	0.000
T_{on}*T_{off}	4	110.38	27.60	14.59	0.000
T_{on}*Ip	4	1498.81	374.70	198.08	0.000
Error	64	121.07	1.89		
Total	80	13821.60			

S = 1.37538 R-Sq = 99.12% R-Sq(adj) = 98.91%

Table 6: Analysis of Variance for MRR, using Mean Data

Source	DF	Seq SS	Adj MS	F	P
Ton	2	4704.71	2352.36	1243.54	0.000
T_{off}	2	2828.09	1414.04	747.52	0.000
Ip	2	2495.32	1247.66	659.56	0.000
SV	2	2063.22	1031.61	545.35	0.000
T_{on}*T_{off}	4	110.38	27.60	14.59	0.000
T_{on}*Ip	4	1498.81	374.70	198.08	0.000
Error	64	121.07	1.89		
Total	80	13821.60			

S = 1.37538 R-Sq = 99.12% R-Sq(adj) = 98.91%

Table 7: Analysis of Variance for Surface Roughness (Mean Data)

Source	DF	Seq SS	Adj MS	F	P
T_{on}	2	5.94699	2.97350	73.22	0.000
T_{off}	2	0.38217	0.19109	4.71	0.012

Ip	2	1.93474	0.96737	23.82	0.000
SV	2	1.67640	0.83820	20.64	0.000
$T_{on}*T_{off}$	4	0.28574	0.07143	1.76	0.148
$T_{on}*Ip$	4	1.16430	0.29107	7.17	0.000
Error	64	2.59920	0.04061		
Total	80	13.98955			

S = 0.201526 R-Sq = 81.42% R-Sq(adj) = 76.78%

Table 8: Predicted Optimal Values, Confidence Intervals and Results of Confirmation Experiments

Performance Measures/ Responses	Optimal Parameter Combination	Predicted Optimal Value	Predicted Confidence Interval (CI_{CE}) at 95% Confidence Level	Experimental value
Material Removal Rate (MRR)	$A_3B_1C_3D_1$	69.67mm^2/min	67.84≤MRR≤ 71.50	69.317 mm^2/min
Surface Roughness(SR)	$A_1B_3C_1D_3$	1.926μm	1.658 ≤ SR ≤2.194	1.692μm

5. Multi-response optimization using Grey Relation Analysis

An efficient solution of the uncertainty, multiple input and discrete data problem can be find out byGrey Relation Analysis (GRA).Grey relational analysis provides the relation between machining parameters and machining performance. In the present study, a linear normalization of the experimental results (S/N ratios) for Material Removal Rate (MRR) and Surface Roughness (SR) were performed in the range of 0 and 1, which is also called the grey relational generating [1].

In order to optimize the MRR and SR simultaneously using grey relational analysis (GRA), the following steps were followed [1]:

1. By using Taguchi Method Convert the experimental data into S/N values,
2. Normalize the S/N ratio,
3. Grey relational generating is perform and then calculate the grey relational coefficient,
4. By using the concept of weighing factor Calculate the grey relational grade for the performance characteristics,
5. Analyse the experimental results using the grey relational grade and statistical analysis of variance (ANOVA),
6. Select the optimal levels of process parameters,
7. Perform the confirmation experiment to verify the optimal process parameter settings.

In grey relational analysis experimental results were normalized first and then the grey relational coefficient has been calculated from the normalized experimental data. Data pre- processing was performed on raw data (Mean data) for all experimental results (Table 9). Data pre-processing is a technique that involves transforming raw data into an understandable format. Raw data (Mean data) can be normalized by using the following expressions: For MRR Higher the better (Eq. (2))

$$x_i^*(k) = \frac{x_i^0(k) - \min x_i^0(k)}{\max x_i^0(k) - \min x_i^0(k)} \text{ .... for MRR} \qquad (2)$$

$$x_i^*(k) = \frac{\max x_i^0(k) - x_i^0(k)}{\max x_i^0(k) - \min x_i^0(k)} \text{ ....for SR} \qquad (3)$$

Where $x_i^*(k)$ is the sequence after the data processing; $x_i^0(k)$is the original sequence of S/N ratio, i=1, 2, 3…., m and k =1, 2,...., n *with* m=27 and n=2; max $x_i^0(k)$ is the largest value of $x_i^0(k)$; min$x_i^0(k)$is the smallest value of $x_i^0(k)$. Table 9shows the normalized S/N ratio for the MRR and SR. The outcomes are denoted as $x_0^*(k)$ and $x_i^*(k)$ for reference sequence and comparability sequence, respectively. Basically, the larger normalized S/N ratio corresponds to the better performance and the best-normalized S/N ratio is equal to unity [1].

6. Grey Relational Grade Generation

The grey relational coefficient is calculated to express the relationship between the best (reference) and the actual normalized S/N ratio. The grey relational coefficient is expressed as follows:

$$£_{ij}(k) = \frac{(\Delta)\min - \zeta(\Delta)\max}{(\Delta)oi(k) - \zeta(\Delta)\max} \qquad (4)$$

Where, $\Delta_{oi}(k)$ is the deviation sequence of reference sequence

Δ_{min} = min [min $\Delta_{oi}(k)$]

Δ_{max} = max [max $\Delta_{oi}(k)$]

$£_{ij}$The grey relational coefficient, ζ is is the distinguishing coefficient which is defined in the range $0 \le \zeta \le 1$.ζ is set as 0.5 in this study.

And hence to calculate the grey relational coefficient from the above equationthe deviation sequence must be calculated. The deviation sequence $(\Delta)oi$ can be calculated as follow:

$(\Delta MRR)o1$ = |*1-0.5769* | = *0.4231*

$(\Delta SR)o1$= |*1-0.6133* | = *0.3867*

The same calculating method was performed for i=1-27, and the results of all $(\Delta)oi$ for i=1-27 are listed in Table (9).

On investigating the data in table (9) we can find the following:

$\Delta_{max} = \Delta_7 (MRR) = \Delta_{24} (SR) = 1.000$,

$\Delta_{min} = \Delta_{20} (MRR) = \Delta_7 (SR) = 0.000$,

Next, the grey relational coefficient is calculated from the equation no. (4) and the values are calculated as follows:

$£_{1(MRR)}$ = [0.000 + (0.5 × 1.000)] / [0.4231 + (0.5×1.000) = 0.5416

$£_{1(SR)}$ = [0.000 + (0.5 × 1.000)] / [0.3867 + (0.5×1.000)] = 0.5638

And grey relational grade is computed by averaging the grey relational coefficient corresponding to each performance characteristic. The overall evaluation of the multiple performance characteristics is based on the grey relational grade, that is:

$$Y_j = \frac{1}{m}\sum_{i=1}^{m} £ \quad (5)$$

Where $\gamma_j$the grey relational grade for the jth experiment and m is the number of performance characteristics. The grey relational grade is a weighting-sum of the grey relational coefficients. The overall evaluation of multiple performance characteristics is based on the grey relational grade. Using the same weighting values of MRR and SR (i.e.$w_1=w_2=0.5$), grey relational grade is calculated.

Table 9: Experimentation data of material removal rate and surface roughness

Sr. no.	Mean (MRR) (Normalized)	Mean (SR) (Normalized)	ΔMRR	ΔSR	GR coeff. (MRR)	GR coeff. (SR)	GR grade S/N ratio	GR grade
1.	0.5769	0.6133	0.4231	0.3867	0.5416	0.5638	-5.15021	0.5527
2.	0.4981	0.5915	0.5019	0.4085	0.4991	0.5503	-5.60178	0.5247
3.	0.7644	0.6596	0.2356	0.3405	0.6797	0.5948	-3.91448	0.6372
4.	0.2923	0.4347	0.7077	0.5653	0.4140	0.4693	-7.09942	0.4416
5.	0.1837	0.0938	0.8163	0.9062	0.3799	0.3555	-8.69013	0.3677
6.	0.8743	0.7711	0.1257	0.2289	0.7991	0.6859	-2.58607	0.7425
7.	0.0000	1.0000	1.0000	0.0000	0.3333	1.0000	-3.52139	0.6667
8.	0.3541	0.4846	0.6459	0.5154	0.4363	0.4924	-6.66403	0.4643
9.	0.6564	0.8569	0.3436	0.1431	0.5927	0.3043	-6.96475	0.4485
10.	0.8915	0.7602	0.1085	0.2398	0.8217	0.6758	-2.51384	0.7487
11.	0.8261	0.7645	0.1739	0.2355	0.7420	0.6798	-2.96383	0.7109
12.	0.7875	0.8543	0.2125	0.1457	0.7018	0.7743	-2.63887	0.7380
13.	0.6556	0.7036	0.3444	0.2964	0.5921	0.6278	-4.29483	0.6099
14.	0.5456	0.6093	0.4545	0.3904	0.5238	0.5615	-5.31040	0.5426
15.	0.8500	0.8657	0.1500	0.1343	0.7692	0.7882	-2.17260	0.7787
16.	0.4049	0.5682	0.5951	0.4318	0.4566	0.5366	-6.07987	0.4966
17.	0.6485	0.7568	0.3515	0.2432	0.5872	0.6727	-4.01457	0.6299
18.	0.6545	0.8316	0.3455	0.1684	0.5852	0.7480	-3.52269	0.6666
19.	0.9046	0.9475	0.0954	0.0525	0.8398	0.9049	-1.18668	0.8723
20.	1.0000	0.8558	0.0000	0.1442	1.0000	0.7761	-1.03076	0.8881
21.	0.8964	0.8789	0.1036	0.1211	0.8283	0.8050	-1.75981	0.8166
22.	0.6828	0.8581	0.3172	0.1419	0.6119	0.7789	-3.15531	0.6954
23.	0.7801	0.9151	0.2199	0.0849	0.6945	0.8548	-2.21285	0.7751
24.	0.9531	0.0000	0.0469	1.0000	0.9142	0.3333	-4.10049	0.6237
25.	0.4483	0.6975	0.5517	0.3025	0.4754	0.6230	-5.20539	0.5492
26.	0.8673	0.9684	0.1327	0.0316	0.7903	0.9405	-1.25566	0.8654
27.	0.7752	0.7755	0.2248	0.2245	0.6898	0.6901	-3.22428	0.6899

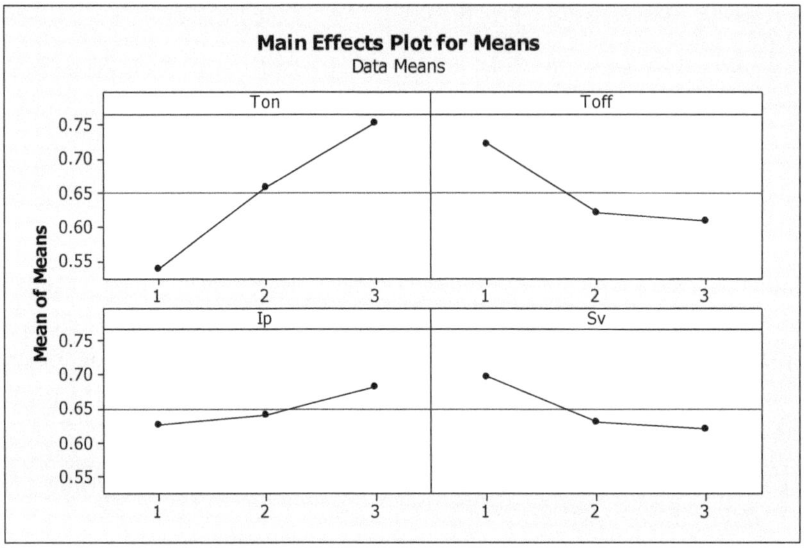

Figure 8: Main Effect plots for Means of G R Grade

7. Results and Discussion

Fig. 4and fig 5 represents the effect of pulse on time on MRR at the chosen machining conditions. It can be observed from the figure that the MRR increased when the pulse on (T_{on}) time was increased due to the number of discharges within a given periodof time increase. Since the amount of material removed from the work specimen highly depends on the discharge energy, it is also seen that with an increase of pulse on time the surface quality of the machined surfaces were decreased because under longer pulse on time the electrical sparks generate bigger craters on the surface of work piece. Fig 4 showsincreasing the peak current results in the increases of number of electrons striking the work surface in a single discharge, which further causes eroding out more material from the work surface per discharge, thereby increase in material removal rate. Fig.4 and fig.5 shows that, the metal removal rate decreases almost constantly with increase in pulse off time (T_{off}) and spark gap set voltage (SV) as less number of sparks are produced at higher value of pulse off time and spark gap set voltage. Fig 6 and fig 7 shows when parameter pulse on time increased the value of Surface Roughness decreases. The pulse off time parameter has direct effect on the Surface roughness, as we increase the pulse off time the value of Surface Roughness also increases.

Table 10: Summary and comparison of results

Method	Optimization technique	Optimal parameters	Experimental values
Single characteristic	Taguchi Method	A3 B1 C3 D1 (MRR)	MRR = 69.317 mm^2/min
Multi-characteristic optimization	Grey relational analysis	A3 B1 C3 D1	MRR = 69.317mm^2/min SR = 2.086 μm

The value of Surface roughness also increases by increasing servo voltage (fig.6). Fig 6 shows that when parameter peak current increases the value of Surface Roughness decreases which is in agreement with the findings of Kumar A., Kumar V., Kumar J. 2012 [2]. Analysis of Variance for MRR and SR is done using Mean data shown in Table 6 and 7 respectively.Figure 8 shows the main effect plots for means of G R Grade. Table 10 shows the comparison of results between single response optimization and multi response optimization using grey relational analysis.

8. Conclusion

1. MRR and SR were optimized individually by the help of Taguchi methodology. Two different optimal settings of process parameters were found for MRR (A3, B1, C3, and D1) and SR (A1, B3, C1 and D3). The material removal rate (MRR) is mostly affected by the pulse-on time, peak current pulse off-time, and servo voltage. The surface roughness values (SR) are influenced mostly by pulse-on time, peak current, pulse off-time, and servo voltage.
2. In case of GRA, grey relational grade was used as a performance index to determine the optimal combination of process parameters for multiple machining characteristics. Both the machining characteristics were given equal importance by assigning equal weights in calculating the grey relational grade. However, with a different set of weights, a different set of optimal parameters for machining characteristics will result. The optimal set predicted will be closer to the optimal set predicted for single characteristic with the largest weight.
3. A single set of parametric combination can never produce the highest productivity (within the possible range) with the best surface finish at high material removal rate. A proper trade-off becomes inevitable to satisfy all the above-mentioned two machining criteria simultaneously.
4. The optimal values obtained using the multi-characteristic optimization model has been validated by confirmation experiments. The optimal setting of process parameters is A3=123 μs, B1 =50μs, C3 =13ampere, D1=30 volt.

Reference

1. Datta S., Mahapatra S.S, Modeling, simulation and parametric optimization of wire EDM process using response surface methodology coupled with grey-Taguchi technique, International Journal of Engineering, Science and Technology, 2010; 2(5): 162-183.
2. Kumar A., Kumar V., Kumar J., Prediction of Surface Roughness in Wire Electric Discharge Machining (WEDM) Process based on Response Surface Methodology, International Journal of Engineering and Technology, 2012; 2(4).
3. Mahapatra S. S. &Patnaik A., Optimization of wire electrical discharge machining (WEDM) process parameters using Taguchi method, International Journal of Advance Manufacturing Technology, 2007; 34: 911–925.
4. NihatTosun , Can Cogun& Gul Tosun, A study on kerf and material removal rate in wire electrical discharge machining based on Taguchi method, Journal of Materials Processing Technology, 2004; 152: 316–322.
5. Philip J. Ross, Taguchi Techniques for Quality Engineering, McGraw Hill Book Company, New York, 2010.
6. R. Ramakrishnan, L.karunamoothy, Modelling and multi-response optimization of Inconel 718 on machining of CNG WEDM process, Journel of material process technology 207(2008) 343-349,Doi:10.1016/j.jmatprotec.2008.06.040.
7. Rong Tai Yang &ChorngJyhTzeng& Yung Kuang Yang & Ming Hua Hsieh,Optimization of wire electrical discharge machining process parameters for cutting tungsten,Int J AdvManufTechnol (2012) 60:135–147, DOI 10.1007/s00170-011-3576-z.
8. Roy, R.K., A primer on Taguchi method, Van Nostrand Reinhold, New York, 1990.

9. Sarkar S., Mitra S. & Bhattacharyya B., Parametric analysis and optimization of wire electrical discharge machining of γ-titanium aluminide alloy, Journal of Materials Processing Technology, 2005; 159: 286–294.

10. Scott F. Miller, Albert J. Shih, Jun Qu,Investigation of the spark cycle on material removal rate in wire electrical discharge machining of advanced materials, International Journal of Machine Tools & Manufacture 44 (2004) 391–400,Doi:10.1016/j.ijmachtools.2003.10.005

11. Singh H. & Garg R., Effects of process parameters on material removal rate in WEDM, Journal of Achievements in Materials and Manufacturing Engineering, 2009; 32(1).

12. Tarang Y. S., S. C. MA and Chung L. K., Determination of Optimal cutting Parameters in Wire Electrical Discharge Machining, International Journal of Machine Tools Manufacturing, 1995; 35(12): 1693-170.

FIGURE LEGEND

Figure1: Set up of Electric discharge machining
Figure 2: Monel K- 500 Plate Mounted on WEDM Machine
Figure 3: MitutoyoSurftest SJ-210 (Surface Roughness Tester)
Figure 4: Main Effect Plot for Means (for Material Removal Rate)
Figure 5: Main Effect Plot for S/N ratio (for Material Removal Rate)

Figure 6: Main Effect Plot for Means (Surface Roughness)

Figure 7: Main Effect Plot of Surface Roughness (S/N ratio)

Figure 8: Main Effect plots for Means of G R Grade

TABLE LEGEND

Table 1: Process parameters and their symbols
Table 2: Levels for various control factors
Table 3: Limiting Chemical Composition, %
Table 4: Taguchi's L_{27} Standard Orthogonal Array
Table 5: Experimental Results of Material Removal Rate and Surface Roughness

Table 6: Analysis of Variance for MRR, using Mean Data
Table 7: Analysis of Variance for Surface Roughness (Mean Data)

Table 8: Predicted Optimal Values, Confidence Intervals and Results of Confirmation experiments
Table 9: Experimentation data of material removal rate and surface roughness
Table 10: Summary and comparison of results

SPECIAL PURPOSE MACHINE FOR CUTTING CHASES IN WALL FOR CONCEALED ELECTRICAL FITTING

Mr.Swapnil prakash magar[1], Dr.S.P.Trikal[2]

[1]S.S.G.M.C.E.Shegaon,Maharashtra,Swapnilmagar24@gmail.com
[2]Dept. of Mechanical engineering,S.S.G.M.C.E.,Shegaon,Maharashtra. Sptrikal1971@yahoo.com

Abstract

The construction industry worldwide having solid growth due to industrialization, urbanization over the last decade. In survey of these construction sites we found that for concealed electrical fitting the chases cut in the wall is done with help of hand tools like hammer, chisel, grinders, demolition hammer which is very tedious work. Therefore we have designed the wall engraving machine which is based on engraving techniques. With the use of wall engraving machine it is possible to remove the material from wall with ease and giving appropriate straight and clean cut to the wall at specified height for conceal electrical fitting. Our aim of designing this machine is making chases in wall of 30mm deep and 40mm wide in single operation without use of any hand tools so that electrical wires can easily accommodate space. The various parts of this machine like motor, cutter, gears are designed and selected by considering the bricks and concrete strength.

Keywords: Engraving; Conventional tools; brick strength; concrete strength; cutter design.

1. Introduction

The construction industry worldwide is having a total growth of 30% from 2010 to 2014. In survey of these construction sites we found that for concealed electrical fitting the chases cut in the wall is done with help of hand tools like hammer, chisel, grinders, demolition hammer which is very tedious work. We designed wall engraving machine which is based on engraving techniques. **Engraving** is the practice of incising a design onto a hard, usually flat surface, by cutting grooves into it. Engravers in ancient times used engraving for make or decorate things that were needed in everyday life.[1]It is one of the processes which may be used by an artist when creating a sculpture. It is an activity where pieces of rough natural stone are shaped by the controlled removal of stone.[2] Our aim to design wall engraving machine is to get rid of all hand tools used in present date and make process simple and accurate.

The various parts used in wall engraving machine.

i. MOTOR
ii. CUTTER BLADE
iii. GEAR BOX
iv. DUST COVER

We used motor of 2200 watt, 3 cutter blade of 125mm diameter cutter blades are zigzag type width of blade is 9mm so that they can cut groove of 40 mm wide and 20 to 30 mm deep. Bevel gear is used to transmit the power from motor shaft to cutters. Total weight of machine is 7.5 kg which is easy to handle during operation.

2. CONVENTIONAL TOOLS AND TECHNIQUES

Cutting a chase is a simple job and most electricians will own a chasing tool. The chasing tool seen in the image below is basically an angle grinder. [1]

Fig.1 Conventional Tools

Chasing tools in the above image:

1. Angle grinder
2. Chisel
3. Lump hammer
4. Large bolster
5. Medium bolster
6. Demolition hammer

2.1 WORKING METHOD

- Mark lines on the wall with a pencil so that you can follow them.
- Use grinder to cut chase lines.
- Finally hammer bolster used to chop out middle part.

2.2 NEED OF EVOLUTION

The construction industry worldwide having solid growth due to industrialization, urbanization over the last decade. To cope up with this growth we need more faster, accurate and easy techniques to cut chases in wall for concealed electrical fittings. There are various drawbacks with conventional techniques like:

1) Time consuming

2) Less accurate

3) Three operation required to complete task

3. PROPOSED SOLUTION

Special purpose machine can be a solution for all problems occur. Our aim is design a special purpose machine is to get rid of all hand tools used in present date and make process less time consuming.

3.1 OBJECTIVE OF SPECIAL PURPOSE MACHINE

- With the use of engraving techniques cut chases in walls.
- To design a machine this is portable and can handle with ease.
- To make the process faster and accurate.
- To get rid of hand tools.
- To make the operation simple.

4. PROPOSED DESIGN

Fig.3 Conceptual Design

We are going to use 3 cutter blade of 125mm diameter. Cutter blades are zigzag type and width of blade is 9mm so that they can cut groove of 40 mm wide and 20 to 30 mm deep.

5. DESIGN PROCEDURE

5.1 WALL STRENGTH

Walls are normally constructed of brick, mortar, and concrete column so we focus on strength of bricks and concrete columns to design our machine.[4]

Bricks compressive strength:

Speciation of Common Clay Building Bricks

Table 1.

Size of bricks (mm)	compressive load (T)	Compressive strength N/mm^2
220.5×100.5	27	11.430
230.00×110	26	10.4

Calculations:

$$\text{Compressive strength} = \frac{\text{Maximum load at failure (N)}}{\text{Average area of bed face } (mm^2)}$$

Result:

Average compressive strength of the given bricks = 10.91 N/mm^2

5.2 CONCRETE STRENGTH

Concrete specimens are tested using the ASTM39 Test Standard Test Method for Compressive Strength of Cylindrical Concrete Specimens by using two samples made from the same test at the same age, usually at strength of 28 days old. Design engineers use the specified strength $f'c$ to design structural elements. This specified strength is incorporated in the job contract documents. The concrete mixture is designed to produce an average strength ($f'cr$) higher than the specified strength, such that the risk of not complying with the strength specification is minimized. Cylindrical specimens for acceptance testing should be 6 x 12 inch (150 x 300 mm) size or 4 x 8 inch

(100 x 200 mm) when specified. The Concrete strength is calculated by dividing the maximum load at failure by the average cross sectional area. [5]

Table 2. Concrete strength per day

Age	Strength percent
1 day	16%
3 days	40%
7 days	65%
14 days	90%
28 days	99%

Result:

i. The compressive strength of structural concrete begins at 17 mpa and can be commercially produce at 138 mpa.
ii. Residential and light commercial building concrete strength ranges from 17 mpa to 24 mpa.
iii. Heavy commercial and special structure typically required concrete strength 28 mpa.

5.3 MOTOR SELECTION

The more HP will give more efficient cut. Selection of motor is done on the basis of following points:[8]

a. concrete strength
b. speed
c. depth of cut (20 to 30 mm)
d. constant torque

By taking all these points under consideration we select motor of 2200 watt.

a) RPM CALCULATIONS

$$RPM = \frac{SFPM \times 12}{Diameter \times \pi}$$

$$\pi = 3.14$$

Blade diameter should be in inches

Surface feet per minute (SFPM) is the speed at which a diamond particle on the perimeter of the blade is traveling. For a given blade diameter, surface feet per minute can be converted to revolutions per minute

Result:

RPM=6115

b) TORQUE CALCULATIONS

HP = Torque x RPM ÷ 5252

Torque = 2.87 Nm

5.4 CUTTER SELECTION

Factors Involving Concrete Cutting

When cutting concrete, several factors influence your choice of blades. These include:

-Compressive strength
-Hardness of the aggregate
-Size of the a aggregate
-Type of sand
-Steel reinforcing (rebar)
-Green or cured concrete

Concrete slabs may vary greatly in compressive strength, measured in pounds per square inch (PSI). Compressive strength in concrete is a measurement of the load carrying capability of concrete.

Aggregate hardness is one important factor when cutting concrete. Because hard aggregate dulls diamond more quickly, segment bonds on diamond saw blades generally need to be softer when cutting hard aggregate. This allows the segment to wear normally and bring new, sharp diamond grit to the surface. Softer aggregate will not dull diamond grit as quickly, so harder segment bonds are needed to hold the diamonds in place on the saw blade long enough to use their full potential. Diamond saw blades are widely used in cutting concrete materials in civil engineering industry. Concrete is a kind of difficult-to-cut composite material. By taking all these factor under consideration we select three types of blade for cutting concrete.[4,5]

i. Diamond saw blades
ii. Tungsten carbides blades
iii. Gemeni concrete blades

Out of these we select diamond saw blades for cutting chases in walls.We made our cutter blades zigzag so that it can cut chases of 40 mm wide in the wall.

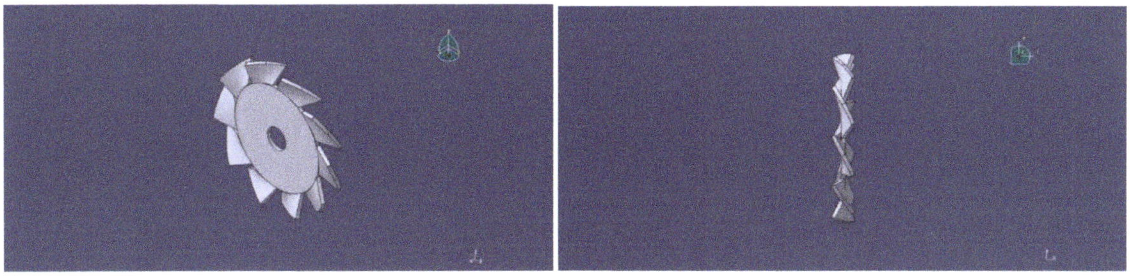

Fig.4 Cutters

5.5 DUST COVER

Dust cover is made of mild steel material. Which is act as safeguard for user from dust while using machine.

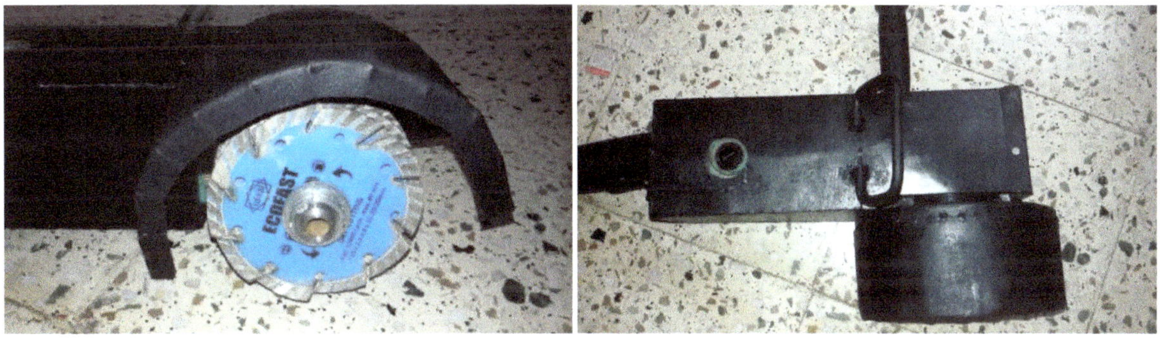

Fig 5.Dust Cover

ADVANTAGES

- Great time saving for cngraving.
- Simple and faster operation.
- Get rid of hand tools.

CONCLUSION

By using this special purpose machine we can have efficient cutting in wall with no other hand tools in less time. operation is going to be a very simple with less efforts. As we uses diamond saw blades for cutting purpose theirs is scope to use other material cutter blades in future.

ACKNOWLEDGEMENTS

The author wish to thanks Mr.D.B.Gaikwad (M.D.) Puja Engineerig works,Aurangabad for his contribution in the design process.

REFERENCES

[1]Gamble, William (1923). *Music Engraving and Printing: Historical and Technical Treatise*. London: Pitman.

[2].Fleming, S.J., Roman Glass; reflections on cultural change. 1999, Philadelphia: University of Pennsylvania Museum of Archaeology and Anthropology.

[3]J. Konstanty, "Theoretical analysis of stone sawing with diamonds", *Journal of MaterialsProcessing Technology*, Vol.123, pp.146-154, 2002.

[4] B. Unver, "A statistical method for practical assessment of sawability of rocks", Proc. of ISRM Int. Symp.Eurock 96', 1996, Barla (Edt.), Rotterdam, Balkema, pp.59-65.

[5]Malhotra, V. M., "Making Concrete Greener with Fly Ash,"*Concrete International*, V. 21, No. , May 1999, pp. 61-66.

[6] Mehta, P. K., "Concrete Technology for Sustainable Development," *Concrete International,* V. 21, No. 11, Nov. 1999, pp. 47-53.

[7]Annual Book of ASTM Standards, ASTM International, West Conshohocken, PA 2006:
Volume 04.02 – Concrete and Aggregate
ASTM C 88 Test Method for Soundness of Aggregates by Use of Sodium Sulfate
ASTM C 418 Test Method for Abrasion Resistance of Concrete by Sandblasting

[8]CEE Motors & Motor Systems Initiative Resources: CEE Summary of Motors and Motors Systems Programs, CEE Guidance Specification for 250-500hp, Low Voltage, General-Purpose Motors, CEE Premium Efficiency Motors List.

[9]. "New Developments in Sawing and Drilling with Diamond Tools," Superabrasives '85 Proceedings, Society of Manufacturing Engineers, Box 930, Dearborn, Michigan 48121

[10] "Application and Safety in the Use of Dry Cutting Diamond Blades," Superabrasives '85 Proceedings

Swapnil P. Magar- Perusing M.E in Advance manufacturing and Mechanical System Design, S.S.G.M.C.E. Segaon,Maharashtra.B.E. in Mechanical K.D.K.C.E. Nagpur,Maharashtra.

Dr.S.P. Trikal- Dept. of Mechanical engineering,S.S.G.M.C.E.,Shegaon,Maharashtra.

COOK OFF STUDY OF A 120 MM GUN BARREL USING FINITE ELEMENT METHOD

Vikas [1],Dr.Nirmal Kant Singh[2], Manoj kumar[3]

1. M.Tech (Thermal Engg.) student, Mechanical Engineering Department, NIT Kurukshetra
2. Associate Professor, Department of Mechanical Engineering, NIT Kurukshetra
3. M.Tech (Thermal Engg.) student, Mechanical Engineering Department, NIT Kurukshetra
Email: vikassoni4you@gmail.com

Abstract

Cook off study is very important to calculate firing rate of a gun. In this paper using analytical method heat transfer to the gun barrel for given ammunition parameters was calculated. Cook off study was done analytically and subsequently finite element method was employed for cook off study using ANSYS 14.0 software .the simulation was run for 521.9 seconds and the result was found to match satisfactorily to the corresponding analytical result. Proposed scheme can accurately study cook off of a gun barrel so accurate firing rate can be determine The cook-off temperature was assumed to 180 ^{0}C for the propellant[1]. The results presented in this report are related to service condition only at 20 ^{0}C.

Keywords: cook-off, gun barrel, wear, finite element method, firing rate.

1. Introduction

When there is firing from a gun, there is large amount of heat input to the bore surface of a gun barrel, and the heat transfer to the barrel is mainly due to forced convection from the hot gases generated inside the barrel due to combustion of propellant. Generally, after firing the barrel is naturally cooled by convection and radiation at its outer surface but natural cooling is inefficient and only a fraction of the total heat input is transferred to the external environment. Hence, during continuous firing at a high rate of fire, the temperature of the gun barrel keeps on rising to ultimately equal to the cook-off temperature. At cook-off temperature, the self ignition of propellant takes place. This premature self-ignition may result in serious damage to the gun barrel and injury to crew members.

Thus, for finding out the maximum firing rate cook-off study is of utmost significance. In this work, a scheme for cook off study of a gun barrel using FEM was proposed and accuracy of the scheme was checked with the analytical results. Present work includes barrel cook off study divided into two parts:

a. Determination of the total heat transfer during a gun fire from hot propellant gases to the gun bore surface
b. Cook off study of the barrel resulting from the heat transfer.

2. Formulation Of Gun Tube Heating Problem

Heat transfer to the gun bore surface can be approximated as an exponentially decaying heat flux [8], [6].

Figure 1 shows a schematic of the heat transfer in a gun barrel.

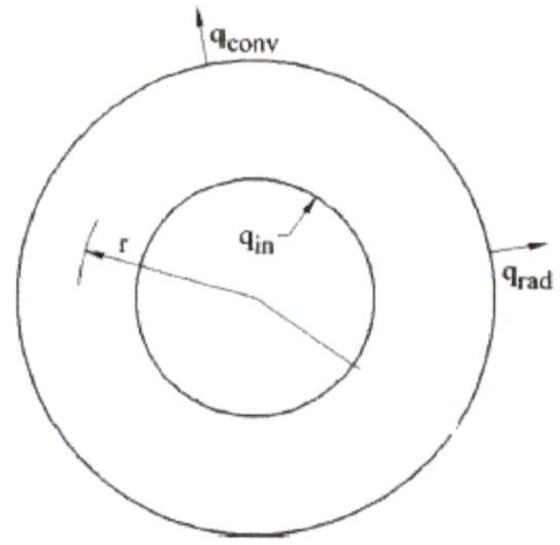

Fig.1. Schematic of heat transfer to gun barrel

Following assumptions and consideration have been made during the present study:

a. The heat conduction in the axial direction is negligible in comparison to conduction in the radial direction [2] and hence, ignored.
b. Friction heating between the projectile driving band and the bore surface was neglected [3].
c. Any effect of gravitation on convection heat transfer was also neglected.
d. The barrel was assumed to be uniformly thick at any transverse cross section along its length.
e. The mathematical formulation is presented using the cylindrical coordinate system.

Based on the above assumptions, any possibility of azimuthal variation of temperature was removed and the problem was reduced to a one-dimensional axisymmetric case. The governing equation, Eq. (1) is the diffusion equation, which is Fourier's conduction equation, combined with the energy equation in cylindrical coordinates [7]

$$\frac{\partial^2 T}{\partial r^2}+\frac{1}{r}\frac{\partial T}{\partial r}=\frac{1}{K}\frac{\partial T}{\partial t} \qquad (1)$$

Where, r is the radius of barrel at chamber, varying from R_i to $R_{0.}$ k and K is the thermal conductivity and thermal diffusitivity of gun steel, assuming constant thermo-physical properties. The initial uniform barrel temperature is taken as 293 K. The boundary conditions at the bore surface and the outside barrel surface are given by Eq. (2). and Eq. (3). respectively

$$-k\frac{\partial T}{\partial r} = h_g\left(T_g - T\right) \qquad (2)$$

$$-k\frac{\partial T}{\partial r} = h_\infty(T - T_\infty) \qquad (3)$$

Where, h_g is the combined convective heat transfer coefficient between the bore surface and hot propellant gases based on exponentially decaying heat flux, the following relation between heat transfer per round and maximum bore temperature has been derived in [4],

$$H_{\infty} = \frac{\kappa*(T_{max}-T_i)}{1.082} * \sqrt{\frac{\pi*t_0}{K}}$$

Where time constant t_0 is:

$$t_0 = \frac{0.8(m_r V_m)}{(d^2 P_{max})}$$

Table.1

The parameter used and heat input result

Notation	Parameters	Value
K	Thermal diffusivity	9 e -6 m^2/s
κ	Thermal conductivity of the barrel material	35 W/mK
m_r	Mass of the shot	6.8 kg
P_{max}	Maximum bore pressure	510 MPa
t_0	Time constant	0.00123 s
H_{∞}	Heat input per round	656833.6138J/m^2

For calculating the time for cook-off some simplifying assumptions [5], are taken into consideration the assumptions are: Constant heat input at the bore surface (constant fire rate). Only convection at the outer surface, barrel temperature is constant across the barrel wall thickness then by using energy balance for the gun barrel

$$mC_v \frac{dT}{dt} = HA_i R_f - h_{\infty} A_0 (T - T_0) \qquad (4)$$

Where

$$m = \rho \frac{\pi}{4}(D^2 - d^2)L \;, A_i = \pi dl \;, \quad A_0 = \pi DL$$

Eq. (4) shows that the higher will be the heat transfer coefficient, lower will be the steady state temperature of the barrel, and lower be the firing rate, lower be the steady state temperature of the barrel.

Let $\theta = T - T_{\infty}$

$= temperature\ above\ ambiant\ temperature$

And $\frac{d\theta}{dt} = \frac{dT}{dt}$

Eq. (4) can be re written as:

$$\frac{d\theta}{dt} + \frac{\theta}{t_0} = \frac{\theta_\infty}{t_0} \qquad (5)$$

Where

$$t_0 = \frac{m\,C_v}{h_\infty A_0}$$

t_0 is the time constant, and it measure of how fast barrel temperature responds to heat flow and

$$\theta_\infty = \frac{HA_iR_f}{h_\infty A_0}$$

θ_∞ is finite temperature rise if firing continued for an infinite time, the solution of Eq. (5) with initial condition

$\theta=0$ at $t=0$ is:

$$\frac{\theta}{\theta_\infty} = 1 - \exp\left(\frac{-t}{t_\infty}\right) \qquad (6)$$

Let ' t_c' is the cook-off time for the barrel and 'T_c' be the cook off temperature of the barrel then $\theta = \theta_c$

This gives:

$$\frac{t_c}{t_0} = -\ln\left(1 - \frac{\theta_c}{\theta_\infty}\right) \qquad (7)$$

The cook-off time was calculated analytically by using Eq. (7). Assuming the thermal properties of the barrel material remains constant, at the elevated temperature, and consider a bore element of .01 mm for the simplicity of analysis because heat input was per m^2 so there was no effect of barrel length on the result. The values of the parameters in above equation to calculate cook-off time are listed in table. 2.

Notation	**Parameter**	**Value**
T_c	Barrel cook-off temperature	180^0 C
A_i	Inner surface area of bore	.0075398 m^2
A_0	Outer surface of the bore	0.012566 m^2
H	Heat transfer per unit area per round	656833 j/m^2
D	Outer diameter of bore	0.2 m

D	Inner diameter of bore	0.120 m
h_∞	Heat transfer coefficient at outer surface of barrel	6.5 W/m^{2}K
C_v	Specific heat	490 j/kg K
L	Length of the bore barrel element	0.01mm
ρ	Density of barrel	7860 kg/m^3
t_0	Time constant	18960.73

Table. 2: Parameters used to calculate cook-off time

a. when firing rate is six rounds per minute:

$$t_0 = \frac{m\,C_v}{h_\infty A_0}$$

$$= \frac{7860 * \pi * (0.2^2 - 0.12^2) * 0.02}{6.5 * \pi * 0.2 * 0.02 * 4} = 18960.73$$

and $$\theta_\infty = \frac{H A_i R_f}{h_\infty A_0}$$

$$= \frac{656833 * \pi * 0.12 * 0.02 * 6}{6.5 * \pi * 0.2 * 0.02 * 60} = 6030.07$$

By putting the value of θ_∞ and t_0 in Eq. (7) we get t_c =507 s so barrel will cook-off after 507 s at a firing rate of six rounds per minute.

3. Finite Element Model Analysis

A one-dimensional, transient thermal analysis was performed with the FEA package ANSYS 14.0 to study cook off of a gun barrel. Chamber of only 0.01 mm length was modelled to avoid heat transfer in longitudinal direction and to reduce the calculation time. The length of the chamber will
not have any effect on the heat transfer and temperature variation along the radius as the heat input to the model in ANSYS is given as heat flux (W/m^2) at the inner surface of barrel and at outer surface of barrel convection was applied with a heat transfer coefficient of 6.5 W/m^{2}K other surface of barrel were perfectly insulated to prevent the heat transfer in axial direction.

The rate of fire for the given 120 mm gun is 6 rounds per min and the cycle may be repeated after 10 s , the heat flux per unit time for firing rate of six rounds per minute can be calculated as following,

$$Q = \frac{H_\infty * R_f}{t}$$

Where
'H_∞' is heat input per unit round per unit area in J/m^2.
'R_f' is rate of fire.

't' is duration of firing.

So

$$Q = \frac{656833.6138 * 6}{60} = 65683.36\ W/m^2$$

In the present study case this heat flux was applied for 60 s with a break of 6s, for 521.9 s. with a time step of 0.01 s and convergence was also checked at different time step and element size. This induces very acute temperature gradients near the bore surface; hence a very fine mesh was used in that zone. Figure. 2 shows meshed model of bore surface.

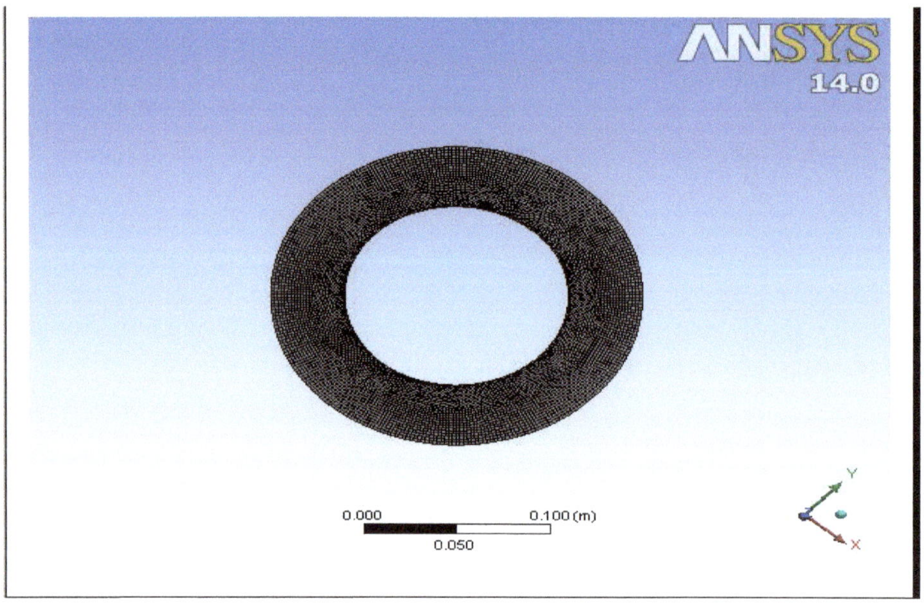

Fig. 2. Meshed model of bore surface

The bore temperature, after a number of cycles, and the corresponding time was noted down. The following table shows the heat input for analysis. Value of heat input is slightly more than the value calculated above, for sixty seconds because in software we cannot apply two values at same time e.g. maximum and zero value at same point sixty second so to compensate the decrease in heat input due to software limitation some extra value of heat input was applied as shown in table. 3.

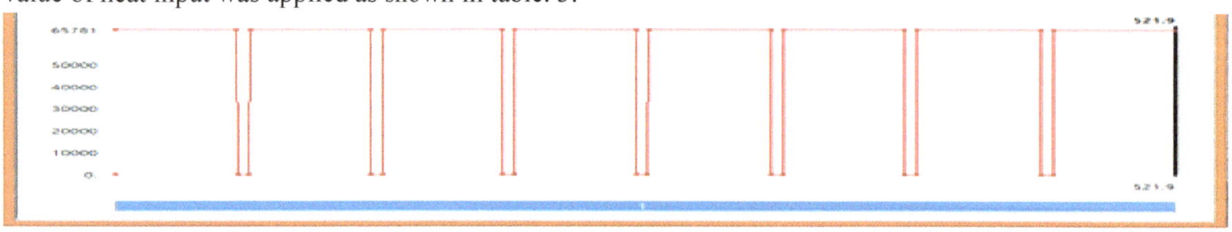

Table 3: Heat input applied for constant heat supply for sixty seconds for firing rate of six rounds per minute

4. Results And Discussion

when firing rate of six rounds per minute

(a.) The cook-off temperature was 507 s when calculated analytically by using Eq. (7). Therefore, the bore temperature will not reach the cook-off temperature for 50 cycles i.e. gun can fire 50 rounds without reaching cook of limit at a time.

(b.) Finite Element Method results:

The cook-off temperature was reached in 476.8 s when heat was supplied continuously for sixty seconds with six seconds gap. Therefore, the bore temperature will not reach the cook-off temperature for 44 cycles i.e. gun can fire 44 rounds without reaching cook of limit. The variation of temperature with time is shown in the Fig. 3. Variation of temperature in radial direction is shown in Fig. 4.

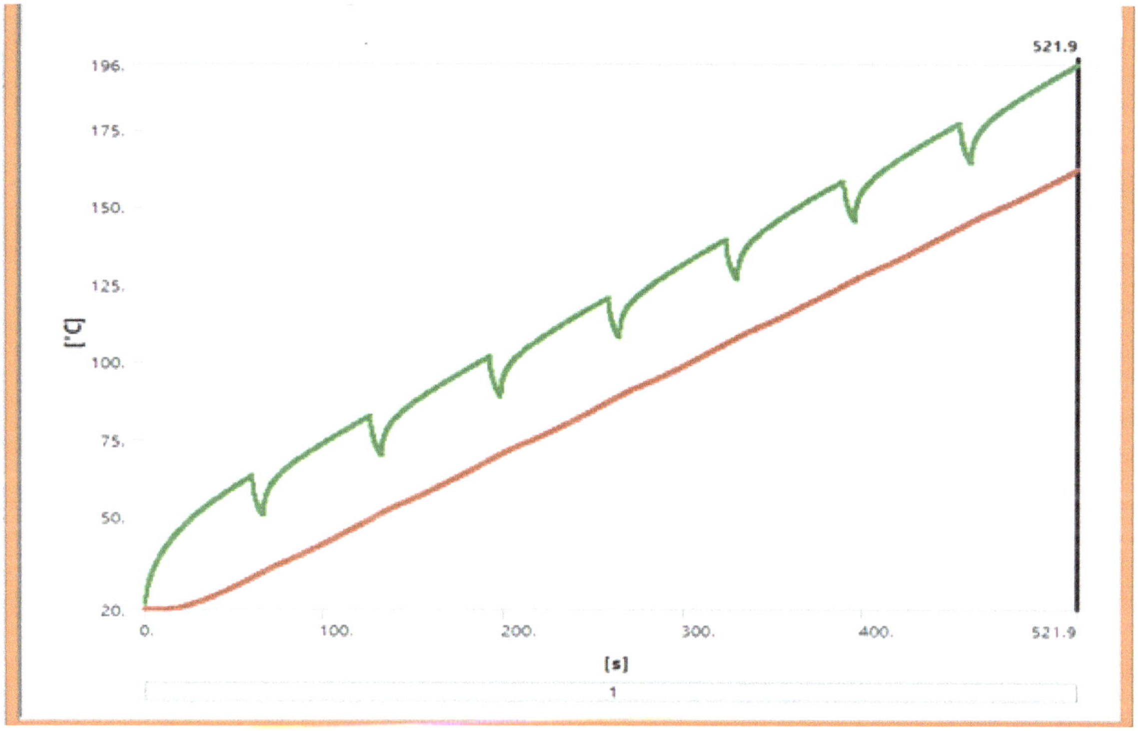

Figure. 3: Variation of temperature with time firing rate of six rounds per minute with constant heat supplied for sixty seconds

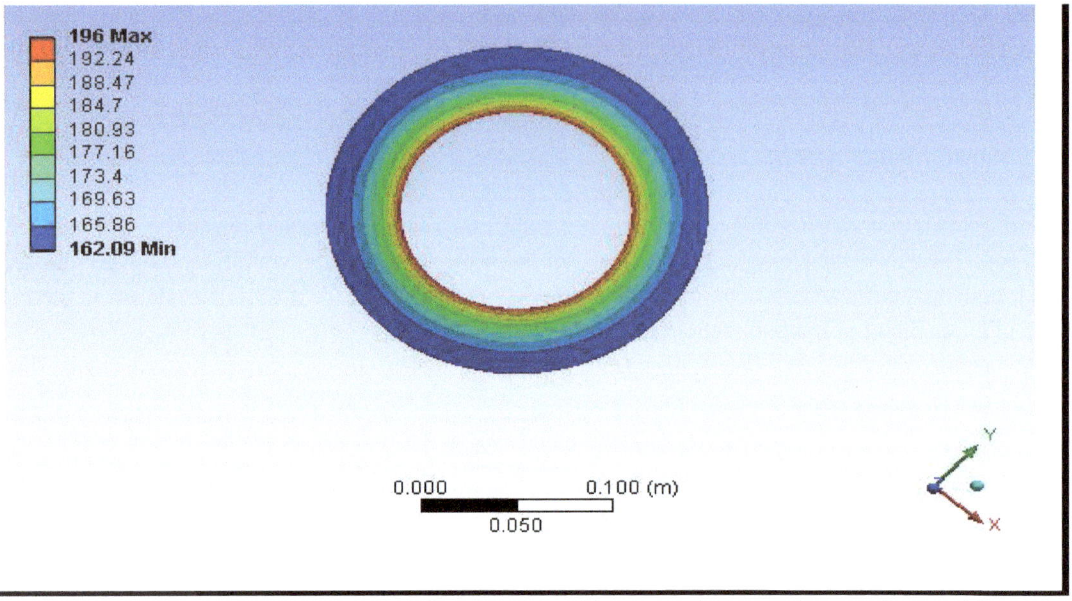

Figure. 4: Temperature variation in radial direction for a firing rate of six rounds per minute with constant heat supplied for sixty seconds

The percentage variation with finite element method is approximately 5.9 %. The difference in two methods may have occurred due to the 3D model used to represent the portion of barrel chamber. Due to the 3D model, the assumption of one-dimensional axisymmetric case was invalidated and there was some heat transfer in the longitudinal direction as well. However, it is the limitation of ANSYS that in 2D analysis, the heat flux in W/m^2 is not recognized by the software. Whereas, the results in the present study are in good agreement (only 5.9 % variation) and no further improvement was felt. However, the results may further be improved by reducing the length of the barrel modeled.

.

5. Conclusion

Present scheme of cook off study of the gun barrel is analytical valid. with the help of Proposed scheme we can find no of shots fired before reaching the cook off of a gun barrel at a given firing rate. The small variation in the results was also discussed and remedy for further improvement in the results was suggested. In this direction, an important advantage is offered by proposed scheme, as it couples internal ballistic parameters with the inputs for finite element methods.

6. Acknowledgment

The authors are grateful to the Director ARDE, Pune for providing the important data necessary for calculation of heat input for given ammunition

7. Nomenclature

D	Bore diameter
h_∞	Heat transfer coefficient
H_∞	Heat input per round
κ	Thermal diffusivity
K	Thermal conductivity of the material
m_c	Mass of propellant
m_r	Mass of the shot
P_{max}	Maximum bore pressure
t_0	Time constant
T_f	Flame temperature of propellant
T_i	Initial temperature
V_m	Muzzle velocity
t	Duration of firing
R_f	Rate of fire
t_c'	Cook-off time for the barrel
T_c	Cook off temperature of the barrel
θ_∞	Finite temperature

8. References

[1] Dr. B Lawton," *Heat Transfer and Erosion In Gun Barrel*‖", Report No. SMMCE/TE/BL/6-94, DACMT, Cranflield University pp. 04-05, pp. 1, pp. 13, pp. 11, pp. 10 of Section 8.

[2] Gerber, N., and Bundy, M. L., 1991. Heating of a Tank Gun Barrel: Numerical Study, Army Ballistic Research Laboratory Report No. BRL-MR-3932.

[3] Gerber, N., and Bundy, M., 1992. Effect of Variable Thermal Properties on Gun Tube Heating, Army Ballistic Research Laboratory Report No. BRLMR-3984.

[4] Wu, B., Chen, G., and Xia, W., 2008. Heat Transfer in a 155 mm Compound Gun Barrel With Full Length Integral Midwall Cooling Channels, Appl. Therm. Eng., 28 [8–9], pp. 881–888

[5] Dr. B Lawton and G. Klingenberg," *Transient Temperature in Engineering and Science* ", Oxford Science Publication, pp. 452-453, 453-454, 61-62, 455, 455- 456, 450, 1995.

[6] Conroy, P. J., 1991. Gun Tube Heating, Army Ballistic Research Laboratory Report No. BRL-TR-3300.

[7] " *Book On Internal Ballistics* ".

[8] Lawton, B., and Klingenberg, G., 1995. Transient Temperature in Engineering and Science, Oxford University Press, Oxford, Chap. 2, pp. 61–63.

Authors Brief Biography

1. ***Vikas-*** M.Tech (Thermal Engg.) student, Mechanical Engineering Department, NIT Kurukshetra, *vikassoni4you@gmail.com*, Interested in thermal analysis and management of gun barrel and in area of CFD.

2. ***Dr.Nirmal Kant Singh-*** Associate Professor, Department of Mechanical Engineering, NIT Kurukshetra, *kant.nirmal@gmail.com* .

3. ***Manoj Kumar-*** M.Tech (Thermal Engg.) student, Mechanical Engineering Department, NIT Kurukshetra, *mksainisingh@gmail.com*, Interested in research in the field of CFD.

FABRICATION, TESTING AND ANALYSIS OF ALUMINIUM 2024 METAL MATRIX COMPOSITE

Marlon Jones Louis

Research Scholar, Department of Mechanical Engineering, Government College of Engineering, Salem 636 011, Tamilnadu, India, mecharlon@yahoo.co.in

Abstract

In the field of material science and engineering, there is a great impact ever since the invention of composites materials. High strength and lightweight remain the winning combination that propels composite materials into new arenas. The composite materials replace conventional materials like steel, cast iron and aluminum alloys by its superficial properties. As literatures were collected, it could be found that metal matrix composites are under serious consideration as potential candidate materials to replace conventional materials in aerospace and automotive applications. In this project, composite material based on aluminum alloy (Al 2024) reinforced with 10% volume fraction of Silicon Carbide Particulates (SiC) and 5 % volume fraction of Graphite particles is produced by stir casting method. The fabricated composite is tested in order to find tensile strength and hardness and analysis is carried out using Ansys software version 10.0. to study on cracking behaviour.

Keywords: Al 2024, Metal Matrix Composite, Stir casting, Ansys

1. Introduction

In an advanced society like ours we all depend on composite materials in some aspect of our lives. Fiber glass, developed in the late 1940s, was the first modern composite and is still the most common. It makes up about 65 per cent of all the composites produced today and is used for boat hulls, surfboards, sporting goods, swimming pool linings, building panels and car bodies. Composites exist in nature. A piece of wood is a composite, with long fibres of cellulose (a very complex form of starch) held together by a much weaker substance called lignin. Cellulose is also found in cotton and linen, but it is the binding power of the lignin that makes a piece of timber much stronger than a bundle of cotton fibres. In engineering materials, composites are formed by coatings, internal adhesives and laminating. An important metal composite is clad metals. Thermostatic controls are made by roll-bonding a high expansion alloy such as copper to a low expansion alloy like steel. When the composite is heated it will deflect to open electrical contacts. Ply wood is a similarly common composite. Since wood is weaker in its transverse direction than its long direction, the alternating grain in plywood overcomes the transverse deficiency.

Composite materials are formed by combining two or more materials that have quite different properties. The different materials work together to give the composite unique properties, but within the composite the materials can be differentiated since they do not dissolve or blend into each other. Composites are made up of two materials namely **matrix** and **reinforcement**. The matrix or binder surrounds and binds together a cluster of fibres or fragments of the stronger material (reinforcement). In Metal Matrix Composites (MMCs), ceramics or metals in form of fibres, whiskers or particles used to reinforce in a metal matrix. Most commonly used matrixes are aluminum, magnesium, copper, titanium and zinc. The most commonly used reinforcements are silicon carbide, alumina, boron, graphite and fly ash. The strengthening effect of the reinforcements in composites depends on the orientation of the reinforcements to the direction of the loads.

1.1 OBJECTIVE

The objectives of this project are

- To fabricate Metal matrix composites with the base metal as Aluminum reinforced with a Volume of 10 % of Silicon carbide particulates and 5 % of graphite particulates by Stir casting method.
- To study the cracks using ANSYS version 10.0 software

2. FABICATION METHOD

2.1 STIR CASTING

The stir casting technique was used to fabricate the composite specimen as it ensures a more uniform distribution of the reinforcing particles. This method is most economical to fabricate composites with discontinuous fibers or particulates. In this process, matrix alloy (Al 2024) was first superheated above its melting temperature and then temperature is lowered gradually below the liquidus temperature to keep the matrix alloy in the semisolid state. At this temperature, the preheated Sic particles of 10 % (by weight) and graphite particle of average size of 23 μm and 45 μm respectively were introduced into the slurry and mixed using a graphite stirrer. The composite slurry temperature was increased to fully liquid state and automatic stirring was continued to about five minutes at an average stirring speed of 300-350 rpm under protected organ gas. The SiCp particles help in distributing the graphite particles uniformly throughout the matrix alloy. The melt was then superheated above liquidus temperature and finally poured into the cast iron permanent mould for testing specimen. The specification of the fabricated billet composite is150 mm length and 50 mm width and a thickness of 20 mm .The composite metal after been ejected from the mold is then rolled in a hot rolling machine up to 15 passes. This is done in order to help in distributing the silicon carbide particulates in the metal matrix and thereby improving the mechanical properties.

Figure 1: Experimental setup for fabricating composite material

Figure 2: Mould setup

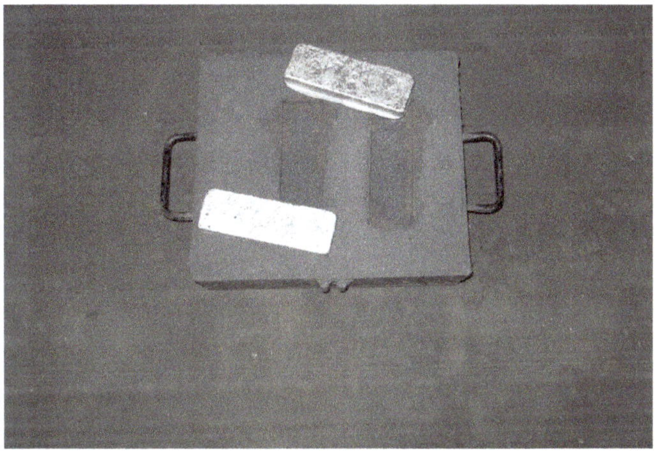

Figure 3: Fabricated composite metal ejected from the mould

3. COMPARISONS OF THE OBTAINED RESULTS

3.1 MECHANICAL PROPERTIES

3.1.1 TENSILE TEST:

As far as the tensile test is concerned the comparison is given below in the table,

S.No	Composition	Yield Strength in KN	Tensile Strength in KN
1	Al-2024	76	93
2	Al-2024 10%SiC+5%Gr	95.10	98.34

Table: 1 Comparison of Tensile Test

3.1.2 STRENGTH:

The yield and tensile strengths of Al 2024/SiC+Gr composite are comparatively higher than the conventional Al 2024 metal.

3.1.3EFFECT OF AL MATRIX ALLOY:

The Al matrix used for the Al 2024/SiC+Gr composite was the most important factor affecting yield strength and ultimate tensile strength of these Al 2024/SiC+Gr composite. Tests showed that Al 2024/SiC+Gr composite had higher strength but lower ductility.

3.1.4 HARDNESS TEST:

Based on the readings obtained in the hardness test under 'B' scale reading the results are shown in tabulation

S.No	Composition	Hardness Number
1	Al-2024	47
2	Al-2024 10%SiC+5%Gr	57

Table: 2 Comparison of Hardness Number

As shown in the tabulation, the hardness of the material as compared to Al-2024 is relatively higher in Al-2024 10%SiC+5%Gr composite.

4. STRESS INTENSITY FACTOR:

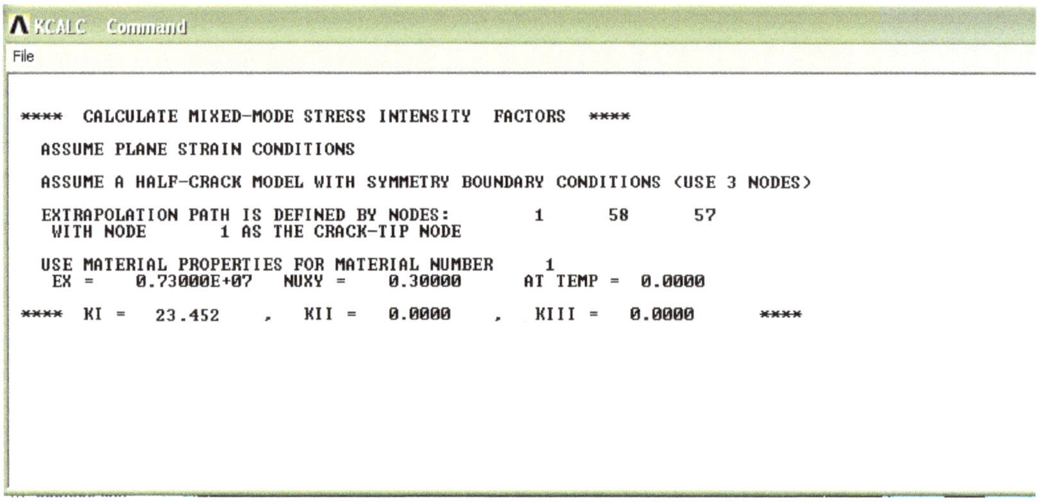

Figure 4: Stress Intensity factor of Aluminium 2024

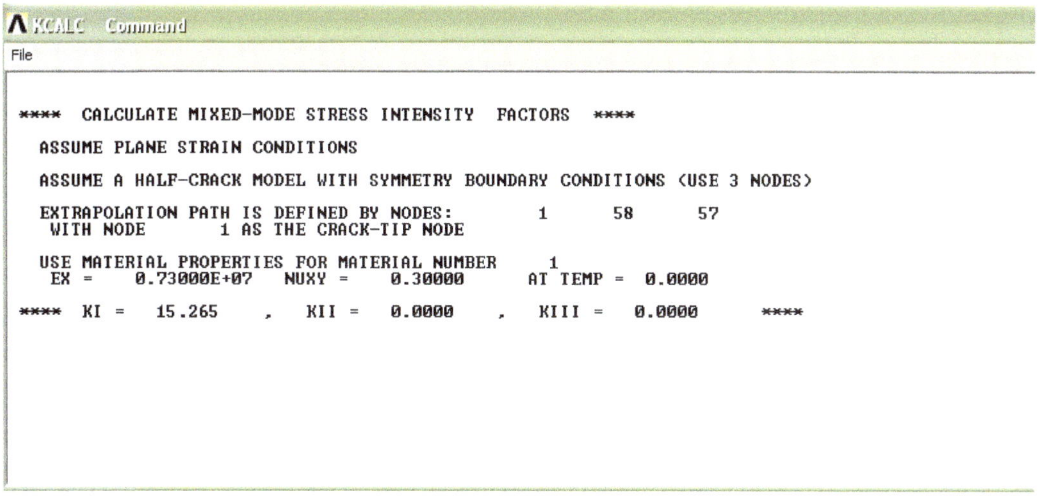

Figure 5: Stress Intensity factor of composite material

From the above figures the Stress intensity factor for Al 2024 is higher when compared with the composite material fabricated and hence this reveals that the stress distribution across the crack tip plastic zone is smaller than the conventional Al 2024 metal leading to minimal stress at the crack tip of composite material which is fabricated.

4.1 DYNAMIC ANALYSIS:

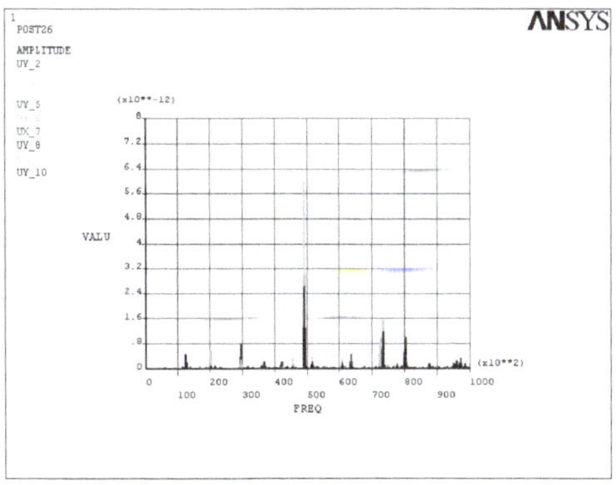

Figure 6: Analysis of aluminum 2024

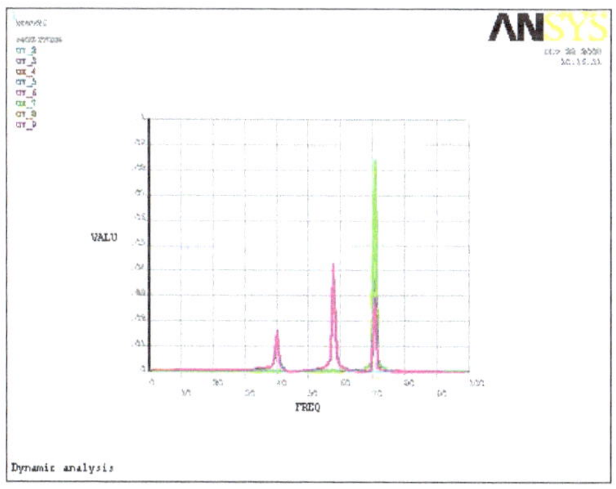

Figure 7: Analysis of composite

From the graph, it shows that the dominant peak value in Al 2024 is 49 KHz whereas in Al-2024 10%SiC+5%Gr the peak value is about 79 KHz. This proves that the life expectancy of the composite involving in crack can withstand more than that compared to the life of the Al 2024.

4.2 STRESS – STRAIN CURVE

The Stress – Strain curve is depicted below,

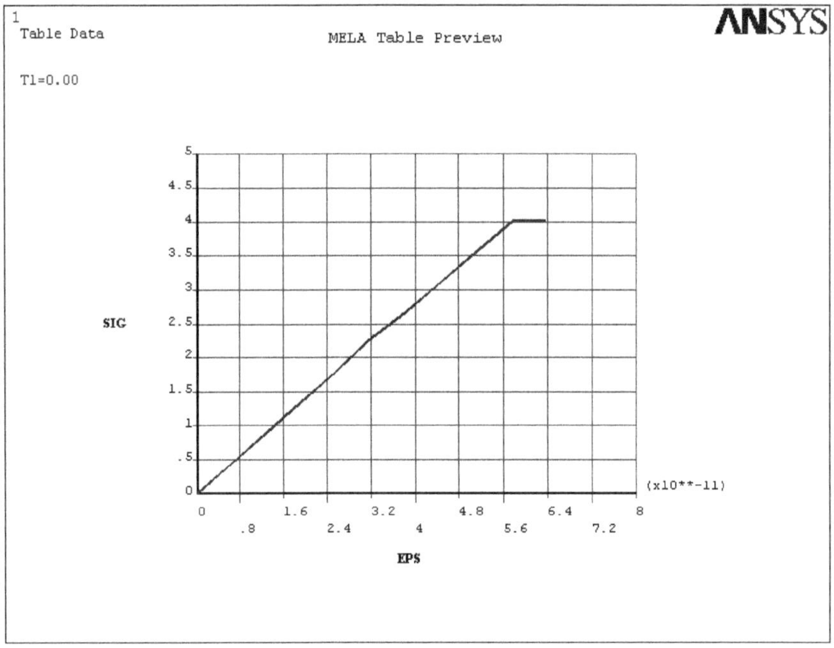

Figure 8: Stress – Strain Curve

From the above stress-strain graph the material is linear up to 2 N/m^2, from then it is nonlinear in nature. This is expected at near crack tip region. But the ultimate tensile strength of Aluminium 2024 + Sic 10% and Gr 5% is about 98 N/m^2. So even though there is a nonlinear material there is no further cracking in the crack tip region for this loading until the stress near the crack tip exceeds or equals to maximum tensile strength of the material.

5. CONCLUSIONS

The tensile test and the hardness test were conducted and the mechanical properties were determined. The experimental results that were found are been compared with the conventional Aluminium 2024 and with the Composite material were one can see that the composite material plays a dominant role than the Aluminium 2024 with respect to its strength, ductility and hardness.

Dynamic analysis is a very important investigation when it comes to the composite materials, where these can exhibit diversity in material properties as well as shapes. The main idea of this work is to perform analysis which gives the information about cracks and the locations of the damages on composite materials.

In real life situations doing a dynamic analysis on structures requires great skill and experience, because the excitation force which is dynamically given on the structure should be chosen very carefully and the excitation point is also plays very important role in dynamic analysis. By doing the computer simulation of dynamic analysis gives us the variety in choosing excitation signal as well as the excitation points.

In this work the modal analysis is done for finding modal frequencies which are very useful to do the frequency analysis which requires the range of frequencies in the system vibrates more (natural frequencies). After doing the frequency analysis the next step is to perform a static analysis or dynamic analysis depending on the requirement. Finding the damages in the materials is of course a very important non-destructive testing which dealt with ease here with computer simulation by using dynamic analysis. Not only finding the damages but the location of damages is always required in process.

6. FURTHER APPROACH TO THE WORK

This work leads to very appealing thoughts about dynamic analysis. This work is carried out to find the crack location and the life expectancy of the composite material model. This work will certainly help in those types of situations where a small part of the structure is to be examined. But doing analysis for the whole structure is often required. There are many factors concerned with cracks but those factors should be accounted for higher studies in this perspective. Further dynamic analysis can be done such as transient dynamic analysis through which velocity – time, Displacement – time plot graphs can help more about to learn cracks in deeper sections.

REFERENCES

[1] American Foundry men's Society, Cast metals handbook, 4th ed., Desplaines (1957)

[2] Byung-Chul ko and Yeon Chul yoo, "Deformation Map of Hot working of AA2024 Composites reinforced with SiC particles",Metals and Materails Vol. 5, No. 6 (1999), pp. 555-558

[3] R.E.Rowlands,Y.Yang and Y.Peng "Metal Matrix Composites",pp.2-37

[4] Walter D. pilkey "peterson's stress concentration factors", 2nd edition. Wiley Interscience (1997)

[5] Vikram Singh and R.C Prasad, 2004 "Tensile and Fracture Behavior of 6061 $AlSic_p$ Metal Matrix Composite", International symposium of research on materials science and engineering

[6] Vidya Sagar Avadutala, 2005, "Dynamic Analysis of Cracks in Composite Materials" Wiley Interscience

[7] P. Rohatgi, 1991, "Cast Aluminum Matrix Composites for Automotive Applications," Journal of Metals.

[8] Abis, S., (1989), "Characteristics of an aluminium alloy/Alumina Metal Matrix composite", Composites science and technology, Vol.35, pp.1-11.

[9] Brian Terry and Jones Glyn.,(1990), Metal matrix composite, Elsevier Science Publishers Ltd., England, pp.41-47.

Permissions

All chapters in this book were first published in IJRAME, by IJRAME; hereby published with permission under the Creative Commons Attribution License or equivalent. Every chapter published in this book has been scrutinized by our experts. Their significance has been extensively debated. The topics covered herein carry significant findings which will fuel the growth of the discipline. They may even be implemented as practical applications or may be referred to as a beginning point for another development.

The contributors of this book come from diverse backgrounds, making this book a truly international effort. This book will bring forth new frontiers with its revolutionizing research information and detailed analysis of the nascent developments around the world.

We would like to thank all the contributing authors for lending their expertise to make the book truly unique. They have played a crucial role in the development of this book. Without their invaluable contributions this book wouldn't have been possible. They have made vital efforts to compile up to date information on the varied aspects of this subject to make this book a valuable addition to the collection of many professionals and students.

This book was conceptualized with the vision of imparting up-to-date information and advanced data in this field. To ensure the same, a matchless editorial board was set up. Every individual on the board went through rigorous rounds of assessment to prove their worth. After which they invested a large part of their time researching and compiling the most relevant data for our readers.

The editorial board has been involved in producing this book since its inception. They have spent rigorous hours researching and exploring the diverse topics which have resulted in the successful publishing of this book. They have passed on their knowledge of decades through this book. To expedite this challenging task, the publisher supported the team at every step. A small team of assistant editors was also appointed to further simplify the editing procedure and attain best results for the readers.

Apart from the editorial board, the designing team has also invested a significant amount of their time in understanding the subject and creating the most relevant covers. They scrutinized every image to scout for the most suitable representation of the subject and create an appropriate cover for the book.

The publishing team has been an ardent support to the editorial, designing and production team. Their endless efforts to recruit the best for this project, has resulted in the accomplishment of this book. They are a veteran in the field of academics and their pool of knowledge is as vast as their experience in printing. Their expertise and guidance has proved useful at every step. Their uncompromising quality standards have made this book an exceptional effort. Their encouragement from time to time has been an inspiration for everyone.

The publisher and the editorial board hope that this book will prove to be a valuable piece of knowledge for researchers, students, practitioners and scholars across the globe.

List of Contributors

Tejas G. Patil
Deprtment of Mechanical Engineering, Godavari College of Engineering, India
Kolhe Na gar, Jalgaon

Atul A. Patil
Deprtment of Mechanical Engineering, Godavari College of Engineering, India
Kolhe Na gar, Jalgaon

Mukund Y Pande
Department of Mechanical Engineering, Godavari College of Engineering, India
1GF'S Godavari college of Engineering Jalgaon Maharashtra. India

Atul Patil
Department of Mechanical Engineering, Godavari College of Engineering, India
1GF'S Godavari college of Engineering Jalgaon Maharashtra. India

K. Esakkiraja
Assistant Professor, Adithya Institute of Technology, Coimbatore

B. Sam Antony
Assistant Professor, Ranganathan Engineering College, Coimbatore

A. Kannan
Assistant Professor, SVS College of Engineering, Coimbatore

Prof. Dr. S. H. Sawant
Professor, Mechanical Engg.Dept, Dr.J. J. Magdum college of Engineering,Jaysingpur, Kolhapur, India

D. Muruganandam
Assistant Professor, Dept. of production engineering, Sairam Engineering college, Chennai, India

D. Raguraman
Assistant Professor, Dept. of production engineering, Sairam Engineering college, Chennai, India

S. Briolin Nayagam

Karan Singla
Department of Mechanical Engineering MIET, Shahabad India

Dr. Deepak Bhardwaj
Department of Mechanical Engineering, BITS, Bhiwani India

Vikas Burman
Department of Mechanical Engineering , RPIIT, Karnal India

Shahrukh Shamim
Department of Material Science and Metallurgical Engineering, Maulana Azad National Institute of Technology, Bhopal, M.P., India.-462051

Jamil Anwer
Department of Material Science and Metallurgical Engineering, Maulana Azad National Institute of Technology, Bhopal, M.P., India.-462051

PRASHANT SINGH
PG SCHOLAR (DEPARTMENT OF MECHANICAL ENGINEERING, MATS University, RAIPUR)

PARIWESH THAKARE
PG S SCHOLAR (DEPARTMENT OF MECHANICAL ENGINEERING, MATS University, RAIPUR)

DHANANJAY SAHU
PG SCHOLAR (DEPARTMENT OF MECHANICAL ENGINEERING, MATS University, RAIPUR)

PRADEEP NAGESH
RESEARCH SCHOLOR, UIT, RGPV, BHOPAL (M.P.)

PRASHANT SHARMA
PROF. & HEAD (CMWS), UIT, RGPV, BHOPAL (M.P.)

Er. Rinku Kumar
Er. Rinku Kumar (M-Tech student at RPIIT Technical Campus, Karnal)

Er. Naveen Hooda
Er. Naveen Hooda (Assistant Professor at RPIIT Technical Campus, Karnal)

Mr. Zeeshan Khan
Dr. D. Y. Patil Institute of Engg. & Tech. Ambi, Talegaon Dabhade, Pune University

Prof. R. J. Patil
Dr. D.Y. Patil Institute of Engg. & Tech. Ambi, Talegaon Dabhade, Pune University

Vinayak D. Yadav
PG Student ME CAD/CAM/CAE

S. G. Bhatwadekar
Associate Professor in Production Engineering

P. PADMAVATHI
M.Tech Student (Thermal Engineering), Aditya Institute of Technology and Management, Tekkali

C. J. RAO
Professor &HOD. Mechanical Engg. Dept. Aditya Institute of Technology and Management, Tekkali

Neeraj kumar
National institute of Technology (NIT) Kurukshetra, India, 136119

Dilbagh Singh
NILLM UNIVERSITY, KAITHAL, 136027

Harry Garg
Central Scientific Instruments Organisation (CSIR-CSIO), Chandigarh, India, 160030

Nirmal Kant Singh
National institute of Technology (NIT) Kurukshetra, India, 136119

Sunil Gaikwad
PG-Student, Mech. Engg. Dept. SCOE, Kharghar, Navi Mumbai, Maharashtra

S. N. Teli
Associate Professor & Head, Mech. Engg. Dept. SCOE, Kharghar, Navi Mumbai, Maharashtra

L. M. Gaikwad
Asst. Professor, Mech. Engg. Dept. BVCOE, Belpada, Navi Mumbai, Maharashtra

Avinash D Jadhav
Department of Mechanical Engineering, GF's Godavari College of Engineering Jalgaon

Tushar A Koli
Department of Mechanical Engineering, GF's Godavari College of Engineering Jalgaon

Vithal Rao Chavan
Asst. Professor, Department of Mechanical engineering, Rao Bahadur Y Mahabaleshwarappa Engineering College, Bellary, India

Dr. K R Dinesh
Principal and Professor, Department of Mechanical Engineering, Government Engineering College, Raichur, India

Dr. K Veeresh
Principal and Professor, Department of Mechanical engineering, Rao Bahadur Y Mahabaleshwarappa Engineering College, Bellary, India

Veerabhadrappa Algur
Asst. Professor, Department of Industrial Production Engineering, Rao Bahadur Y Mahabaleshwarappa Engineering College, Bellary, India

S. P. Jagadish
Asst. Professor, Department of Mechanical engineering, Rao Bahadur Y Mahabaleshwarappa Engineering College, Bellary, India

C. Madhu Mohan
P.G. Scholar, Department of Industrial & Production engineering, Rao Bahadur Y. Mahabaleshwarappa Engineering College, Bellary, India

Sagar W. Matte
P. G. Scholar, Mechanical Engineering, BIT, Ballarpur-India

Umesh G. Telote
P.G guide & Assistant professor, Mechanical Engineering, Ballarpur Institute of Technology, Ballarpur
C/o. W. S. Matte, Wadgaon road, near dhanoje kunbi samaj mandir, Chandrapur- 442401,

Mr. Chetan Rane
PG Scholar, Mechanical Engineering Department, SCOE, Kharghar, India

Prof. M. B. Sorte
HOD and Assistant Professor, Mechanical Engineering Department, MGMCET, Kamothe, India

Abhishek Verma
Student, Mechanical Engg. Deptt. M.I.E.T, KUK

Vishal Kumar
Assistant Professor in Mechanical Engg. Deptt. MIET, KUK

P. Balakrishnan
PG Scholar, Department Of Mechanical Engineering, Regional Centre Anna University: Tirunelveli Region, Tirunelveli -627 007

Dr. K. Karuppasamy
Assistant Professor, Department Of Mechanical Engineering, Regional Centre Anna University: Tirunelveli Region, Tirunelveli -627 007

J. Ramkumar
Teaching Faculty, Department Of Mechanical Engineering, Regional Centre Anna University: Tirunelveli Region, Tirunelveli-627007

P. Anu Nair
PG Scholar, Department Of Mechanical Engineering, Regional Centre Anna University: Tirunelveli Region, Tirunelveli -627 007

Dr. K. Karuppasamy
Assistant Professor, Department Of Mechanical Engineering, Regional Centre Anna University: Tirunelveli Region, Tirunelveli -627 007
Madhusadhanam, Thrikkodithanam P.O, Changanacherry, Kottayam, Kerala

R. Arularasan
Faculty, Students Dept. of Mechanical Engineering, University College of Engineering Arni, Thatchur, Tiruvannamali District- 632326, India

K. Sivasakthivel
Faculty,Students Dept. of Mechanical Engineering, University College of Engineering Arni, Thatchur, Tiruvannamali District- 632326, India

M. Alagesan
Students Dept. of Mechanical Engineering, University College of Engineering Arni, Thatchur, Tiruvannamali District- 632326, India

R. Anand
Students Dept. of Mechanical Engineering, University College of Engineering Arni, Thatchur, Tiruvannamali District- 632326, India

K. Arshad Ahmed
Students Dept. of Mechanical Engineering, University College of Engineering Arni, Thatchur, Tiruvannamali District- 632326, India

S. Vimalan
Students Dept. of Mechanical Engineering, University College of Engineering Arni, Thatchur, Tiruvannamali District- 632326, India

Abhishek Tripathi
Post Graduate (PG) student, Mechanical Engineering Department, Radharaman Institute of Technology and Science, Bhopal (M.P.), India

Ajay Kumar Singh
Associate professor & HOD, Mechanical Engineering Department, Radharaman Institute of Technology and Science, Bhopal (M.P.), India

Amit kumar

Vishal Gulati

Amitesh Goswami
Mr. Swapnil prakash magar
S.S.G.M.C.E. Shegaon, Maharashtra

Dr. S. P. Trikal
Dept. of Mechanical engineering, S.S.G.M.C.E., Shegaon, Maharashtra

Vikas
M.Tech (Thermal Engg.) student, Mechanical Engineering Department, NIT Kurukshetra

Dr. Nirmal Kant Singh
Associate Professor, Department of Mechanical Engineering, NIT Kurukshetra

Manoj kumar
M.Tech (Thermal Engg.) student, Mechanical Engineering Department, NIT Kurukshetra